Current Controversies in the Biological Sciences

Basic Bioethics

Glenn McGee and Arthur Caplan, editors

Current Controversies in the Biological Sciences

Case Studies of Policy Challenges from New Technologies

Karen F. Greif and Jon F. Merz

The MIT Press
Cambridge, Massachusetts
London, England

MIT Press books may be purchased at special quantity discounts for business or sales promotional use. For information, please e-mail <special_sales@mitpress.mit.edu> or write to Special Sales Department, The MIT Press, 55 Hayward Street, Cambridge, MA 02142.

This book was set in Sabon by SNP Best-set Typesetter Ltd., Hong Kong. Printed on recycled paper and bound in the United States of America.

Library of Congress Cataloging-in-Publication Data

Greif, Karen F. (Karen Faye), 1952–
Current controversies in the biological sciences : case studies of policy challenges from new technologies / Karen F. Greif and Jon F. Merz.
    p. cm. — (Basic bioethics)
 Includes bibliographical references and index.
 ISBN-13: 978-0-262-07280-9 (hardcover : alk. paper)
 ISBN-13: 978-0-262-57239-2 (pbk. : alk. paper)
    1. Medicine—Research—Government policy. 2. Medicine—Research—Moral and ethical aspects. I. Merz, Jon F. (Jon Frederick), 1956– II. Title.
R850.G74   2007
610.72—dc22

                                                                    2006027119

10 9 8 7 6 5 4 3 2 1

For our parents

Contents

Series Foreword

We are pleased to present the twentieth book in the series Basic Bioethics. The series presents innovative works in bioethics to a broad audience, and introduces seminal scholarly manuscripts, state-of-the-art reference works, and textbooks. Such broad areas as the philosophy of medicine, advancing genetics and biotechnology, end-of-life care, health and social policy, and the empirical study of biomedical life are engaged.

Glenn McGee
Arthur Caplan

Basic Bioethics Series Editorial Board
Tod S. Chambers
Susan Dorr Goold
Mark Kuczewski
Herman Saatkamp

Preface

Why write this book? Technological advances in biology have enormous potential impacts on our lives, and there is a need for citizens and policymakers alike to understand the scientific, political, and ethical issues that underlie the policy decisions that might be made. Many issues arising from new advances in biology do not have simple answers. Conflicting goals and beliefs influence the discussions that are associated with the development of policy, and compromise between conflicting views may be difficult. This book is intended as an introduction to science policy. Additional readings are suggested for those who wish to dig deeper into their areas of interest.

This book examines those interactions between science and government in which policy (and political) decisions are made, using examples drawn from the biological sciences. These examples are selected to demonstrate the different ways in which science and politics intersect. Policy may take various forms, such as international laws and treaties, federal or state laws, court decisions, and specific regulatory measures (to control but not proscribe particular behaviors). Policy may be reflected too in the lack of adoption of any formal regulations. What causes government to step in and regulate a biological technology? Does the public have a right—or the ability—to control science? We attempt to show the limits of scientific knowledge and its ability to contribute to effective policymaking.

Writing any book on government policy inevitably results in a sea of acronyms. We have tried to limit our use of acronyms to common ones, and redefine them frequently. Nevertheless, acronyms are unavoidable, and we encourage readers to use patience in their first perusals of the text. It takes a little time to become conversant with the alphabet soup of government.

The issues we discuss are controversial, and our goal is to present them in a balanced manner, reflecting the range of current viewpoints. We try to avoid polemics—but it is almost impossible to avoid the insertion of at least some of our own opinions. We urge readers to draw their own conclusions from the information presented.

Acknowledgments

This book is an outgrowth of a course titled "Biology and Public Policy" that one of us (Greif) has been teaching at Bryn Mawr College since 1991. Student demands for reading materials inspired us to provide a book that would be accessible to a broad audience with relatively limited biology backgrounds who are concerned about the place of science in our society. The actual writing was helped by countless interactions with our colleagues at Bryn Mawr and the Center for Bioethics at the University of Pennsylvania, especially Margaret Hollyday, Xenia Morin, Janet Ceglowski, Paul Grobstein, Arthur Caplan, Sheldon Zink, and Kenneth Foster. Thanks to Chris Frey (North Carolina State University) and Sheri Alpert (Notre Dame) for their feedback on draft chapters. Greif also thanks the Center for Bioethics at the University of Pennsylvania for providing office space and support for the sabbatical during which much of this book was drafted, and Bryn Mawr College for additional research support.

Chapter 3 is a substantially modified version of J. F. Merz, "Disease Gene Patents," in *Encyclopedia of Medical Genomics and Proteomics*, ed. J. Fuchs and M. Podda (New York: Marcel Dekker, 2004); J. F. Merz and M. K. Cho, "What Are Gene Patents and Why Are People Worried about Them?" *Community Genetics* 8 (2005): 203 (with permission).

Chapter 3, section 2 has been abstracted with permission from a longer chapter titled "Discoveries: Are There Limits on What May Be Patented?" in *Who Owns Life?* ed. D. Magnus, A. Caplan, and G. McGee (Amherst, NY: Prometheus Books, 2002).

1

An Uneasy Balance: Science Advising and the Politicization of Science

Science and technology-related issues are pervasive in today's society. Science contributes in many ways to our lives, whether directly in health-related matters or more indirectly through effects on the environment, economic development, and international relationships. What is science and technology policy? While difficult to define, one author described it as "a governmental course of action intended to support, apply, or regulate scientific knowledge or technological innovation" ([1], 12). As we will explore, policy sometimes takes the form of governmental action, but occasionally inaction results because of political considerations. Nongovernmental actors also affect public policies, including nonprofit advocacy organizations, educational institutions, and businesses. Policies may be divided into two types: decisions affecting the funding or direction of science ("policy for science"), and decisions that draw on scientific data to inform policy debate ("science in policy") [2]. Issues regarding the funding and direction of science are obvious examples of science and technology policy, but we will demonstrate that the latter (science in policy) are as important for the population at large.

How are policy decisions made? An abstract description includes five stages: the perception and definition of a problem by the public and policymakers, the formulation of possible solutions by policymakers, the adoption of a policy, its implementation, and then an evaluation of the outcome of the decision [1]. It is rare that the political policymaking process follows this tidy description—a field of social science, policy science, attempts to develop a rational framework for understanding, predicting, and directing the policymaking process [3].

Several features of science policy issues distinguish them from more general policy questions [1]. Particularly in the life sciences, the pace of technological change is rapid, and issues arising from new developments are novel. The technologies are complex, and difficult for both policymakers and the general public to grasp. New developments may carry irreversible consequences, and once in use, it may be difficult to stop their application. New technologies may raise strong public worries about threats to health and safety, the environment, or other areas of concern. Finally, many developments challenge deeply held social, moral, and religious values. All these factors may contribute to the difficulty in establishing effective policy. As will be demonstrated by the case studies presented in the upcoming chapters, how a question is formulated—by whom, and under what time and political constraints—can have an enormous impact on the decisions that are made.

Although scientific input is only one factor in policymaking, having accurate, timely, and accessible information is valuable for developing appropriate responses. Policy is made by all branches of the federal government—executive (including regulatory agencies), legislative, and judiciary—and state governments. Foreign governments also make policy, and treaties are often used to secure consistent international policies on far-reaching issues, such as in the domains of the environment (e.g., global warming and biodiversity), trade, and human rights. Given the range of policy challenges facing governments, how does scientific understanding and knowledge contribute to the decision-making process? This chapter provides an overview of the ways in which scientific information may be used by the federal government to develop policy. It then goes on to discuss the inherent conflict between science and politics, and how this leads to the apparent politicization of science.

Science Policy and Government

In the United States, science may contribute to policy discussions on several levels. There are close to a thousand advisory committees in the federal government; about half of these deal directly or indirectly with scientific or technological matters [4, 5]. Scientists may advise the president and other members of the executive branch on establishing directions for research and setting the agenda for future development through cabinet-level positions. Scientists offer testimony to Congress, adding their expertise and opinions to the debate. They also contribute to the development of regulations by the numerous regulatory agencies given responsibility for the oversight of different science-related activities. The courts influence policy by an array of decisions; some rule directly on matters regarding science (patents, etc.), and some reverse policy decisions made by other branches of government. Judicial rulings informed by "expert testimony" may alter existing policies or drive the development of new ones. The courts might also determine that a new law or regulation violates the Constitution or statutes, requiring reevaluation by the body creating or instituting the policy. The government may request that studies be conducted by independent nonpartisan organizations such as the National Academy of Sciences (NAS) to provide information to aid the policy process.

In a democratic society, policy decisions are rarely made without some consideration of public opinion—unpopular decisions might be rebuked at the ballot box. As citizens, scientists may seek to influence politicians to support their views. Scientists and their employer institutions (corporate or academic) and professional societies may actively lobby for specific policy decisions; one major focus of such lobbying is research funding. Scientists may also work as advisers to organizations that take activist roles in influencing public opinion and driving policy decisions. Individuals with scientific experience or interests may work as journalists to help inform the public on new issues. At the same time, the public's understanding, or lack thereof, of new scientific developments may lead to calls for governmental action. If not tempered by sound advising, poorly conceived or nonsensical policies may result.

Science in the Executive Branch

The president appoints individuals to a number of senior-level advisory positions in science and technology; these advisers wield significant power in determining the influence of science in government [6]. Since World War II, the highest ranking of these is the assistant to the president for science and technology (APST); the same individual usually (but not always) serves as the director of the Office of Science and Technology Policy (OSTP), an advisory group created in 1976 by an act of Congress. The president is not required to name such an assistant; President George W. Bush's senior science adviser, John Marburger, was named head of the OSTP only [7]. The National Science and Technology Council was established in 1993 by an executive order, and includes the president, vice president, APST (if there is one), cabinet secretaries, and agency heads with significant science and technology responsibilities [8]. The council's main objective is setting clear national goals for investments in science and technology. Other high-level advisory groups are the President's Council on Science and Technology, which examines a broad range of topics, and the President's Council on Bioethics, which provides input on ethical issues arising from developments in the life sciences. Most cabinet departments include directors with direct responsibility for science and technology policy; among these are the departments of agriculture, commerce, defense, energy, health and human services, interior, and labor. Other independent agencies with directors or administrators named by the president include the National Science Foundation (NSF) and the Environmental Protection Agency (EPA) [4].

Senior-level advisers must have the president's ear if they are to contribute to the policymaking process. Advisers who have only limited access to the president or key deputies will have little impact. The effectiveness of science advising in the White House is tied to the president's interest in scientific issues. In addition, senior-level appointments generally reflect the ideology of the president; these individuals serve to translate the president's viewpoints into policy directions for the agencies that fall under the aegis of the directors [1]. It is therefore not surprising that regulatory agencies appear to make U-turns in overarching policy with each change in administration.

An increasing concern is the growing delay in the appointment of individuals to these important advisory positions [4, 9, 10]. Two factors contribute to the problem: the growing number of presidential appointments overall (with over five hundred senior-level positions alone), and the requirement that many nominees must be confirmed by the Senate. Identifying qualified candidates who are interested in taking a government position may be difficult, particularly in view of the amount of paperwork involved in the review process, the heavy workload, and the comparatively low salaries. Candidates must undergo rigorous background checks that may take months. Finally, Senate confirmation hearings may be delayed if an influential member disapproves of a candidate. As a result, the amount of time for a nominee to be approved has increased from just over two months during the Kennedy administration nearly nine months during the first term of President George W. Bush [4]. In addition, as of January 2002, halfway through President Bush's first term, there were close to one hundred positions for which candidates had not even been named [9].

Delays in filling senior-level appointments may have a chilling effect on policy development, leaving agencies without strong leaders to direct policy. Delaying appointments may be used deliberately to slow the development of new regulations, as was done by President Reagan as part of his generally antiregulatory stance [9].

Federal Advisory Committees

Federal advisory committees play important roles in shaping public policy. There are hundreds of advisory committees focusing on science and technology issues; some advise regulatory agencies, and others serve to advise the president or Congress. Committees may be created specifically to address controversial issues for which the government feels it needs expert advice. In 1972, concerned at the ever-growing number of committees, Congress enacted the Federal Advisory Committee Act (FACA, 5 U.S.C. App.), aimed at limiting the creation of new committees, and establishing standards for committee membership and operations. FACA also mandated "transparency" in committee deliberations; membership on most committees is published, and at least some meetings are open to the public. Central to the law's mission is a mandate that membership on the committees should be balanced, and viewpoints should be represented by accomplished individuals in the policy area. For science and technology, committee members should be chosen for their expertise in the relevant scientific area and their respect within the professional community [4]. Creating effective and unbiased committees is a major challenge, and will be discussed further below.

Federal Agencies

Federal agencies—part of the executive branch, but often referred to as the fourth branch of the federal government because of their unique powers—are the operational arm for many executive and congressional science policy mandates; they create functional policy in response to law. Science oversight is highly fragmented within many agencies and departments. Among those most involved with the biological sciences are the Department of Health and Human Services (DHHS), which includes the Food and Drug Administration (FDA), the Centers for Disease Control and Prevention (CDC), and the National Institutes of Health (NIH); the NSF; the Department of Agriculture; and the EPA. Almost all other departments also contribute to issues in the biological sciences, including the Department of Defense (for example, on bioterrorism policy), the Department of the Interior (the Fish and Wildlife Service), the Department of Labor (the Occupational Safety and Health Administration), and the Consumer Products Safety Commission.

When Congress passes a law, it is the responsibility of agencies to develop regulations that define and enforce the legislation's mandates. Developing regulatory policy again requires the input of science: What is practical? What limits should be set? For example, if a statute mandates that new drugs must be safe and effective, it is the FDA's responsibility to develop and enforce regulations to achieve that goal. Public or congressional objections to new regulations may impel the agency to revise the regulation.

Although many agencies, such as the NIH and the FDA, have quasi-independent status, this freedom is tempered by the strong role that the president plays in determining overall policy direction through senior-level appointments. Congress may also act to limit agencies' ability to enact or enforce policy by controlling appropriations; actions unpopular with Congress may lead to reductions in operating funds or substantive changes in the agency's enabling legislation.

Science and Congress

Congress exerts enormous influence on the direction of science through the appropriations process. The scientific community has a major stake in the congressional determination of levels of research funding, sometimes termed policy for science. Although the purse is arguably the biggest tool wielded by the federal government, other kinds of policy decisions regarding innovation, intellectual property, and trade also fall under this rubric. Intense lobbying by scientists and scientific organizations for funding in specific areas is common. Critics suggest that such lobbying is self-serving; the goal is to gain research funds for one's own projects rather than make choices in the best interests of the nation [11]. Scientists counter that the products of research may not be predicted and broad general support is needed. Congress also provides funding for federal agencies by passing budgetary bills. Both the president, who proposes a budget, and Congress have considerable impact on agency activities, from conducting research to enforcing regulations, through their control of funding.

Science also widely informs policymaking by Congress. Since few politicians have scientific training, they may turn to congressional staffers or outside experts to provide guidance [1, 7, 12]. Congressional committees hold hearings on selected topics, and the invited scientists may offer testimony relating to issues of science and technology. Congressional hearings may also serve to put regulatory agencies on notice that a response is needed—or Congress may act. Congress, however, may pass laws that cannot be implemented by the relevant regulatory agency for either practical or political reasons; carrying out the mandate may be too expensive, too complex, too unpopular, or simply impractical. Congress has even prohibited agencies from spending money on specific regulatory activities.

Congress may turn to government support agencies for a thorough study of the issues. The General Accounting Office (renamed the Government Accountability Office in 2004, or the GAO), established in 1921, frequently provides reports on the possible effects or results of legislation or regulations. It may explore the economic costs of action or the effectiveness of certain approaches to a problem, and then makes recommendations for change or improvement. From 1972 until its closure in 1995, the Office of Technology Assessment (OTA) provided hundreds of comprehensive reports to Congress on a wide range of scientific and technological issues [13, 14, 15]. These reports, intended for congressional committees, served a much wider community. The OTA, however, was criticized for its slow response to requests for information. The new Republican-dominated Congress in 1995 closed the OTA purportedly for primarily budgetary reasons. Yet, a perceived "liberal" bias in its reports contributed to its demise [15]. Without the OTA, Congress now turns more to an independent organization, the NAS, for advice (see below).

Science and the Courts

The judiciary branch of government plays an active role in science policy. Far from being purely reactive, often the courts step in to resolve controversies for which policy has yet to be developed [2]. Court decisions may interpret the impacts of science and technology, generate "authority" for scientific knowledge, and place limits on certain scientific activities. Judicial review of federal agencies is mandated by the Administrative Procedure Act of 1946 and later legislation, which authorizes the courts to invalidate decisions if they are not based on sound evidence [2]. Nevertheless, judicial decisions also can produce an incoherent set of policies when conflicts are resolved on a case-by-case basis, and bring up fundamental questions about the competence of courts to make social policy in light of the practical constraints on fact finding and jurisdiction raised by cases and controversies presented to the courts [16]. Many of the case studies described in the upcoming chapters are influenced by judicial decisions.

Outside Advisory Groups

The federal government may seek advice from outside groups in shaping science policy. Depending on their membership, outside advisory groups may provide independent nonpartisan or highly skewed advice. The National Research Council (NRC) is the operational arm of the NAS, the National Academy of Engineering (NAE), and the Institute of Medicine (IOM). President Abraham Lincoln asked Congress to establish the NAS in 1863 as an independent organization to provide scientific advice to the government; membership was offered to the leading scientists of the day. In 1916, the NRC was founded to carry out the research and advisement activities of the NAS, leaving the NAS and its affiliates as largely honorary societies. Membership in the NAS is highly prestigious. There are around eighteen hundred living members; an additional thirty-one hundred individuals are members of the NAE and IOM [17].

Each year, select committees formed at the NRC research topics requested by Congress, federal agencies, or other groups. The NRC has an internal system of assuring broad representation on committees, and members must reveal any potential bias or conflict of interest. The final reports, like those of the OTA, are widely read and cited. Yet the NRC is also criticized for the slow appearance of reports [18, 19]. Because the NRC is comparatively independent of political pressure, it may produce reports that run contrary to what the agency requesting the study anticipated. The NRC's recommendations are not binding, so the government or other critics may choose to ignore the study's conclusions or seek to discredit them.

Other "think tank" organizations that may conduct research under contract with the government include the more politically liberal Brookings Institution, the politically conservative Heritage Foundation, and the libertarian Cato Institute as well as more neutral policy research institutes such as the RAND and MITRE Corporations. Such consultants—of which there are many—are generally referred to as "Beltway bandits" for their proximity to the main highway that loops around Washington, DC. The federal government may also assemble its own study groups to explore issues; these panels are often criticized as reflecting the bias of the administration.

The Politicization of Science: Conflicting Goals of Science and Politics

Whatever the means of input, there is a constant tension between science and politics. From the perspective of science, policies should reflect careful consideration of the scientific data, and should be in line with the findings and recommendations of science. Scientists who offer advice to policymakers, however, often complain that their input is ignored or distorted during the policymaking process. Political values and necessities may conflict sharply with the data presented by scientists. A policy may be developed that represents a compromise between the criteria determined by science and the pragmatic needs of politics. An effective policy should be cost-effective and fair, place limited demands on government, and provide assurance to the public that the goals will be met [20]. If an administration's position is not supported by the data, it may ask for further studies rather than accept what is offered. In extreme cases, scientific data might be buried in the face of the apparent demands of politics.

The selective use of scientific advice and information has received heavy media coverage in recent years. This strategy is not new, though, President Richard Nixon removed all science advising from the White House during his tenure because he objected to reports with recommendations against his own projects; he also expressed strong irritation toward the apparent left-leaning political viewpoints of many leading scientists [1, 21]. Examples of policies that either ignored or ran contrary to scientific input are common in the physical sciences—for instance, the cancellation of the Superconducting Supercollider for budgetary reasons in the 1980s despite strong support from physicists.

Science advice is subject to harsh criticism from both the left and right wings of the political spectrum. Advocates for more regulation might argue that scientific evidence is distorted in order to avoid establishing regulations, while those opposed to regulation contend that science is distorted in order to promulgate intrusive and inappropriate regulation [1, 6, 22]. Critics label advisers as incompetent or biased, committees as unbalanced or unduly influenced by certain positions, and supporting science as flawed and incomplete. Because scientific information is rarely clear-cut, science policy recommendations remain vulnerable to criticism. In addition, critics may seize on reports of scientific misconduct as justification for discounting all work in a controversial area [15]. Finally, because many leading scientists are also recipients of federal funding, critics charge that their advice is tainted by the desire to obtain more research funding.

The level of concern over suppression of scientific information and manipulation of committees reached new heights during the presidency of George W. Bush [23]. For example, shortly after taking office in 2001, the Bush administration rescinded the new limits on arsenic levels in drinking water introduced late in the Clinton administration; arsenic is known to cause cancer. The mining industry strongly opposed and lobbied against the new regulations. Christine Todd Whitman, the new EPA director, argued that the scientific data supporting the lowered limits were uncertain [24, 25]. After a storm of protest from both environmental groups and members of Congress, the EPA asked the NRC, which had issued a comprehensive report in 1999, to review the scientific evidence on the effects of arsenic again. The NRC report, released in September 2001, found that even its previous recommended standards were probably too high [26]. In November 2001, the EPA agreed to adopt the standards proposed by the Clinton administration starting in 2006 [27]. The Bush administration's proindustry position on environmental and health issues continues to draw criticism from advocates of strong regulation.

Beginning in 2003, a growing chorus of critics maintained that the Bush administration sought to suppress science and stack membership on advisory committees by selecting only those representatives who express the administration's preferred viewpoints [5, 28, 29, 30, 31]. Critics argued that biasing scientific analysis inherently subverts the advisory committee process [21]. One example was the failure to reappoint two members of the President's Council on Bioethics who expressed strong support for human cloning and stem cell research during their first term on the committee, contrary to the more limited support expressed by the administration and the council's chair [32]. Marburger, the OSTP head, responded that such attacks were a significant distortion of the administration's actions and a reflection of partisan politics leading up to the national election of 2004 [33]. He also reminded the critics that science is but one input into the policy process.

A second criticism leveled at the Bush administration is that it subjects candidates for committees to questions regarding their political views and affiliations that are inappropriate given the FACA guidelines and other legislation [30, 34]. The administration even asked potential committee members if they had voted for the president. A GAO report in April 2004 recommended that additional guidelines be developed to assure that advisory committees are both independent and balanced [35]. A follow-up response by the GAO, requested by Congress, indicated that while existing law prohibits discrimination in federal hiring based on political affiliation, the applicability of such antidiscrimination regulations to federal advisory committees must be determined on a case-by-case basis [36]. Thus, although the scientific community and other critics may find such political litmus tests distasteful, they are not necessarily illegal. Nevertheless, creating committees whose scientists do not represent the range of expertise relevant to the difficult issues under discussion does not appear to achieve the goals of the advisory process. Some members of Congress, however, argue that many "scientific" issues have, at their heart, nonscientific controversies. Asking about political affiliations and positions is therefore appropriate in order to best represent differing points of view [37]. In February 2005, Representative Henry A. Waxman (D-CA) introduced the Restore Scientific Integrity to Federal Research and Policymaking Act (HR 839) to block political litmus tests and other interference for federal scientists. In October 2005, Senator Richard Durbin (D-IL) attached a similar amendment to the appropriations legislation for the DHHS, the Department of Education, and the Department of Labor. President Bush signed the appropriations bill into law on December 30, 2005 [38].

Given the heightened partisan rhetoric over science advising in recent years, is it possible to find a balance? A number of suggestions have been made: reestablish the OTA to improve the quality of scientific advice to Congress and reduce the dominance of advising in the executive branch, regularize science policy in the executive branch, and involve the public more in deliberations so that citizens feel more invested in the decisions [7, 15, 39]. To imagine that scientific advising will ever be free of politics is both naive and self-defeating. The challenge remains to find ways to insulate scientific advising from political ideology so that differing interpretations of scientific data are represented and considered when making new policy.

While most people would agree that advances in scientific knowledge, particularly in biomedical areas, have improved their lives, scientific discoveries may also give rise to contentious and sometimes alarming developments. Science is not seen as a universal good. Particularly in recent decades, many now view both scientists and science with suspicion and distrust. Nevertheless, both government and the public must find ways to make decisions on applications of new knowledge.

Case Studies in Science Policy

Eleven chapters of this book use case studies to explore mechanisms of scientific input into policy decisions and examine the issues raised here. Each chapter includes background information on the biology underlying the issue as well as an exploration of policy.

Chapter 2 explores policy for science using the Human Genome Project (HGP) to discuss the federal funding of research. It compares the "big science" of the HGP to more typical investigator-initiated research projects, and looks at the potential impact of big projects on the focus and direction of research in biomedical areas. Two short sections explore peer review and the alternative approach to funding using congressional earmarking, and congressional influence on the direction of science.

Chapter 3 examines aspects of information sharing, and the conflict between the public and private support of research, through the history and impact of gene patenting. The effects of patenting on access to information are discussed. The broader impact of patenting the genome is also explored. Two sections examine cases in which human tissues and DNA were exploited by researchers, raising questions about fairness and commercialization in biotechnology.

Chapter 4 explores issues of self-regulation by the scientific community using assisted reproductive technologies as a case study—asking, When should government step in to control the directions of research and clinical medicine? The development of regulation in the United Kingdom is compared with the absence of oversight in the United States. The two sections in this chapter discuss the recent push to ban human cloning and its potential impact on stem cell research, and the early history of recombinant DNA research as an exemplar of self-regulation.

Chapter 5 uses the development of new drugs to treat AIDS to introduce the role of federal agencies in regulating science. The conflict between public demand, the interests of industry, and safety concerns is explored. The two sections provide a perspective on how regulations protecting human and animal subjects were developed. The appropriateness of certain kinds of human experimentation is discussed.

Chapter 6 addresses the role of scientific input into court cases, and the contrast between scientific evidence and public perception. Silicon breast implants are used to illustrate how misperception about the risks led to huge settlements in the absence of any scientific evidence showing that the implants caused the medical problems. The sections here describe the current guidelines for scientific evidence in the courts, and also touch on continuing controversy concerning the use of DNA testing in forensics.

Chapter 7 explores the role of the media in influencing public opinion about science using coverage of new treatments in the "war" on cancer. Coverage can have impact on public perceptions, decisions by policymakers, and the stock value of companies conducting research. Media coverage can mislead the public and artificially raise hopes. The responsibility of journalists in informing the public is discussed in a section about the risks of electromagnetic fields (power lines and cell phones).

Chapter 8 looks at the complex relationship between free enterprise and scientific responsibility. The tobacco industry is used as a case study to explore why government may be reluctant to regulate, even in the face of clear evidence that a product is unhealthy. The concealment of evidence from the public is also discussed. Two sections address conflicts of interest and scientific misconduct.

Chapter 9 examines the emerging area of bioterrorism, provides a brief history of biological weapons, and discusses the 2001 attack involving anthrax-laden letters. The government public health response is scrutinized in the broad context of civil liberties. The two sections here use the recent SARS epidemic to assess public health responses, and explore moves to censor science and classify some forms of research.

Chapter 10 examines international policy issues involving science, and looks at the differing responses to genetically modified organisms in the United States and abroad, exploring how public opinion can impact policymaking internationally. A section examines the international impacts of mad cow disease.

Chapter 11 explores the complexities of environmental policymaking using air pollution as its case study. The challenges of competing interests are discussed and the difficulties of developing rational policy are outlined. One section examines lead poisoning and the challenges of generating effective policy even when the risks are known. A second section offers insights into risk assessment and how it is used in policymaking.

Chapter 12 examines situations in which scientists are asked to weigh in on issues that do not have a scientific basis. The shortage of organs for transplantation places pressures on physicians to develop rational approaches to the distribution of organs. The current situation for organ transplantation in the United States is described. Proposals on how to increase the rate of donation are discussed. Two sections address the possibility of using animal organs for transplant along with end-of-life issues.

Chapter 13 provides a synthesis of and conclusions about science policy drawn from the case studies. It presents continuing challenges and unresolved questions.

References

1. Barke, R. *Science, technology, and public policy.* Washington, DC: CQ Press, 1986.

2. Jasanoff, S. *Science at the bar: Law, science, and technology in America.* Cambridge, MA: Harvard University Press, 1995.

3. The policy sciences. Available at <http://www.policysciences.org> (accessed December 28, 2005).

4. Committee on Science Engineering and Public Policy. *Science and technology in the national interest: Ensuring the best presidential and federal advisory committee science and technology appointments.* Washington, DC: National Academy Press, 2004.

5. Steinbrook, R. Science, politics, and federal advisory committees. *New England Journal of Medicine* 350, no. 14 (April 1, 2004): 1454–1460.

6. Jasanoff, S. *The fifth branch: Science advisors as policymakers.* Cambridge, MA: Harvard University Press, 1990.

7. Kelly, H., I. Oelrich, S. Aftergood, and B. H. Tannenbaum. *Flying blind: The rise, fall, and possible resurrection of science policy advice in the United States.* Federation of American Scientists, occasional paper No. 2, December 2004.

8. Office of Science and Technology Policy, National Science and Technology Council. Available at <http://www.ostp.gov> (accessed January 13, 2005).

9. Light, P. C. Our tottering confirmation process—presidential appointment process. *Public Interest* 147 (Spring 2002). Available at <www.findarticles.com/p/articles/mi_m0377/is_2002_spring/ai_84557329> (accessed August 2, 2006).

10. Lee, C. Confirmations fail to reach Light's speed: Initiative fell short, its director says. *Washington Post*, June 20, 2003, A23.

11. Greenberg, D. S. *Science, money, and politics: Political triumph and ethical erosion.* Chicago: University of Chicago Press, 2001.

12. Morgan, M. G., and J. Peha, eds. *Science and technology advice to the Congress.* Washington, DC: RFF Press, 2003.

13. Morgan, M. G., A. Houghton, and J. H. Gibbons. Improving science and technology advice for Congress. *Science* 293 (September 14, 2001): 1999–1920.

14. Leary, W. E. Congress's science agency prepares to close its doors. *New York Times*, September 24, 1995, A26.

15. Keiper, A. Science and Congress. *New Atlantis* 7 (Fall 2004/Winter 2005): 19–50.

16. Horowitz, D. J. *The courts and social policy.* Washington, DC: Brookings, 1977.

17. National Academies. History of the national academies. 2004. Available at <http://www.nationalacademies.org/about/history.html> (accessed January 7, 2005).

18. Lawler, A. Is the NRC ready for reform? *Science* 276(1997): 900.

19. Lawler, A. New report triggers changes in the NRC. *Science* 289(2000): 1443.

20. Office of Technology Assessment. *Environmental policy tools: A user's guide.* Washington, DC: Office of Technology Assessment, September 1995.

21. Branscomb, L. M. Science, politics, and U.S. democracy. *Issues in Science and Technology* 21, no. 1 (Fall 2004): 53–59.

22. Gough, M., ed. *Politicizing science: The alchemy of policymaking.* Stanford, CA: Hoover Institution Press, 2003.

23. Mooney, C. *The Republican war on science.* New York: Basic Books, 2005.

24. Jehl, D. E.P.A. to abandon new arsenic limits for water supply. *New York Times*, March 21, 2001, A4.

25. Kaiser, J. Science only one part of arsenic standards. *Science* 291(2001): 2533.

26. National Research Council. *Arsenic in drinking water: 2001 update.* Washington, DC: National Academy Press, 2001.

27. Seelye, K. Q. E.P.A. to adopt Clinton arsenic standard. *New York Times*, November 1, 2001, A18.

28. Malakoff, D. Democrats accuse Bush of letting politics distort science. *Science* 301 (August 15, 2003): 901.

29. The Union of Concerned Scientists. *Scientific integrity in policy making.* Washington, DC: Union of Concerned Scientists, February 18, 2004.

30. The Union of Concerned Scientists. *Scientific integrity in policy making: Further investigation of the Bush administration's misuses of science.* Washington, DC: Union of Concerned Scientists, July 2004.

31. Hadley, C. Science policy in the USA. *EMBO Reports 5*, no. 10 (2004): 932–935.

32. Holden, C. Researchers blast U.S. bioethics panel shuffle. *Science* 303 (March 5, 2004): 1447.

33. Marburger, J. H. I. *Statement of the honorable John H. Marburger, III on scientific integrity in the Bush administration.* Washington, DC: Office for Science and Technology Policy, April 2, 2004.

34. Lawler, A., and J. Kaiser. Report accuses Bush administration, again, of "politicizing" science. *Science* 305 (July 16, 2004): 323–325.

35. U.S. General Accounting Office. *Federal advisory committees: Additional guidance could help agencies better assure independence and balance.* 04–328. Washington, DC: General Accounting Office, April 2004.

36. U.S. General Accounting Office. *Legal principles applicable to selection of federal advisory committee members.* B–303767. Washington, DC: General Accounting Office, October 18, 2004.

37. Mervis, J. Congressmen clash on politics and scientific advisory committees. *Science* 305 (July 30, 2004): 593.

38. Union of Concerned Scientists. Legislative information: Update on scientific integrity legislation and amendments. January 3, 2006. Available at <http://www.ucsusa.org/scientific_integrity/restoring/scientific-integrity-legislative-information-page.html> (accessed January 16, 2006).

39. Guston, D. H. Forget politicizing science. Let's democratize science! *Issues in Science and Technology* 21, no. 1 (Fall 2004): 25–28.

Further Reading

Kelly, H., I. Oelrich, S. Aftergood, and B. H. Tannenbaum. *Flying blind: The rise, fall, and possible resurrection of science policy advice in the United States.* Federation of American Scientists, occasional paper no. 2. December 2004. Available at <http://www.fas.org>.

• A lucid overview of science advising in the U.S. government as well as its weaknesses and proposals for improvement

Morgan, M. G., and J. Peha, eds. Science and technology advice to the Congress. Washington, DC: RFF Press, 2003.

• A thorough overview of the congressional use of scientific and technological information in formulating policy

2

Big Science: The Human Genome Project and the Public Funding of Science

Government support of scientific research has strong influences on the direction of that research; this policy for science significantly affects what research is done and what knowledge might be gained. This chapter explores the relationship between the government funding of basic research in biology and the researchers themselves. Who should decide what research will be supported? What criteria should be used to determine what research will be funded? Can scientists influence the decision-making process? The Human Genome Project (HGP) set out to map the entire genetic complement of human beings—a total of three billion base pairs of DNA. Accomplishing this task meant a major commitment of funds by the federal government, beyond what was already being granted to support research. How did scientists persuade the government that the project was worthwhile?

Federal Funding of Research and the HGP

The history of the HGP stands in contrast to the typical investigator-initiated funding of research in the biological sciences (see "Funding Biomedical Research" section below). We are familiar with big science projects in the physical sciences, such the Manhattan Project to develop the atomic bomb, the international space station, and a series of multibillion dollar physics projects on nuclear fusion and particle accelerators (some of which were canceled after expenditures of billions of dollars due to overruns and changing budget priorities). The HGP is the first genuine example of big science in the biological sciences, with an estimated original price tag of $3 billion. It should be noted, however, that $3 billion spread over ten to twenty years represents only a tiny fraction of the federal research and development budget for the biological sciences. The development of the HGP demonstrates both the power that politically astute scientists have in directing government decisions about funding and the effect that major funding in an area of biology can have on the direction of new research.

The federal government began to fund scientific research at significant levels in the latter half of the twentieth century, after Vannevar Bush, head of the Office of Scientific Research and Development in the Roosevelt administration, developed a plan to continue support of scientific research after the end of World War II [1]. Bush's recommendations laid the foundation for the establishment of the NSF, the NIH, and other federal agencies supporting scientific research. Central to Bush's recommendations were that agencies granting funds would be autonomous entities, run by scientists, not career administrators, and that the direction of research would be determined by the scientists themselves.

The budgets of federal granting agencies have grown enormously in fifty years. In fiscal year 2005, the projected support for research in the life sciences was $25.5 billion at the DHHS, $578 million at the NSF, $289 million at the Department of Energy (DOE), $1.4 billion at the U.S. Department of Agriculture (USDA), and $695 million at the Department of Defense (DOD), most of which was targeted for bioterrorism research (see chapter 9) [2]. This amount represents only a tiny fraction of the country's total budget of several trillion dollars. The availability of funds for research in areas determined by federal agencies influences the directions and limits of research in the biological sciences [3]. Funding through the NIH has been particularly generous in the biomedical sciences, with the stated justification that such research will ultimately benefit the health and welfare of Americans. For researchers in non-health-related areas of biology, funding has been more limited, and is awarded primarily through the NSF and the USDA.

Until the 1980s, the federal government funded the majority of research in the biological sciences. The proportion of research dollars coming from industry has increased dramatically, however. Recent estimates are that over 60 percent of biomedical research is supported by industry [4]. The implications of such support will be discussed later.

Beginnings of the HGP

The seeds of the HGP were sown in the mid-1980s by scientists working independently [5, 6]. The first was Robert Sinsheimer, then chancellor of the University of California at Santa Cruz (UCSC). Sinsheimer proposed that UCSC might develop an Institute on the Human Genome and thereby bring the biology program at UCSC into greater prominence. He convened a meeting at UCSC in 1985 of leading molecular biologists. The group agreed that a project to develop a large-scale genetic linkage map, a physical map, and the capacity for large-scale DNA sequencing was both appropriate and feasible. Wholesale sequencing of the entire human genome was deemed not technically possible in the view of the gathered scientists; they proposed that sequencing targeted regions would be of interest. Sinsheimer explored sources of funding for such a project (including going directly to Congress [see "Funding Biomedical Research" section below]) but was not successful. One of the attendees at the Santa Cruz meeting, Walter Gilbert, a 1980 Nobel Prize winner for his work in molecular biology, became the HGP's strongest proponent in the years that followed.

The second champion of an HGP was Charles Delisi, who became director of the Office of Health and Environmental Research (OHER) at the DOE in spring 1985. OHER had funded projects to investigate the effects of radiation on Japanese survivors of the bombing of Hiroshima and Nagasaki. Delisi reasoned that the search for DNA mutations might be extended into a project to map and sequence the human genome. He proposed that the national laboratories that had grown as a result of the Manhattan Project and other weapons programs could be redirected to study the human genome. Delisi convened a meeting in Santa Fe in early 1986. The consensus again was that genetic linkage and physical mapping was feasible.

Unlike Sinsheimer, Delisi was in a good position to influence political decisions concerning funding. He had easy access to government officials with control of funding and also managed considerable funds himself within OHER. In spring 1986, he submitted a proposal to the director of the DOE for initial funding for the project, stressing that the DOE was well situated to provide leadership for a major, multiyear endeavor. His proposal, for $78 million from 1987 to 1991, was passed on to the White House Office of Management and Budget, where it also gained support. With this, Delisi redirected $5.5 million of his 1987 budget toward human genome research.

The third scientist who had an early influence on the HGP was Nobel laureate Renato Dulbecco, then president of the Salk Institute for Biological Research in California. He published a commentary in *Science* magazine in 1986, suggesting that cancer research would be aided by detailed knowledge of the human genome [7]. This article raised awareness among a broader scientific audience of the possibility of sequencing the human genome. In summer 1986, a conference was held at Cold Spring Harbor, New York, attended by over three hundred molecular biologists. At an open session, Gilbert reported on past meetings and suggested that the genome might be sequenced for $3 billion. The idea of mapping the genome was viewed favorably in principle. Strong objections were voiced concerning the potential intellectual value of a complete sequence, however, given that only a small fraction of the genome coded for actual genes. The project represented the worst characteristics of "discovery science," being repetitive, tedious, and without underlying hypotheses. Many feared that the high cost of a project would undermine investigator-initiated research by redirecting limited resources to the sequencing project. Others expressed concerns about the appropriateness of having the project managed by the DOE, since expertise in molecular biology appeared to lie elsewhere.

In 1987, Gilbert announced plans to form a company to carry out the genome project. He had already helped found the Swiss company, Biogen, in 1978. Biogen was one of the first companies established to pursue commercial goals in biotechnology. Gilbert's proposed new company, Genome Corporation, would carry out mapping and sequencing activities, and market sequence information, clones, and services. Information would be gathered more efficiently and economically than by individual labs working independently. His proposal to commercialize the genome appalled many molecular biologists. Still, because of the stock market downturn in October 1987, Gilbert was unable to raise the money needed to establish the company. He continued to champion a government-funded project instead.

To help resolve some of the issues, the NRC was asked by leading molecular biologists to conduct a study assessing the feasibility and value of an HGP. The NRC obtained funding from the James S. McDonnell Foundation to conduct its study. The report, released in February 1988, recommended a fifteen-year project, funded at $200 million per year, to develop linkage and physical maps of the human genome, develop faster methods for sequencing, and ultimately sequence the entire human genome once the technology was available to allow it [8]. In order to be able to place genetic information in context, the mapping and sequencing of the genomes of other species was also necessary. The report argued that the potential value of this knowledge merited a major commitment by the federal government. Funding for the project should come from new sources, so as not to negatively impact other investigator-initiated research. Funds should be awarded to individuals, collaborative groups, and academic centers using peer-review criteria. The report recommended against a small number of centralized sequencing facilities, in contrast to the plan envisioned by the DOE, in order to broadly draw on existing expertise and develop a stronger scientific workforce. The report also stressed the need to develop means to store and disseminate the large amount of data that would be generated by the project [8]. Finally, the report recommended that oversight be provided by a single agency with a scientific advisory panel.

With the endorsement of the NRC, Congress began to explore ways to support the HGP. Leading scientists gave testimony to congressional committees, offering broad visions of future applications and improvements in human health. Two competing agencies sought to gain oversight of the project, the DOE and the NIH. Eventually, an agreement was reached to give the NIH lead responsibility, but to allow substantial funding of research through the DOE. The NIH Office of Human Genome Research was created in October 1988, with Nobel laureate James Watson as director. In 1989, it became the National Center for Human Genome Research (NCHGR) with a budget of $59.3 million. The HGP formally began in October 1990, although elements of the project had begun earlier.

The Biology behind the HGP

The HGP grew out of expanding knowledge about the nature of DNA, and the development of tools to manipulate it [9]. Two major areas of biological research—genetics and molecular biology—provided the basic information needed to make the HGP a reality.

The discussion below describes the strategies used to apply basic information to the goals of the HGP. These goals include developing linkage and physical maps of all chromosomes, locating genes within the genome, developing better technology for genetic analysis, and ultimately determining the sequence of all three billion base pairs of DNA.

Linkage Mapping, Physical Mapping, and Gene Discovery

Recognizing an inheritance pattern is the first step in determining the association of a gene with a disease or physical trait. The gene must be located within the genome (the total complement of DNA in the organism) and its function determined before there is much possibility of developing gene-based therapies for the disease. Given the immense size of the human genome, how might this be accomplished? Geneticists, beginning with Gregor Johann Mendel (1822–1884), recognized that certain traits tended to be inherited together long before scientists in the mid-twentieth century determined that genetic information was carried in the form of DNA. This pattern of coinheritance of traits is termed linkage. For example, white domestic cats with blue eyes are often deaf. Linkage suggests that the genes for the two traits are located fairly close to each other on a piece of DNA. Geneticists were eager to discover ways to determine whether a given individual (or fetus) might be carrying an allele (a version of a gene that recognizably affects its function) that causes a potentially devastating genetic disease. Given that variations in DNA sequence occur, was it possible to use these as a means to predict disease? If variations in DNA sequence could be found that were linked to a given disease, even if the variations themselves were not part of the gene itself, they might be used as a diagnostic tool.

Human gene mapping did not begin until the 1960s, when mouse-human cell fusion, or somatic cell hybridization, was used to associate certain gene products with identified chromosomes. The development of fluorescent dyes that labeled banding patterns in human chromosomes allowed further genetic mapping. The banding patterns in human chromosomes were so distinctive that deletions, translocations, inversions, or other changes could be recognized easily. In 1980, the development of in situ hybridization allowed for more detailed localization. A piece of DNA may be synthesized with a radioactive label, and this serves as a "probe" for the gene in the chromosome. The DNA within the chromosome is treated to separate the strands and the probe is allowed to bind to its complementary sequence. The radioactivity is detected using X-ray film. In a successful experiment, a spot of radioactivity is found on a particular site on a particular chromosome, identifying that region as the gene location.

The resolution of mapping using in situ hybridization may narrow the location of a gene to a region of several million base pairs. To increase the resolution of a map, other techniques are used. This approach, to search for a gene within a fairly large region of DNA, is called "positional cloning" [10]. This approach takes advantage of variations in DNA sequences to associate the presence of a given allele of the gene to an identified marker. A marker is simply a sequence of DNA whose location is known. Its function (if any) is not known; its usefulness lies in its close linkage to the unknown gene of interest. A region of DNA is treated with a restriction enzyme that cuts the DNA into several fragments at a specific sequence of bases. If there are allelic differences in the gene sequence, a site for the restriction enzyme may be gained or lost, leading to production of different-size fragments, known as restriction fragment length polymorphisms (RFLP). The next step is to determine if a particular pattern of fragments is reliably associated with the disease. If so, the marker might be used to diagnosis the presence of a defective allele. The marker may also be used to narrow the region of the chromosome that contains the gene of interest. RFLP mapping helped to locate the genes for Huntington's disease and Duchenne muscular dystrophy in 1983. As the number of markers increased, so did the pace of new gene discovery.

RFLP technology was expanded in the HGP to develop an array of markers called sequence tagged sites (STSs). These are known sequences of DNA regularly spaced on the chromosomes. They serve two purposes: to facilitate the localization of genes by their proximity (linkage) to given STSs, and to help align pieces of DNA in a physical map.

A second approach to gene localization was to identify DNA sequences from genes, or expressed sequence tags (ESTs). These sequences could be made from messenger RNA (mRNA) using reverse transcriptase, generating complementary DNA sequences (cDNAs). Fragments of these cDNAs could be used to hybridize to genomic DNA, thereby marking a region as containing a gene. It is important to note that ESTs do not determine the function of the gene but only its location. The development of ESTs caused controversy in 1991, when Craig Venter revealed that the NIH was filing patent applications on thousands of ESTs, although nothing was known about the genes of which they were fragments (see below).

Since human somatic cells contain two copies of each chromosome, there is considerable interest in recognizing allelic variations of genes and their potential link to disease or other traits. Differences in gene sequences are frequently the result of changing a single base pair. These single nucleotide polymorphisms (SNPs) tend to be inherited in blocks on a single chromosome. These blocks of SNPs are termed a haplotype and can be recognized using extensions of RFLP analysis. Over twenty-seven million human SNPs had been identified as of November 2005 [11]. Research is necessary to determine which of these chromosomal variants are relevant as markers for disease. An alternative approach is to use highly variable sequences called microsatellites as markers in linkage studies. Whether either of these approaches may be used to identify disease genes requires the detailed study of defined human populations.

One of the HGP's goals was the construction of a detailed linkage map, consisting of markers separated by ever-decreasing distances, as described above. Another goal was the development of a physical map. A physical map consists of fragments of DNA that are aligned in their linear sequences. This is made possible by the cloning of many fragments of DNA produced by treatment with different restriction enzymes. The fragments are inserted into plasmids, which are circular strands of DNA from bacteria. Plasmids may be rapidly and cheaply reproduced, or cloned, providing many copies of the DNA fragment. These plasmids are collected into a library of DNA fragments covering the extent of human DNA. Fragments may then be aligned using STSs or by sequencing the ends of these fragments to determine areas of overlap. The order of the overlapping fragments (or contigs) along a chromosome is determined as a giant jigsaw puzzle, using sequence overlap and the markers developed in linkage mapping. Each new gene may then be assigned first to a given chromosome and then to a smaller region within it. The two techniques combined narrow down the region in which a given gene might be positioned, reducing the time spent combing the genome for its location.

The final stage of the HGP is to sequence all the aligned fragments of DNA. This yields the sequence of the human genome. The completion of this part of the project is hampered by long stretches of repeated DNA in noncoding regions. Researchers are challenged to identify how many repeats are present. Other stretches of DNA prove difficult to clone for a variety of technical reasons. Therefore, it is not surprising that the "complete" sequence still contains many gaps. It now appears that less than 5 percent of the total sequence codes for genes.

Bioinformatics

The planners of the human genome project recognized that handling the vast amount of data generated would present a major challenge. Data come in several forms: markers and map information (both linkage and physical), DNA sequences, DNA fragments in a variety of vectors (DNA libraries), and identified genes. How might all this information be managed? The NIH established a gene sequence repository in 1982 called GenBank that would allow retrieval of gene sequences using newly developed computer programs. Investigators were expected to submit gene sequences at the time of publication of their research; each new gene was given a unique identification number. By the mid-1980s, however, it was clear that the pace of discovery of new sequences was overwhelming GenBank's ability to manage them and a more extensive effort was required. The late Senator Claude Pepper recognized the importance of computerized information-processing methods for biomedical research and sponsored legislation that established the National Center for Biotechnology Information (NCBI) in November 1988 as a division of the National Library of Medicine at the NIH. The NCBI now maintains GenBank (which now contains over forty million sequences [11]), other databases such as RefSeq (a collection of sequences from several species) plus numerous other resources for molecular biologists. The NIH is also constructing a library of clones of all human genes, called the Mammalian Gene Collection. Databases are also maintained in Europe and Japan. Researchers and members of the public may access databases at no charge via the Web.

Having repositories of DNA sequences is not useful unless there are means to extract information from them. This process, called gene "mining," required the development of computer algorithms that permit comparison of sequences and recognition of similarities [12]. David Lipman, Eugene Myers, and colleagues at the NCBI developed the first truly successful algorithm, called BLAST, in 1990 [13]. BLAST allows researchers to compare newly discovered sequences with those already in the databases. Sequence alignment and similarity comparisons allow researchers to place new genes among functional families, and to recognize homologies between sequences from different species. BLAST analysis proved enormously helpful in gene identification in a broad range of applications beyond the genome project itself.

As map and sequence information is generated, algorithms are needed to order fragments in physical maps. Two programs, named phrap and phred, developed by Phil Green and Brent Ewing at the University of Washington and Washington University at St. Louis, have been heavily used for these purposes. Phred, published in 1998, is particularly useful in automatically interpreting sequence data [14]. This proved useful for Venter's "shotgun" approach to sequencing the human genome (see below). Additional programs allow for alignment of the many cloned DNA fragments within chromosomes.

A particularly difficult challenge is identifying genes in the finished sequence of human DNA [12, 15]. Surprisingly, researchers cannot agree on how many genes are contained within the genome. Original estimates prior to the HGP were in the range of a hundred thousand genes; current estimates range from twenty-five to forty-five thousand genes, with most researchers predicting numbers at the low end. This is only about twice that found in *C. elegans* (roundworm) and *Drosophila* (fruit fly), two model organisms whose genomes have been sequenced. BLAST analysis helped researchers discover many families of related genes by identifying sequence homologies.

Many genes are not members of gene families. How might they be identified? ESTs are powerful tools in that they are fragments of expressed genes. Genes may be missed because of their small size, however, or because the genes do not code for protein but rather for RNA. Comparing sequences with another species is a particularly powerful approach, since most of our genes are shared with other organisms.

Another approach is to search for common regulatory sequences of genes (promoters) that might signal that a gene is nearby. The search for genes is complicated by the presence of pseudogenes, sequences that share similarities with actual genes, but represent nonexpressed evolutionary dead ends. Gene prediction programs such as Ensembl, Genie, and GenomeScan all have limitations, being either prone to over- or underestimate the number of actual genes in model systems.

The HGP and Global Activities

An important feature of the HGP is that it included a mandate that 3 percent of the budget should be used to study the ethical, legal, and societal aspects of the research. James Watson was a strong proponent of such research. One of his first acts as director of the Office of Human Genome Research was to establish a working group on the ethical, legal, and social implications (ELSI) of the HGP, and name Nancy Wexler, a leading researcher of human genetic diseases (and who was at risk for developing Huntington's disease), as its chair. Congressional concern about privacy issues also strengthened the role of the ELSI program as it developed. Congress mandated specific ELSI funds in appropriation legislation for NCHGR beginning in 1991. By 1993, the ELSI budget portion had risen to 5 percent. The ELSI program provides grant funds to explore a variety of bioethical and policy issues associated with genetic information, among them privacy, discrimination, and prenatal genetic testing. The establishment of ELSI was unprecedented in the history of big science, and was in stark contrast to the absence of consideration of the potential social effects of the atomic bomb until after it was used in World War II.

Discussions about HGPs were also taking place in many countries around the world. Several European countries, notably Great Britain and France, had long histories of genetics research. In Asia, Japan had developed DNA sequencing capabilities in the early 1980s, and had a modest but growing research program already in place by the mid-1980s. As individual countries explored ways to increase their genome efforts, discussions were held at a conference in Cold Spring Harbor, New York, in spring 1988 to develop a coordinated international program. The Human Genome Organization (HUGO) was intended to foster international cooperation in genome research [16]. Funding was provided initially by private sources, including the Howard Hughes Medical Institute in the United States and the Wellcome Trust in the United Kingdom. HUGO remains active as a largely privately funded entity, with a number of international advisory committees that focus on ethical issues and the dissemination of information. Although competition has not been eliminated, some degree of coordination in research efforts has been maintained. HUGO's direct influence on the direction of the HGP is limited, however, and it remains somewhat on the fringes.

The ramping up of the HGP took several years, as funds were awarded to establish centers for research and technology development to achieve the project's goals. Watson argued strongly before Congress that funding should be maintained and set a series of goals to be achieved in shorter time periods. He was sensitive to the fact the most legislators are not able to wait fifteen years for outcomes. For example, he set a goal of developing a genetic map consisting of six hundred to fifteen hundred markers in five years [17, 18]. As it turned out, the genome was mapped with three thousand markers by the target date of 1994. Five centers were established that conducted much of the sequencing: three NIH-funded centers at the Whitehead Institute at MIT, Baylor College of Medicine, and Washington University at Saint Louis; the DOE-funded Joint Genome Institute in California; and the Sanger Institute in Great Britain. The U.S. centers were awarded funding using peer-review criteria, and continued funding was based on success at achieving goals. In Great Britain, the Sanger Institute received the bulk of its funding from the Wellcome Trust, not the British government [19]. Ultimately, twenty centers in six countries participated in sequencing. Guidelines were established for standards and quality control. Most important, principles for information sharing were laid out at a conference in Bermuda in 1996 that required the rapid release of genome information into the public domain. President Clinton and Britain's prime minister Tony Blair endorsed these "Bermuda principles" in 2000 [17, 19]. Controversy over information sharing proved to be one of the greatest challenges in the HGP.

Emerging Controversies

The project was shaken in July 1991, when Venter, then at the NIH, testified before Congress and announced that the NIH had filed for patents on thousands of ESTs, the short sequences of cDNA associated with genes of unknown function [6, 17, 18]. Bernadine Healy, then the NIH director, argued that patenting partial genes would benefit society by stimulating further work to develop diagnostic tools and other applications (see also chapter 3). Watson strongly disagreed, and resigned as the NCHGR director in April 1992. The NIH eventually backed off the notion of patenting ESTs, after Harold Varmus became the NIH director in 1993. Venter left the NIH in 1991 to form the nonprofit Institute for Genome Research, and continued mapping the genome using ESTs. Francis Collins, known for his work in identifying the genes for Huntington's disease and cystic fibrosis, was named director of the NCHGR in 1993. Collins proved an effective advocate of the HGP before Congress by stressing the potential medical benefits of genomic information.

By 1992, physical maps of chromosome 21 and the Y chromosome, among the smallest human chromosomes, were developed [18]. Rough genetic maps of the entire genome were developed, and these continued to be refined over the ensuing years. While mapping efforts continued smoothly, sequencing lagged behind until the development of high-speed DNA sequencers in the mid-1990s. In 1995, Venter and his colleagues published the first complete sequence of a bacterium, *Haemophilus influenzae*, whose genome contained 1.8 million base pairs. This was achieved using a new approach called shotgun sequencing. Rather than working from detailed physical and genetic maps, Venter and his colleagues simply fragmented the entire genome, cloned each fragment, and then sequenced it. The assembly of the sequences was done using new computer programs to align the fragments.

HGP researchers were skeptical that the approach would be effective for larger genomes containing high proportions of repetitive DNA. Yet in 1998, Venter announced that he intended to use the shotgun approach to sequence the human genome, bypassing the publicly funded project, and formed a company, Celera Genomics, to do so. Celera was funded in large part by the Perkin Elmer Corporation, a manufacturer of state-of-the-art DNA sequencing machines. Venter intended to use large numbers of high-speed sequencers to achieve the goal of a complete sequence by 2001. He announced that the sequence information would be made available, but that the company intended to put limits on how the information might be accessed.

The race was on. Scientists in the HGP resented Venter's grab for glory, arguing that he had free access to map information already developed by the HGP, but that he had no intention of sharing his own information [17, 19]. They feared that the sequence of the human genome would be privatized, in direct opposition to the guidelines laid out in the Bermuda principles. They also worried that if the privately funded sequencing effort were successful, Congress would reduce or cut off funding for the HGP. Finally, researchers contended that the shotgun approach would produce a sequence with many gaps, and would be incomplete.

Collins responded by setting new goals for the HGP, including a rough draft of the genome by 2001 [17]. As milestones were reached—including the sequence of the roundworm, *C. elegans*, in 1998 by the HGP, and a draft sequence of the fruit fly, *Drosophila melanogaster*, in 2000 by Celera—the controversy between the two sequencing projects persisted. Despite a joint announcement of rough drafts of the genome in 2000 and the simultaneous publication of "complete" working drafts in 2001 [20, 21], issues concerning the availability of genome information continued. Celera's refusal to release its data at the time of publication in February 2001 created additional conflict. Both teams announced completion of the sequencing effort in 2003, with more than 98 percent of the gene-containing regions of the genome sequenced. Data from the publicly funded project were freely available, while Celera's remained subject to licensing controls depending on the user. In May 2005, Celera announced it was no longer limiting access to sequence information. As more and more genetic sequences were published, Celera's opportunity to make money from its databases diminished [22].

Assessing the Success of the HGP

Did the HGP achieve its goals? The project is unprecedented in the history of big science in that it achieved more than its stated goals, finishing two years ahead of schedule and underbudget [23]. In addition to the mapping and sequencing of the human genome, researchers sequenced the genomes of a number of other species, including the bacterium *E. coli*, yeast (*S. cerevisiae*), the roundworm *C. elegans*, the plants *Arabidopsis thaliana* and rice, the parasite that causes malaria and its mosquito host, and the mouse. The project spurred technological developments, including high-speed DNA sequencers, DNA microarrays, and data-analysis software. The project also led directly and indirectly to the formation of thousands of biotech companies involved in genetics research and development. This technology transfer provided a core of support that was leveraged by industry into a well-endowed basic research enterprise. Nevertheless, as will be seen in later chapters, applications of genome information carry with them considerable bioethical challenges.

Continuing Scientific Challenges

The complete sequencing of the genome is only the first step in understanding the information contained within it [12, 24]. Even identifying all the genes in the genome does not explain how these genes work together to produce the human organism or what roles genetic errors play in causing many diseases. The complexities of gene interactions represent a major challenge for the future. Researchers are beginning a systematic analysis of the functional roles of gene products, and how genes and their products influence the expression of other genes. Proteomics—the identification of the function of all proteins encoded by the genome—and functional proteomics—how these gene products interact—will require scientific efforts that may prove to be greater than for the genome project itself.

Researchers also wish to learn what subsets of genes are responsible for distinguishing humans from other animals. We share about 98.8 percent of our gene sequences with our closest relative, the pygmy chimpanzee. Understanding both the similarities and differences between humans and other organisms may help to define both the uniqueness of humanity and its commonality with other species. The differences between humans and other animals may turn out to be quantitative rather than qualitative; differences in gene regulation may determine some of our human-specific traits.

An immediate focus is to identify genes associated with diseases, either by causing the disease itself or increasing the risk for developing the disease. Another application is the use of SNPs to predict whether a given individual will respond well to a drug treatment or not, since the "genetic background" appears to influence whether a treatment is effective. The NIH and its international partners conducted a large-scale "HapMap" project to identify haplotypes associated with disease or drug responses; the first map was published in October 2005 [25, 26]. Yet the increased use of genetic screening raises difficult questions about access to genetic information and the possibility of discrimination on the basis of one's genome. For example, despite considerable protest, a company named DeCODE Genetics was authorized in 1998 to conduct a genetic analysis of Icelanders and compare gene patterns with medical records [27].

Haplotype analysis may also challenge common views about racial differences between human populations. Initial studies suggest that virtually all human genomes are about 99.9 percent identical [28], and there is little indication that different races differ in particular alleles. "Race" appears to be more of a cultural construct than a genetic one [29]. A proposal to study anthropological differences between populations of humans, the Human Genome Diversity Project, was rejected because of concerns about the potential for exploitation of indigenous populations and the intrusion into cultural beliefs about ethnic origins.

Lessons from the HGP

What can be learned from the HGP? The project would not have happened without the efforts of scientists to promote it and argue for continued funding before Congress. With less politically astute advocates, it appears likely that the research may have proceeded in a highly fragmented manner. In other words, research into the human genome would have happened, but with much less focus and coordinated activity. Sir John Sulston and Georgina Ferry [19] suggest that without the HGP, the human genome would be "owned" by private companies, instead of being available to all. Indeed, the race that occurred between Celera Genomics and the HGP showed this to be the case, resulting in the rather unique situation of government-funded researchers competing with commercial firms to ensure public access to data.

The HGP demonstrates that big science projects in the biological sciences can be successful and that coordination of the efforts of thousands of researchers at different sites, even internationally, is possible. The project also shows that "discovery-based" science can yield information that may prove invaluable in supporting the hypothesis-driven research projects that develop from it. Moreover, the project stimulated the development of new technologies that will be useful in many molecular biological applications. Some of these developments made it possible to avoid the drudgery that critics feared would be the hallmark of sequencing efforts. New technologies afford opportunities in business to apply the knowledge in areas of new drug development, disease diagnosis, and information technology as well. The technology also has an impact on DNA forensics, the use of DNA evidence in the courts.

The HGP has undoubtedly influenced the direction of future research in biomedical science by making available tools to answer questions about genetic diseases and risk factors as well as the inheritance of other characteristics. Genetic information is also changing the nature of drug discovery, where drugs can now be designed or chosen to target disease-causing molecular problems. The information also allows a different level of analysis as to how genes interact with each other. In addition, the data stimulate research into the evolutionary roots of humans and the relationship of humans with other species. Genome information on other species will have an enormous impact on agriculture, issues of biodiversity, and other environmental challenges.

It is important to realize that while the HGP was in progress, many advances in biology were made independent of the project. The HGP did not prosper at the expense of other areas of biology, especially other areas of biomedical research. For example, the 1990s were designated the "Decade of the Brain" by President George H. W. Bush in response to a joint resolution of Congress [30], and enhanced support for research in neuroscience produced a wealth of new information about brain development and function. Whether congressional decisions subtly altered funding for other biology endeavors is difficult to assess. Whether other big science projects will be funded in the future is unclear—few other questions will challenge the public imagination so effectively. The immediate future of the HGP is also uncertain. The future of the centers as leaders of additional efforts remains to be determined. In fall 2003, Elias Zerhouni, the NIH director, announced a new set of directives, the "NIH Roadmap" [31]. The directives include initiatives for more interdisciplinary work, an increased focus on emerging technologies, and more directed efforts to "translate" basic research into clinical applications. Traditionally, pharmaceutical and biotechnology companies have conducted most "translational research," so this shift in focus by the NIH is a major change [32]. The HGP's success provides an impetus to keep some of the work building on the findings of the HGP in the public sector. In December 2005, the NIH announced the Cancer Genome Atlas, an effort to identify changes in the genome associated with cancer. Samples of different types of cancerous tissue will be analyzed for genetic mutations associated with each specific form of the disease [33].

The HGP starkly reveals the inherent conflict between public and private ventures. At the same time, it serves as a model for the public support and promotion of basic research that yielded the knowledge and intellectual capital to spur private investment and commercial activity. The importance of access to information for scientific progress was also brought into sharp relief during the competition to complete the genome sequence.

The increasing emphasis on genetics resulting from the HGP may strengthen perceptions of "genetic determinism," the view that we are a product of our genes alone. Attitudes about the relationship between genetics and the definition of an individual have far-reaching impacts on many areas of society; these will be discussed in upcoming chapters.

References

1. Barke, R. *Science, technology, and public policy.* Washington, DC: CQ Press, 1986.

2. National Science Foundation, Division of Science Resources Statistics. Federal obligations for research, by field of science and engineering and agency: FY 2005 projected. November

2005. Available at <http://www.nsf.gov/statistics/infbrief/nsf06300> (accessed December 18, 2005).

3. Moore, D. T. Establishing federal priorities in science and technology. In *AAAS science and technology policy yearbook 2002*, ed. A. H. Teich, S. D. Nelson, and S. J. Lita, 273–284. Washington, DC: American Association for the Advancement of Science, 2002.

4. Bekelman, J. E., Y. Li, and C. P. Gross. Scope and impact of financial conflicts of interest in biomedical research. *JAMA* 289, no. 4 (January 22–29, 2003): 454–465.

5. Cook-Deegan, R. *The gene wars: Science, politics, and the human genome.* New York: W. W. Norton and Co., 1994.

6. Kevles, D. J. Out of eugenics: The historical politics of the human genome. In *The code of codes: Scientific and social issues in the human genome project*, ed. D. J. Kevles and L. Hood, 3–36. Cambridge, MA: Harvard University Press, 1992.

7. Dulbecco, R. A turning point in cancer research: Sequencing the human genome. *Science* 231 (1986): 1055–1056.

8. National Research Council, Committee on Mapping and Sequencing the Human Genome. *Mapping and sequencing the human genome.* Washington, DC: National Academy Press, 1988.

9. Judson, H. F. A history of the science and technology behind gene mapping and sequencing. In *The code of codes: Scientific and social issues in the human genome project*, ed. D. J. Kevles and L. Hood, 37–80. Cambridge, MA: Harvard University Press, 1992.

10. Weaver, R. F. *Molecular biology.* 2nd ed. Boston: McGraw-Hill Publishers, 2002.

11. National Center for Biotechnology Information. December 1, 2005. Available at <http://www.ncbi.nlm.nih.gov/> (accessed December 18, 2005).

12. Birney, E., A. Bateman, M. E. Clamp, and T. J. Hubbard. Mining the draft human genome. *Nature* 409, no. 6822 (2001): 827–828.

13. Altschul, S. F., W. Gish, W. Miller, E. Meyers, and D. Lipman. Basic local alignment search tool. *Journal of Molecular Biology* 215, no. 3 (1990): 403–410.

14. Ewing, B., L. Hillier, M. C. Wendl, and P. Green. Base-calling of automated sequence traces using phred: I. Accuracy assessment. *Genome Research* 8 (1998): 175–185.

15. Snyder, M., and M. Gerstein. Defining genes in the genomics era. *Science* 300, no. 5617 (April 11, 2003): 258–260.

16. Human Genome Organization. *HUGO.* November 26, 2002. Available at <http://www.gene.ucl.ac.uk/hugo/> (accessed November 3, 2003).

17. Roberts, L. Controversial from the start. *Science* 291, no. 5507 (February 16, 2001): 1182–1188.

18. Roberts, L. A history of the human genome project. *Science* 291, no. 5507 (February 16, 2001): 1195–1200.

19. Sulston, J., and G. Ferry. *The common thread: A story of science, politics, ethics, and the human genome.* Washington, DC: Joseph Henry Press, 2002.

20. International Human Genome Sequencing Consortium. Initial sequencing and analysis of the human genome. *Nature* 409, no. 6822 (February 15, 2001): 860–921.

21. Venter, J. C., M. D. Adams, E. W. Myers, et al. The sequence of the human genome. *Science* 291, no. 5507 (2001): 1304–1351.

22. Marris, E. Free genome databases finally defeat Celera. *Nature* 435 (May 5, 2005): 6.

23. Collins F. S., M. Morgan, and A. Patrinos. The human genome project: Lessons from large-scale biology. *Science* 300, no. 5617 (April 11, 2003): 286–290.

24. Collins, F. S., E. D. Green, A. E. Guttmacher, and M. S. Guyer. A vision for the future of genomics research: A blueprint for the genomic era. *Nature* 422 (April 24, 2003): 835–847.

25. National Human Genome Research Institute. International HapMap project. Available at <http://www.genome.gov/Pages/Research/HapMap/> (accessed November 3, 2003).

26. The International HapMap Consortium. A haplotype map of the human genome. *Nature* 437 (October 27, 2005): 1299–1320.

27. Merz, J. F., G. McGee, and P. Sankar. "Iceland Inc."?: On the ethics of commercial population genomics. *Social Science and Medicine* 58 (2004): 1201.

28. Pääbo, S. The mosaic that is our genome. *Nature* 421, no. 6921 (January 23, 2003): 409–412.

29. Sankar, P., and M. Cho. Toward a new vocabulary of human genetic variation. *Science* 298 (2002): 1337–1338.

30. Bush, G. H. W. Presidential proclamation 6158. Available at <http://www.loc.gov/loc/brain/proclaim.html> (accessed November 12, 2003).

31. Zerhouni, E. The NIH roadmap. *Science* 302 (October 3, 2003): 63–72.

32. Duyk, G. Attrition and translation. *Science* 302 (October 24, 2003): 603–605.

33. NCI/NHGRI. NIH launches comprehensive effort to explore cancer genomics. December 13, 2005. Available at <http://www.nih.gov/news/pr/dec2005/nci-13a.htm> (accessed December 18, 2005).

Further Reading

Cook-Deegan, R. *The gene wars*. New York: W. W. Norton and Co., 1994.

• A wonderful history of the events leading up to the genome project

International Human Genome Sequencing Consortium. Initial sequencing and analysis of the human genome. *Nature* 409 (2001): 860–921.

• A technical summary of the initial characterization of genome with some history

National Research Council, Committee on Mapping and Sequencing the Human Genome. *Mapping and sequencing the human genome*. Washington, DC: National Academy Press, 1988.

• The NRC report that endorsed the HGP

Sulston, J., and G. Ferry. *The common thread: A story of science, politics, ethics, and the human genome*. Washington, DC: Joseph Henry Press, 2002.

• A personal account of the HGP by the director of the Sanger Institute in Great Britain

Funding Biomedical Research: Peer Review versus Pork Barrel

Large government commitments to new programs in the biological sciences like the HGP have been rare. Most government funding for biological research is awarded to individuals or small groups of researchers through an individual grant application subject to peer review [1]. Individual scientists or small groups of scientists propose projects that are assessed for their quality and feasibility, and are then given priority ratings. These reviews are carried out by panels comprised of working scientists who apply their knowledge of the discipline to select those projects most likely to advance knowledge in the field. A central requirement of most proposals is that they be hypothesis driven—that the research is designed to answer a particular question [2]. This type of approach can be distinguished from so-called discovery science in which research is conducted to see what emerges, without mechanistic models or hypotheses. One of the early objections to the HGP was that it appeared to reflect the worst features of discovery science, being both expensive and without clear goals. If a proposal is not rated highly enough to be funded, investigators may revise and resubmit proposals for reconsideration. There is no guarantee that a revised proposal will be funded, however. There is simply not enough money to fund all strong proposals.

The competition for government-funded grants is stiff, and many worthy proposals are not funded. Despite the doubling of the NIH's budget between 1998 and 2003, the success rate for new grant applications remained at only 20 to 25 percent [3]. The total number of extramural grant applications increased from nine thousand in 1970 to over forty thousand in 2004. Ten thousand new awards were made in 2004, for a total of $3.2 billion; the average award ranged from $300,000 to $400,000 per year. Funds cover the costs of research supplies and equipment, salaries and benefits, and overhead (including providing maintenance services, the costs of heat and electricity, and general operating expenses). The NIH supported over forty-seven thousand projects in 2004, for a total of $19.6 billion [4]. The large number of projects supported results from the continued support of multiyear awards. The funding climate at the NIH will deteriorate at least in the short term, since funding for the NIH increased only by 1 to 2 percent per fiscal year since 2003, despite inflation estimates of over 3 percent [5]. The final budget for the NIH in fiscal year 2006 shows a decrease for the first time in thirty-six years; because of cuts, the NIH 2006 budget is smaller than that of 2003 [6]. The decline in available funds will further heighten competition, and success rates are likely to drop.

Many scientists argue that this process of peer review provides an efficient and equitable means to award research funds. Enormous advances in the understanding of basic biological mechanisms have been made, and ideas are translated into clinical applications. The biomedical research enterprise in the United States is a model of productivity for other countries [7]. Scientists suggest that if problems in the review process develop, then revisions to the process should be made. Such revisions have been instituted. For example, in 1994, a report to the U.S. Senate presented a series of recommendations to expand the pool of reviewers, develop ways to monitor potential discrimination, and develop scoring systems with separate scores for different review criteria [8].

Peer review, however, is criticized as both biased and inefficient [9]. Critics note that about 25 percent of all research funds go to a small number of leading research institutions and suggest that since many reviewers come from these same institutions, the situation is self-perpetuating [10]. Although data are limited, there is a strong perception on the part of researchers, both those who are successful in getting funded and those who are not, that bias exists [9].

One response to these criticisms is the congressionally mandated Experimental Program to Stimulate Competitive Research (EPSCoR). In 1978, Congress mandated that the NSF develop a program to stimulate research in states that traditionally receive small proportions of federal research support. The program is now offered at the NIH and other federal agencies as well. Twenty-five states and territories are eligible for EPSCoR funding from dollars set aside from other research support. One investigator or group of investigators in a state may receive funding with matching funds from the state itself, in competition with applications from other states. The goal is for researchers to "graduate" to regular funding and thereby expand the amount of funding brought into the state. [11] The amount of funding in this set-aside program is modest—only a few hundred thousand dollars. At best,

EPSCoR is viewed as a modest success; no state has achieved a big enough slice of the research funding pie to graduate from the program.

Critics also suggest that the current system of peer review is risk averse—it discourages high-risk or innovative research, since investigators may not be able to demonstrate directly the feasibility of elements of the proposed work. Reviewers instead award funds to straightforward but less interesting projects that are more assured of success. Perhaps in recognition of this criticism, the NIH recently announced a new initiative to establish special grants for high-risk innovative research, and a similar program exists at the NSF [12]. Yet these programs are limited and fund only a small number of researchers.

Alternatives to the current system of peer review generally are aimed at reducing the workload for both applicants and reviewers, and changing the "culture of review." For example, by shortening the length of proposals, more reviews (perhaps twenty to thirty) would be feasible. The ratings might be purely numerical, eliminating the time requirement for writing a lengthy critique. Proponents argue that this approach would eliminate some of the bias that is inevitable when only one or two scientists review a proposal [13]. Another suggestion is to "fund the person, not the proposal." Researchers with strong credentials and productivity would only have to submit a brief résumé of their proposed research. Junior scientists would be funded based on support from mentors [13, 14]. One problem with these approaches is that they inevitably limit feedback to researchers, who then have little opportunity to resubmit an improved proposal.

Another possibility is that approach used in some smaller funding agencies, such as the Defense Advanced Research Projects Agency, which funds highly innovative and speculative research projects. Decisions for funding are made by staff members who are held accountable for the success of funded ventures [15]. Such an approach might work on a small scale for the NIH, although it is unlikely to be manageable given the large number of applications overall.

Despite criticisms, there seems little impetus to change the approach to determining merit in grant funding. Private foundations also use some form of peer review to award funds. For good or ill, peer review remains the general standard and no alternative has been offered. Federal agencies will continue to award funds to individual investigators using peer review. Nevertheless, there is growing concern that limited funding and stiff competition for research funds may discourage bright individuals from pursuing careers in scientific research.

Enter "Pork"

Congress controls the budgets of federal agencies by voting on appropriations bills that provide monetary support. Not all funding is actively debated or reviewed, however. Riders to legislation, known as "earmarking," or more colloquially as pork-barrel funding, may be added. Earmarking is as old as the congressional appropriations system. Representatives and senators attach riders to appropriations legislation for government agencies, directing funds for projects or programs, usually to be performed in their home districts. These funds are often added in conference committee, and may never be discussed on the floor of Congress [16]. The legislator who brings in such funds gains prestige for pulling federal dollars into his/her district with the intended benefit of earning votes for the next election. Memberships on appropriations committees in the House and Senate are highly coveted; the chairs of these committees are particularly adept at adding earmarked funds to legislation.

Academic institutions are now active participants in the earmarking process. In 1983, the Catholic University of America and Columbia University actively lobbied for earmarked funds, and were successful [17]. Academic institutions regularly hire professional lobbyists to influence earmarking decisions, with concomitant increases in total dollars earmarked for science. In 2000, academic institutions and organizations spent over $44.5 million in lobbying efforts [18]. In 1980, $86.2 million was earmarked for academic institutions. In 2000, at least $1.04 billion was directed to academic institutions [17, 19]. This represents around 2 percent of the total support provided by the federal government for research and development outside of national defense [20]. Frequently, funds are to be used for infrastructure improvements such as new buildings or equipment to ostensibly support a scientific project. In some cases, however, the link between funds and research is dubious.

Earmarking is strongly criticized by the scientific community on two grounds: that it undercuts the peer-review process, thereby reducing the quality of funded research; and that it may take funds away from other more worthy programs. Supporters of earmarking counter that the peer-review system tends to stifle innovation and high-risk research; that most research dollars go to a relatively small number of leading institutions, leaving smaller and poorer ones with little support; and that earmarking provides a means to direct research dollars into poorer states. They argue that institutions that receive earmarked funds can then improve their overall research program and become more competitive for peer-reviewed funding. James

D. Savage challenges this last notion, contending there is little evidence that the beneficiaries of earmarking increase their overall quality ranking [19].

An additional incentive for seeking earmarked funds is that the federal government provides limited support for improving or replacing aging infrastructure, the buildings and other facilities needed to conduct high-quality research. The first direct lobbying by Columbia University and the Catholic University was to gain funds to allow the construction of new research buildings. Programs directed at facilities support by the NSF were limited in size and scope, and eliminated almost entirely in the 1990s, and most other funding agencies provide little or no support directed at infrastructure. Earmarked funds for the construction and renovation of research buildings provides a means for academic institutions, regardless of status, to maintain the facilities required for high-tech projects. Anyone who has been to West Virginia will recognize the power of pork, with everything from highways to university research buildings bearing the name of their senatorial benefactor, Robert Byrd.

Federal legislators have used their power and prestige to gain support for pet projects since the establishment of the government. Yet the directing of funds into science is a fairly recent development. Despite the criticism of earmarking, there is little evidence that recipients have any will or desire to decline funds. Even critics acknowledge that their home institutions gratefully receive earmarked funds [19]. Public outrage over pork in the federal budget has had little effect in discouraging Congress from the practice.

Does earmarking take away funds for other research? While it is not possible to directly link earmarking to reduced funding for the NIH and other agencies, earmarking does increase the overall federal budget. In today's budgetary climate of growing deficits and limited revenues, any action that tends to increase overall spending may lead to reductions in appropriations for specified agencies, as Congress attempts to hold down budget increases. Thus, research funding through more traditional federal mechanisms may feel the pinch because of excess spending elsewhere.

References

1. National Institutes of Health. Setting research priorities at the National Institutes of Health. June 18, 2003. Available at <http://www.nih.gov/about/researchpriorities.htm> (accessed November 5, 2003).

2. National Institute of Allergy and Infections Diseases. NIAID funding: How to plan a grant application. Available at <http://www.niaid.gov/ncn/grants/plam/plan_c1.htm> (accessed November 11, 2003).

3. National Institutes of Health. NIH competing research project applications by type of grant, fiscal years 1970–2004. December 7, 2005. Available at <http://grants2.nih.gov/grants/award/success/srbytype7004.htm> (accessed December 18, 2005).

4. National Institutes of Health. NIH awards (competing and noncompeting) by fiscal year and funding mechanism, fiscal years 1994–2004. December 7, 2005. Available at <http://grants2.nih.gov/grants/award/trends/fund9404.htm> (accessed December 18, 2005).

5. Koizumi, K. *Congressional action on research and development in the FY 2005 budget.* Washington, DC: American Association for the Advancement of Science, December 2004.

6. American Association for the Advancement of Science R&D Budget and Policy Program. Congress caps another disappointing year for r&d funding in 2006. January 4, 2006. Available at <http://www.aaas.org/spp/rd/upd1205.htm> (accessed January 9, 2006).

7. Varmus, H. Testimony on revitalization of the NIH before the Senate Committee on Labor and Human Resources. March 6, 1996. Available at <http://www.hhs.gov/asl/testify/t960306c.html> (accessed May 1, 2005).

8. U.S. General Accounting Office. *Peer review: Reforms needed to ensure fairness in federal agency grant selection.* GAO/PEMD–94–1. Washington, DC: General Accounting Office, June 1994.

9. Chubin, D. E., and E. J. Hackett. *Peerless science: Peer review and U.S. science policy.* Albany: State University of New York Press, 1990.

10. Silber, J. Earmarking: The expansion of excellence in scientific research. In *AAAS science and technology policy yearbook 2002*, ed. A. H. Teich, S. D. Nelson, and S. J. Lita, 105–114. Washington, DC: American Association for the Advancement of Science, 2002.

11. Lambright, W. H. *Building state science: The EPSCoR experience.* Washington, DC: American Association for the Advancement of Science, 2000.

12. Zerhouni, E. The NIH Roadmap. *Science* 302 (October 3, 2003): 63–72.

13. Kaplan, D. How to improve peer review at NIH. *Scientist* 19, no. 17 (September 12, 2005): 10.

14. Henneberg, M. Peer review: The holy office of modern science. *naturalSCIENCE.* (February 20, 1997): 1. Available at <http://naturalscience.com/ns/articles/01-02/ns_mh.html> (accessed July 6, 2006).

15. Cook-Deegan, R. Does NIH need a DARPA? *Issues in Science and Technology* 13, no. 2 (Winter 1996): 25–28.

16. Minge, D. The case against academic earmarking. In *AAAS science and technology policy yearbook 2002*, ed. A. H. Teich S. D. Nelson, and S. J. Lita, 115–120. Washington, DC: American Association for the Advancement of Science, 2002.

17. Savage, J. D. *Funding science in America: Congress, universities, and the politics of the academic pork barrel.* Cambridge: Cambridge University Press, 1999.

18. Center for Responsive Politics. Legislative spending: Education. Available at <http://www.opensecrets.org/lobbyists/indusclient.asp?code=W04> (accessed October 3, 2003).

19. Savage, J. D. Twenty years later: The rise of academic earmarking and its effect on academic science. In: *AAAS science and technology policy yearbook 2002*, ed. A. H. Teich, S. D. Nelson, and S. J. Lita, 97–103. Washington, DC: American Association for the Advancement of Science, 2002.

20. National Science Foundation, Division of Science Resources Statistics. Federal obligations for research, by field of science and engineering and agency: FY 2005 projected. November 2005. Available at <http://www.nsf.gov/statistics/infbrief/nsf06300> (accessed December 18, 2005).

Further Reading

Martino, J. Science funding: Politics and porkbarrel. New Brunswick, NJ: Transaction Publishers, 1992.

Minge, D. The case against academic earmarking. In *AAAS science and technology policy yearbook 2002*, ed. A. H. Teich, S. D. Nelson, and S. J. Lita, 115–120. Washington, DC: Amercian Association for the Advancement of Science, 2002.

Savage, J. D. Funding science in America: Congress, universities, and the politics of the academic pork barrel. Cambridge: Cambridge University Press, 1999.

Savage, J. D. Twenty years later: The rise of academic earmarking and its effect on academic science, In *AAAS science and technology policy yearbook 2002*, ed. A. H. Teich, S. D. Nelson, and S. J. Lita, 97–104. Washington, DC: American Association for the Advancement of Science, 2002.

Silber, J. Earmarking: The expansion of excellence in scientific research. In *AAAS science and technology policy yearbook 2002*, ed. A. H. Teich, S. D. Nelson, and S. J. Lita, 105–114. Washington, DC: American Association for the Advancement of Science, 2002.

Choosing Research Directions: Advocacy, the "Disease-of-the-Month," and Congressional Oversight

Even within traditional funding via agencies, Congress frequently attempts to micro-manage the awarding of funds by designating them for specific research purposes. Although scientists may argue that they should take the lead in determining how research funds should be dispersed, Congress maintains an interest in how research funds are used to benefit the public, in response to lobbying by interest groups [1, 2, 3]. For example, in 1985, two new institutes of the NIH were established through congressional legislation: the Institute for Arthritis, Musculoskeletal, and Skin Diseases (NIAMSD), and the Institute for Nursing Research [4]. The rationale for the decision was a perception that insufficient research was conducted in these areas. Included in the legislation for the NIAMSD was a mandate for research on specific disorders along with congressional oversight of research projects. The legislation was strongly supported by a coalition of senior citizens' groups that lobbied for more research into arthritis, a common serious illness in the elderly. The proliferation of new NIH institutes has been criticized for reducing coordination, efficiency, and flexibility in awarding research funds overall; currently, eighteen institutes compete for funding under the NIH umbrella [4].

The large amount of support for the National Cancer Institute (NCI) may be attributed in part to the efforts of Mary Lasker, head of the Lasker Foundation until her death in 1994 [5]. In 2003, the NIH announced it was funding new centers to study the environmental causes of breast cancer; the decision was made in response to both congressional legislation and strong lobbying by advocates of more research on women's health [6]. This response followed earlier funding of research on breast cancer by the DOD.

Funding for research on HIV/AIDS treatment and prevention provides a cogent example of how external forces can influence federal support for research (see also chapter 5). The federal government was slow to allocate research funds after AIDS was initially described in 1981 [7]. Most agree that discomfort with the lifestyle of the first-affected individuals—homosexuals and drug users—and the mode of transmission of the HIV virus contributed to the lack of commitment for funding during the Reagan administration. Still, through strong lobbying by an active and committed gay community, and a growing recognition that AIDS affected a broad range of people, particularly in sub-Saharan Africa, NIH funding for AIDS research and treatment grew rapidly from a few hundred thousand dollars in the mid-1980s to $2.3 billion in 2000. During this period, antiretroviral drugs were developed to slow the progression of the disease, with a decline in the death rate from AIDS in the developed world. The number of new cases per year in the United States has remained stable, however, and new cases are growing explosively internationally [8]. Nevertheless, concern has been raised that funding for AIDS research is disproportionately high given its impact on the U.S. population compared with diseases such as cancer and cardiovascular conditions.

Legislators have suggested that the allocation of funds by the NIH should be determined proportionately to their impact on the population in terms of death and suffering [9]. Others counter that the NIH should not allocate funds using a "dollar-per-death" ratio for two reasons: it puts sufferers of serious illnesses in competition with each other; and basic research can provide information useful for understanding and treating many diseases at once [9, 10]. For example, information gained about viruses through AIDS research has been valuable in understanding other viral diseases. Given that extremely dangerous, and highly contagious, viruses such as Ebola and Marburg exist, this understanding could someday prevent a world pandemic (see also chapter 9).

The Institute of Medicine (IOM) released a report in 1998 that generally supported the criteria by which the NIH sets funding priorities, but recommended that the NIH do a better job of involving interested parties in the process. The IOM was strongly opposed to relying solely on a medical model for determining allocations. Finally, the IOM suggested that while Congress may mandate specific research programs, it should refrain from doing so unless other approaches have proven to be inadequate [11]. Recently, though, there have been calls for directed funding, and it appears that the NIH is prepared to accede to the demands [6].

Legislators persist in proposing new support for research on a given disease as a result of either personal experience or pressure from constituents. An individual legislator who has a family member suffering from a given disease has compelling personal reasons for wanting research into its treatment or cure, and lobbying by single-disease organizations is considerable. Congresspersons repeatedly propose legislation to provide for the more direct micromanagement of the allocation of funds by the NIH. For example, in the 107th session of Congress in 2001–2, ten bills were introduced directed at funding for specific maladies such as muscular dystrophy, blood cancer, and liver disease [12].

A new twist to congressional oversight emerged in 2003, when both legislators and lobbying groups sought to remove funding for NIH-approved grants viewed as morally offensive. In July 2003, two conservative representatives, Patrick Toomey (R-PA) and Chris Chocola (R-IN), proposed an amendment to the $27.6 billion budget authorization bill for the NIH that would remove funding for five projects (mostly studying sexual behavior) the representatives deemed shocking and a waste of taxpayers' money [13]. The amendment failed by a 212–210 vote. In October 2003, a religious lobbying group, the Traditional Values Coalition, asked the House Committee on Energy and Commerce to demand that the NIH investigate the review process of close to two hundred approved or funded projects on controversial subjects involving sexuality and sexual behavior [14, 15]. The NIH reviewed the proposals and reaffirmed its support for them, reporting that all had passed peer review [16]. Once again, in July 2005, the House voted to cancel two NIH-funded psychology grants. Representative Randy Neugebauer (R-TX) argued that the grants— one on marriage, and a second on visual perception in pigeons—exemplify the NIH's failure to focus on serious diseases [17]. Critics suggest that attempts to micromanage the NIH based on religious or political ideology threaten both researchers and the research enterprise [18, 19]. Nonetheless, a recent study suggests that many scientists who work in socially sensitive areas sometimes modify what they do, analyze, and publish in an effort to avoid controversy [20].

When Vannevar Bush proposed mechanisms for the government support of research, he claimed that decisions for setting priorities should lie in the hands of scientists with the expertise to compare relative merits. Nevertheless, Congress is charged with the responsibility to see that taxpayer money is used wisely to benefit the country as a whole. Who should set funding priorities?

References

1. Agnew, B. Decisions, decisions: NIH's disease-by-disease allocations draw new fire. *Scientist* 12, no. 7 (March 30, 1998): 1.

2. Matherlee, K. *The public stake in biomedical research: A policy perspective.* Washington, DC: National Health Policy Forum, George Washington University, November 1999.

3. Malakoff, D. Perfecting the art of the science deal. *Science* 292, no. 5518 (May 4, 2001): 830–835.

4. Varmus, H. Proliferation of National Institutes of Health. *Science* 291 (March 9, 2001): 1903–1905.

5. Martino, J. P. *Science funding: Politics and porkbarrel.* New Brunswick, NJ: Transaction Publishers, 1992.

6. Kaiser, J. In step with advocates, NIH backs targeted research centers. *Science* 302 (October 24, 2003): 547.

7. Shilts, R. *And the band played on: Politics, people, and the AIDS epidemic.* New York: St. Martin's Press, 1987.

8. Subcommittee on Labor, Health, and Human Services, and Education and Related Agencies, Committee on Appropriations. *Acquired Immune Deficiency Syndrome Prevention*, S. Hrg. 108–846, 160th Cong. Washington, DC: U.S. Government Printing Office, 2001.

9. Istook, E., and J. E. Porter. Should disease prevalence determine NIH fund allocation? *FASEB Newsletter.* October 1998. Available at <http://www.fasb.org/opar/newsletter/oct98/10x98x3.html> (accessed September 24, 2003).

10. Nathanson, N. Sustaining the investment in AIDS research. *Scientist* 13, no. 1 (January 4, 1999): 9.

11. Institute of Medicine, Committee on the NIH Research Priority-Setting Process. *Scientific opportunities and public needs: Improving priority setting and public input, and the National Institutes of Health.* Washington, DC: National Academy Press, 1998.

12. U.S. Senate. Search results "NIH." *U.S. Government Printing Office.* Available at <http://frwebgate.access.gpo/cgi-bin/multidb.cgi> (accessed September 5, 2003).

13. Marshall, E. Sex-study grants scrape through. *Science* 301 (July 18, 2003): 289.

14. Weiss, R. NIH faces criticism on grants: Coalition assails "smarmy projects." *Washington Post*, October 30, 2003, A21.

15. Kaiser, J. NIH roiled by inquiries over grants hit list. *Science* 302 (October 31, 2003): 758.

16. Russell, S. Agency stands behind AIDS, sexuality research; NIH ends probe of grants religious right queried. *San Francisco Chronicle*, January 14, 2004, A5.

17. Kaiser, J. House "peer review" kills two NIH grants. *Science* 309 (July 1, 2005): 29–31.

18. Herbert, B. The big chill at the lab. *New York Times*, November 3, 2003, A19.

19. Kaiser, J. Sex studies denounced, NIH's peer-review process defended. *Science* 302 (November 7, 2003): 966–967.

20. Kempner, J., C. S. Perlis, and J. F. Merz. Forbidden knowledge. *Science* 307 (February 11, 2005): 854.

Further Reading

Agnew, B. Decision, decisions: NIH's disease-by-disease allocations draw new fire. *The Scientist* 12, no. 7 (1998): 1. Available at <http://www.thescientist.com/yr1998/mar/agnew_pl_980330.html>.

Institute of Medicine, Committee on the NIH Research Priority-Setting Process. Scientific opportunities and public needs: Improving priority setting and public input at the National Institutes of Health. Washington, DC: National Academy Press, 1998.

3

Who Owns the Genome? The Patenting of Human Genes

Significant advances in genetic technologies since the 1970s have enabled scientists to develop new tools for research and clinical use. These developments also coincided with changes in policy regarding the ownership of inventions made using federal research support (the 1981 Bayh-Dole Act). Since the late 1970s, thousands of patents have been granted covering a broad array of genetic tools as well as human genes themselves. An entire biotechnology industry in the United States and elsewhere has grown into a multibillion dollar enterprise. The U.S. industry, as of 2005, is made up of more than fourteen hundred firms that have drawn investments of nearly $100 billion in the last twenty years; has brought more than 150 drugs or vaccines to market and has over 350 more in clinical trials; and includes publicly traded biotechnology firms with a market capitalization in excess of $300 billion [1]. The industry is strongly dependent on patents, and the Biotechnology Industry Organization cites U.S. Patent and Trademark Office (USPTO) data showing that nearly sixty-five thousand relevant patents had been issued between 1989 and 2002, inclusive [1]. According to Robert Cook-Deegan, this includes patents on some thirty thousand human genes and gene products [2]. Overall, some 20 percent of human genes (excluding proteins and similar gene products) have been patented [3].

There are concerns that gene patents may have negative effects on clinical genetic medicine, and that the patents may inhibit or interfere with basic genetics research as well as the development of drugs and biotechnology products. This chapter describes U.S. patent law, and then examines the kinds of inventions covered by human gene patents, summarizing the body of empirical research looking at the effects of these patents in the United States.

Patents

Patents are government grants of legal rights in certain types of intellectual property. Intellectual property, broadly speaking, is a product of the mind. Not all products of the mind can be protected as property; ideas are not patentable. Most intellectual property begins with ideas, but legal protection is offered only for various embodiments of ideas. The source of legal protections in the United States is the Constitution, which empowers Congress "to promote the Progress of Science and useful Arts, by securing for limited Times to Authors and Inventors the exclusive Right to their respective Writings and Discoveries" [4]. Under this authority, the federal government has enacted laws creating legal rights in copyrights, trademarks, and patents. Copyright protects unique expressions in tangible form, including works of art, writings, music, software, and the like [5]. Trademarks protect unique and valuable expressions or symbols that gain recognition in the commercial marketplace as indicators of quality and genuineness [6]. Patents—called utility patents—protect useful inventions and discoveries [7]. In the United States, special types of patents are also available for bred strains of plants and designs. Finally, trade secrets, which can be anything that has competitive value (such as pricing information, core technologies like the recipe for Coke, or customer lists) that is kept secret or confidential, are protected by various state and federal laws on contract, theft, and industrial espionage. The form of protection granted by each form of intellectual property varies. Authors, firms, and inventors will utilize the full range of rights to retain the strongest legal protections as well as the ability to commercialize and profit from their respective writings and discoveries. This discussion focuses only on utility patents.

In the United States, utility patents are granted by the USPTO for new, nonobvious, and useful inventions and discoveries. Similar standards of patentability are applied around the world. Patents can be granted for different types of inventions, including machines, articles of manufacture, compositions of matter such as chemicals and alloys, methods or processes, and improvements to these. A patent may also be issued on a nonunique product (such as human insulin) made by an otherwise-patentable method (recombinant technologies). As a condition for getting a patent, an inventor must disclose enough details to enable another person who is skilled in the applicable sciences and engineering to understand, reproduce, and use the invention. In return, the patent system rewards the inventor with a period of exclusivity during which time profits may be earned from the invention's commercialization. A patent owner has the right to exclude others from making, using, or selling a patented invention or method for a period of twenty years from the date of filing of the patent application [7].

Throughout the developed world, patents are awarded following an examination by a patent agency (e.g., the European Patent Office and the USPTO). Examination procedures ensure that inventions fulfill the standards for patentability, and that the patent grants protection only for that which has been invented and no more. The patent *claim* defines the scope of patent protection. Typically, there is a negotiation between the inventor and the patent examiner, with the former trying to get broad protections, and the latter seeking to grant a patent narrowly restricted to the technological improvements made by an invention and described in the specification. Broad claims often may be granted for breakthrough inventions such as on the laser or, in biotechnology, those on the polymerase chain reaction (PCR), recombinant DNA techniques, gene knockout methods, and even individual gene sequences (see chapter 2). Because broad claims to inventions such as a sequence or a recombinant protein are so basic, they cannot easily be invented around, and any improvements are likely to require licenses before they can be used commercially. In biotechnology, such licenses may be impossible to secure, since the owners of the dominant patents are likely to depend on them to maintain their market exclusivity [8, 9]. Broad claims, in any technology, may create a disincentive for downstream development because of the need to pay owners for the right to use their patents, and there's always the risk that those owners may not grant licenses, instead using the patents to block competitive products [10]. Of course, competition will arise as third parties attempt to invent around the patent, and make technological and product improvements to the basic invention [11].

There are many similarities between U.S. and foreign patent laws. The major difference, however, is that all countries except for the United States award patents in cases of competing or overlapping inventions to the first to file a patent application. The United States awards the patent to the first to invent. Through a process called interference, the USPTO will determine which of multiple patent applicants was the first to conceive and, exercising due diligence, reduce to practice an invention. Reduction to practice means either making the invention to prove that it works, or describing the invention in sufficient detail (as in a patent application) that it would allow someone skilled in the relevant science or technology to reproduce the invention without undue research or experimentation. Roughly 100 to 150 interferences are declared each year, out of over three hundred thousand patent applications, or four for every ten thousand applications [12]. A recent study of interference cases showed that the applicant who won the race to the patent office is only slightly more likely than the loser to win the inventorship race as well [13]. Given the relative rarity of interferences and the high costs of resolving the interference (estimated to be at least $100,000 and upward of $1 million for each party), it may be time to do away with this anachronism of U.S. patent law.

A patent grants what is called a negative right—the right to prevent others from making, using, or selling the claimed invention without permission. A patent owner may turn to the government—through lawsuits for infringement—to use its judicial and police powers to block others from profiting from the invention, seize and destroy infringing goods, and collect damages from those who infringe. A patent does not grant its owner the positive right to use an invention, which otherwise is subject to legal restraints (e.g., human cloning) or regulatory premarketing licensing requirements (e.g., drugs and medical devices). There is likewise no legal compunction for a patent owner to "work" or license others to use a patented invention, and as a general rule, a patent may be used wholly to keep products from coming to market [14]. Exceptions have been recognized for compulsory licensing of patented inventions when deemed necessary to protect public health and welfare (such as national defense and drugs) [15]. In the United States, the federal government retains "march-in" rights—the right to use or license others to use for governmental purposes—to patents resulting from federally funded research in any case in which the inventions are not developed for practical application, or if deemed necessary to alleviate health or safety needs that are not reasonably satisfied. This right, statutorily created in 1980, has never been exercised [16].

Human Gene Patents

Human gene patents cover discoveries or inventions, depending on one's point of view, that result from the cloning and description of the sequence of a gene, the role or function of which is somewhat understood. As cloning and sequencing capabilities rapidly evolved in the 1980s, patent applications on human genes were filed in increasing numbers. Questions of the wisdom of patenting genes were highlighted by the 1991 patent application filed by the NIH that was subsequently amended to cover thousands of ESTs (see chapter 2). ESTs are unique nucleotide sequences that have no known function other than as a distinctive marker. The NIH applications were ultimately withdrawn, but the concerns over the patenting of life and genes survived. In 1995, Jeremy Rifkin's Foundation on Economic Trends coordinated a joint statement signed by leaders of eighty different religious organizations opposing the patenting of human genes and transgenically modified animal life [17].

This statement unfortunately conflated the patenting of genes, which are chemicals, with the patenting of life itself. This confusion has persisted, largely driven by two related issues. For one, debates swirled around the first patents on genetically modified organisms. In 1980, the U.S. Supreme Court decided *Diamond v. Chakrabarty*, which was a landmark case in U.S. patent law [18]. Chakrabarty's invention was a microbe modified to break down oil, which could be useful for cleaning up oil spills. The USPTO had refused to grant the patent on the living cells, and the Supreme Court decided, in a 5–4 decision, that life-forms are indeed patentable subject matter. This led directly to the patenting of higher life-forms such as the Harvard oncomouse (a mouse model for cancer) and Cre-Lox technology (a method for inserting genes into an organism) for oncomouse production [19, 20, 21], and more recently, modified and cloned mammals like Dolly.

Second, *Chakrabarty* represents the increasing commercialization and commodification of life, characterized by the patenting of not only living things but genes as well. In this sense, and given its timing in the early days of molecular genetics discovery, *Chakrabarty* ushered in the genomics age. Of course, genes are not living but are of the living, unique to the living, and our common inheritance, and the discomfort of patenting a mouse is perhaps loosely akin to that felt by the patenting of human genes (see "Who Owns Life?" section below).

From an intellectual property perspective, however, *Chakrabarty* is not relevant to the patenting of genes. Genes in their natural form, located in our cells, are not patentable. Yet purified and isolated genes are quite patentable because they are chemicals, or in the patent vernacular, compositions of matter. Purified and isolated compositions of matter have been patentable for more than a hundred years. For example, insulin isolated and purified from blood is patentable, as is recombinant insulin, made from human genes. Isolated and purified genetic materials differ from nongenetic compositions of matter only in their source of raw material, and the source is irrelevant for the purposes of U.S. patent law.

But this begs the question of precisely what it is that patents on human genes cover, and what is necessary to secure a patent on what, seemingly, is a naturally occurring thing. Human gene patents cover three distinct types of invention: diagnostics, compositions of matter, and functional uses. Importantly, concerns about the broad scope of gene patents led the USPTO in 2001 to clarify its patentability standard for genes, requiring that a patent applicant make a "credible assertion of specific and substantial utility" of the genetic invention [22, 23].

Each type of gene patent is discussed here in turn, with examples, highlights of areas of concern, and what is known about each. This overview is centered on U.S. patent law and what is known about how gene patents are being used in the United States. Some of the problems discussed have begun to spill over to Europe, Japan, Canada, and Australia. This is not meant to be a comprehensive international review [24] but rather is an attempt to demonstrate the breadth of gene patents, discuss concerns about how they are being used, and summarize relevant empirical data.

Diagnostic Uses

The first type of genetic "invention" covers the testing of genetic differences. These types of patents have been referred to as "disease gene patents" because they claim the characterization of an individual's genetic makeup at a disease-associated gene when performed for the purpose of diagnosis or prognosis [25]. These patents typically cover all known methods of testing, including the use of hybridization, Southern analysis, PCR, and DNA chips. Since the fundamental discovery patented is the statistical observation of a genetic difference and a phenotypic difference (such as the occurrence of disease), then any method for testing for the genetic difference can be covered by the patent [26].

Well-known examples of disease gene patents include those covering genes implicated in breast and ovarian cancers (BRCA1 and 2), colon cancers, cystic fibrosis (CFTR), hemochromatosis (HFE), and a growing number of neurological diseases including late onset Alzheimer's (Apo-E), Canavan disease, Charcot-Marie-Tooth disease, spinal muscular atrophy, spinocerebellar ataxia, and others.

There are several characteristics of genes and disease gene patents that demonstrate how the genome is being divided up by small patent claims to overlapping genetic territory. First, any one gene may have multiple patents claiming the diagnosis of different polymorphisms (variations in gene sequence). Thus, several patents have been issued for the testing of different mutations in the CFTR gene [27]. Further, some diseases (at least the phenotypic expressions of them) are caused by multiple genes, such as Charcot-Marie-Tooth, a muscle disease [28]. Questions about ownership and access get messy when there are many hundreds of known mutations in several genes associated with the same disease, as exemplified by

BRCA1 and BRCA2, for which there are at least a dozen U.S. patents on tests of these two genes [3, 29]. Finally, patents can issue on the same exact molecular test when it is performed for different diagnostic or prognostic purposes. For example, an Apo-E test (for the version of amyloid protein found in nerve cells), in which the number of E2, E3, and E4 alleles carried by an individual is assessed, can be performed for each of the following patented uses: determining whether a patient is at risk of early onset Alzheimer's disease (AD) [30]; assessing an AD patient's prognosis [31]; determining a course of therapy for AD based on the individual's genetic background [32]; and assessing a patient's prostate cancer risk [33]. Apo-E is also used for the assessment of cardiovascular risk, but this use has not (yet) been patented. In these cases, a patent thicket is created that can lead to difficulties in securing licenses and expenses in paying multiple "stacked" royalties to multiple patent owners [34].

As of 2005, the owners of the overwhelming majority of issued genetic diagnostic patents have not aggressively enforced their rights against clinical molecular diagnostics laboratories. Nonetheless, a majority of genetics laboratories across the United States report that their use of one or more of the above disease gene patents has been challenged [35, 36]. In some cases, patent owners have been willing to grant a license to laboratories performing the testing. Per test royalties include $2 for the ΔF508 mutation of CFTR (University of Michigan), $5 for Gaucher's disease (Scripps Institute), $12.50 for Canavan disease (Miami Children's Hospital), and reportedly more than $20 for HFE (Bio-Rad) [37]. In some cases, an up-front license fee (not tied to the number of tests performed) has been demanded as well. While these royalties arguably reduce access and create problems for laboratories, they must be examined in the context of the U.S. commercial, profit-centered health care system.

Clinical as well as research laboratories typically pay royalties for the use of patented technologies. For example, the price of widely used PCR machines and reagents include a premium paid for the use of the patented technologies. In addition, a royalty of about 9 percent is paid for all testing done by licensed laboratories [35]. As discussed in great detail by Dianne Nicol [38], the most recent patents enforced against biotechnology companies and testing laboratories are those that claim the extremely broad uses of DNA sequences not directly part of known disease genes for identifying genetic variations between individuals or populations [39]. Disease gene patents are substantively different from these more typical patented tools, which are instruments and methods that are used by laboratories for testing for a variety of specific disease genes. Since a disease gene patent claims all methods of testing for a specific gene, there is no plausible way of working around these patents and the patents may be used to monopolize a test.

Fortunately, in only a handful of cases have patent owners refused to grant licenses to laboratories to allow them to perform specific tests. In a few cases, patent owners have used the patents to monopolize the testing service, requiring physicians and laboratories to send samples for testing to the owner or its specified licensees. Thus, tests for breast and ovarian cancer genes (Myriad Genetics) and a set of neurological disorders (Athena Diagnostics) are generally available from only these commercial laboratories [14]. Smithkline Beecham Clinical Laboratories made a brief attempt at capturing the testing market for hemochromatosis before the business unit was sold to Quest Diagnostics, which then transferred ownership to Bio-Rad [37]. Myriad has extended its reach beyond the U.S. borders, seeking to enforce its BRCA patents in, among others, France, Canada, and the United Kingdom. The test for Canavan disease, despite being easily included in the panel assays that many laboratories can run, was restricted to selected labs around the United States by the patent owner, Miami Children's Hospital (see the "Canavan Disease" section below) [40].

In these cases, laboratories have been told where patient samples must be sent to have the patented tests performed and how much it will cost. Being compelled to stop providing testing services has serious implications for the ability of molecular pathologists to make use of the newest technologies, treat their patients with comprehensive medical services, train residents and fellows, perform research, and run their labs in an efficient manner. Hospital-based laboratories must often absorb part of the fixed, monopoly costs of the tests that they are compelled to offer patients, but for which health insurance may not cover the full price. Seen in this light, these patents raise the costs of clinical services and restrict physicians' ability to practice medicine [14, 41].

Compositions of Matter

The second broad type of genetic invention relates to compositions of matter (i.e., chemicals and materials), including the isolated and purified gene (cDNA) and all derivative products (e.g., recombinant proteins or drugs, viral vectors and gene transfer "therapies," and transfected cells, cell lines, and higher-order animal models in which the patented gene has been inserted or knocked out).

Patents on human genetic compositions of matter cover a broad array of chemicals and technologies. For example, human insulin, human growth hormone, and many other proteins that can be isolated and purified from, say, human blood or urine, can be patented. Further, synthesized products can be covered by various patent claims, including: claims to the sequences used (both the sequence to be transcribed into RNA and proteins as well as promoter sequences); the virus or other vector containing the claimed sequence; transfected cells, cell lines, and nonhuman organisms created and used in these processes; and perhaps most important, the proteins or other therapeutic products made by these claimed processes. The last, which can be covered by "product by process" claims in patents, allow patent owners to prohibit the use or sale of products made by the claimed processes, regardless of where the product is made [42].

Functional Use

Finally, a third and emerging class of gene patents is that claiming the functional use of a gene. These patents are based on the discovery of the role genes play in disease or other bodily and cellular functions or pathways, and claim methods and compositions of matter (typically called "small molecule" drugs) used to up- or down-regulate the gene. These drugs likely are not gene products themselves but other types of chemicals found to affect gene functioning, and the drugs are likely patentable themselves as unique chemical entities useful as therapy. For instance, a patent that was recently invalidated claimed methods and compositions of matter for the selective inhibition of the Cox-2 gene, which prevents inflammation and pain. The patent was invalidated because the patentee, the University of Rochester, failed to disclose a chemical entity that would perform such selective inhibition [43]. The patent claimed the mechanism by which three drugs that later came to market work: Celebrex, which is comarketed by Pharmacia (of which Searle is a part) and Pfizer; Pfizer's Bextra; and Merck's Vioxx (see also chapter 5). Each one of these chemical entities may be patented as a new, nonobvious, and useful drug for the treatment of inflammation and pain, but the Cox-2 patent attempted to claim all drugs that work by manipulating the function of the target gene.

A similar case to the Cox-2 litigation involves a patent awarded to Harvard and MIT, and exclusively licensed to Ariad Pharmaceuticals. The patent claims the basic regulation of any genes by reducing the intracellular activity of the transcription factor NF-kB [44]. Transcription factors regulate the expression of many different genes. On award of the patent, Ariad sued Eli Lilly for infringement by their osteo-porosis drug Evista and their sepsis drug Xigris, and has asserted the patent against numerous other companies. Lilly's patent applications for these two compounds themselves predate the filing of the NF-kB application [45]. Ariad should have a hard time winning, both because, like the selective Cox-2 inhibition patent, the NF-kB one fails to disclose specific agents for regulating the factor, and because the company is trying to assert its patent in a way that would remove from the market chemical entities already developed when the discovery was made.

Finally, there is the case of Viagra. Pfizer, which has had its erectile dysfunction drug on the market for several years, recently received a patent claiming the molecular pathway by which Viagra works. The patent claims any selective phosphodiesterase (PDE5) inhibitor used to treat impotence [46]. These inhibitors act to increase blood flow to the penis and enable a man to produce an erection. Immediately on allowance of its patent in late 2002, Pfizer sued Bayer and Glaxo-SmithKline for their drug Levitra, and Eli Lilly and their partner Icos for their drug Cialis, both of which drugs were then proceeding toward FDA approval (and have since been approved) [47]. The difference between the Viagra case and the Cox-2 one is that Pfizer actually has and claims a specific class of drugs that work by the claimed functional pathway. Whether this is an adequate basis on which to allow Pfizer to lay claim to all drugs that work by the same molecular mechanism is a fundamental legal question that looms over the pharmaceutical industry.

Arguably, granting exclusive rights to the molecular mechanisms by which drugs work gives away too much. Pharmaceutical firms have historically competed at the product end of the development pipeline, as exemplified by independently patented drugs Viagra, Cialis, and Levitra. Gene patenting, and in particular functional claims, has pushed intellectual property protections back to the basic discovery end of the pipeline. This can make a fundamental change in the ways pharmaceutical firms act and substantially alter the competitive environment. Whether these patents will be held up as valid remains to be seen.

Concerns about Gene Patents and Research

One of the primary concerns about human gene patents is that they will make it more difficult to perform research, thereby delaying or impeding the discovery and development of diagnostics and therapeutics [48]. In the United States, there is no exemption from infringement for research activities written in the patent statute, but the federal courts have defined an extremely narrow exemption for certain research activities. As recently summarized by the Court of Appeals for the Federal Circuit in a lawsuit against Duke University, "Regardless of whether a particular institution or entity is engaged in an endeavor for commercial gain, so long as the act is in furtherance of the alleged infringer's legitimate business and *is not solely for amusement, to satisfy idle curiosity, or for strictly philosophical inquiry*, the act does not qualify for the very narrow and strictly limited experimental use defense" [49]. Duke was not excused from the potential infringement of patents covering laboratory equipment simply because the equipment was used solely for research and educational purposes, which the court found to be the core of Duke's business. A strong argument can be made that the exemption should be much broader, encompassing research aimed at a better understanding of the claimed invention—such as how it works and whether it works as claimed by the patent, how to improve on it, and how to work around it. Indeed, practically speaking, this may in fact be how patents are most commonly used. As a colleague stated it, research *on* the invention should be exempt while research *using* the invention is infringement [50]. As stated earlier, the patent law trades a period of exclusivity for disclosure, and competitors should not have to wait for the period of exclusivity to end before learning from that disclosure and attempting to improve on it. The fact that competition occurs is shown by a simple example, again drawing on the case of Viagra: a U.S. patent search for four different combinations of PDE, or PDE5, or phosphodiesterase and erectile or dysfunction in patent claims yields seventy-six patents assigned to eighteen different companies and two universities [51].

Little is known about how gene patents are being used, and whether they are having a net beneficial or detrimental effect on scientific research and commercial product development. Patents clearly are seen as a necessary stimulus for the infusion of venture and risk capital in the biotechnology industry; less obvious is the role patents play in motivating academic researchers. Some data have been generated about the licensing of biotechnology patents. These studies suggest that most genetic inventions are not patented, but when they are, they are being licensed on exclusive terms [52, 53]. In turn, researchers and firms appear to have developed various strategies to minimize the potential detrimental effects of the patents, including taking licenses when possible, inventing around patent inventions, conducting research in other countries not covered by the patent, and using publicly available resources, litigation, and the deliberate infringement of the patent [54]. Nonetheless, much remains unknown about the effects of these practices on basic research and commercial competition.

Conclusion

Human gene patents cover a broad range of different but related types of inventions. Each type has its own potential uses and marketable products, and each raises potential problems depending on how the patents are used in the relevant marketplace. Much remains unknown, and indeed, the market is still adapting to these patents. Thus, it is extremely important to continue to study and monitor how gene patents are being used, licensed, and enforced in order to ensure that policy interventions can be implemented, if necessary, to achieve the twin goals of public health promotion and economic development.

References

1. Biotechnology Industry Organization. Biotechnology industry facts. <http://www.bio.org/speeches/pubs/er/statistics.asp> (accessed April 30, 2005).

2. Cook-Deegan, R. Personal communication, 2003.

3. Jensen, K., and F. Murray. Intellectual property landscape of the human genome. *Science* 310 (2005): 239–240.

4. U.S. Constitution, art. 1, sec. 8, clause 8.

5. U.S. Code Title 17.

6. U.S. Code Title 15, ch. 22, subch. III, secs. 1051–1127.

7. U.S. Code Title 35.

8. Bar-Shalom, A., and R. Cook-Deegan. Patents and innovation in cancer therapeutics: Lessons from CellPro. *Milbank Quarterly* 80 (2002): 637–676.

9. *Amgen, Inc. v. Hoechst Marion Roussel, Inc.*, 314 F.3d 1313 (C.A.F.C. 2003).

10. Bar-Shalom, A., and R. Cook-Deegan. Patents and innovation in cancer therapeutics: Lessons from CellPro. *Milbank Quarterly* 80 (2002): 637–676.

11. Merges, R. P., and R. R. Nelson. On the complex economics of patent scope. *Columbia Law Review* 90 (1990): 839–916.

12. Merz, J. F., and M. R. Henry. The prevalence of patent interferences in gene technology. *Nature Biotechnology* 22 (2004): 153–154.

13. Lemley, M. A., and C. V. Chien. Are the United States patent priority rules really necessary? *Hastings Law Journal* 54 (2003): 1299–1333.

14. Merz, J. F. Disease gene patents: Overcoming unethical constraints on clinical laboratory medicine. *Clinical Chemistry* 45 (1999): 324–330.

15. Consumer Project on Technology. Health care and intellectual property: Compulsory licensing. Available at <http://www.cptech.org/ip/health/cl> (accessed April 30, 2005).

16. U.S. Code Title 35, § 203 (1999).

17. Kevles, D. J. Patenting life: A historical overview of law, interests, and ethics. 2001. Available at <http://www.yale.edu/law/ltw/papers/ltw-kevles.pdf> (accessed April 11, 2005).

18. 447 U.S. 303 (1980).

19. U.S. Patent No. 4,736,866.

20. U.S. Patent No. 5,087,571.

21. U.S. Patent No. 5,925,803.

22. 66 Federal Register 1092–1099, January 5, 2001.

23. Berkowitz, A., and D. J. Kevles. Patenting human genes. In *Who Owns Life?* ed. D. Magnus, A. Caplan, and G. McGee, 75–98. Amherst, NY: Prometheus, 2002.

24. Katzman, S. What is patent-worthy? Due to differing legal standards for patentability, applications for gene patents face different outcomes in various countries. *EMBO Report* 2, no. 2 (2001): 88–90.

25. Merz, J. F., M. K. Cho, M. A. Robertson, and D. G. B. Leonard. Disease gene patenting is a bad innovation. *Journal of Molecular Diagnostics* 2 (1997): 299–304.

26. Merz, J. F., and M. K. Cho. Disease genes are not patentable: A rebuttal of McGee. *Cambridge Quarterly Healthcare Ethics* 7 (1998): 425–428.

27. U.S. Patent Nos. 5,981,178, 5,407,796, and 5,776,677.

28. U.S. Patent Nos. 5,306,616 and 6,001,576.

29. U.S. Patent Nos. 6,083,698, 5,912,127, 5,756,294, 5,753,441, 5,750,400, 5,709,999, 5,654,155, 6,083,698, 5,622,829, 6,033,857, 6,124,104, and 6,492,109.

30. U.S. Patent No. 5,508,167.

31. U.S. Patent No. 6,027,896.

32. U.S. Patent No. 5,935,781.

33. U.S. Patent No. 5,945,289.

34. Heller, M. A., and R. S. Eisenberg. Can patents deter innovation? The anticommons in biomedical research. *Science* 280 (1998): 698–701.

35. Cho, M. Ethical and legal issues in the 21st Century. In *Preparing for the Millennium: Laboratory Medicine in the 21st Century*, 47–53. Orlando, FL: AACC Press, 1998.

36. Cho, M. K., S. Illangasekare, M. A. Weaver, D. G. B. Leonard, and J. F. Merz. Effects of patents and licenses on the provision of clinical genetic testing services. *Journal of Molecular Diagnostics* 5 (2003): 3–8.

37. Merz, J. F., A. G. Kriss, D. G. B. Leonard, and M. K. Cho. Diagnostic testing fails the test: The pitfalls of patenting are illustrated by the case of haemochromatosis. *Nature* 415 (2002): 577–579.

38. Nicol, D. Balancing innovation and access to healthcare through the patent system: An Australian perspective. *Community Genetics* 8 (2005): 228–234.

39. U.S. Patent Nos. 5,851,762, 5,612,179, 5,192,659, and 5,096,557 are being enforced by Genetic Technologies, Ltd., which acquired the patents from GeneType, AG. Genetic Technologies Ltd. Intellectual Property. Available at <http://www.gtg.com.au/> (accessed April 30, 2005).

40. Merz, J. F. Discoveries: Are there limits on what may be patented? In *Who Owns Life?*, ed. D. Magnus, A. Caplan, and G. McGee, 99–116. Amherst, NY: Prometheus Press, 2002.

41. Leonard, D. G. B. Medical practice and gene patents: A personal perspective. *Academic Medicine* 77 (2002): 1388–1391.

42. Ragavan, S. Can't we all get along? The case for a workable patent model. *Arizona State Law Journal* 35 (2003): 117–185.

43. *University of Rochester v. G. D. Searle, Inc.*, 249 F. Supp. 2d 216–236 (W.D.N.Y. 2003).

44. U.S. Patent No. 6,410,516.

45. Aoki, N. Patenting biological pathways; Ariad suit raises questions on laying claim to processes in treating disease. *Boston Globe*, July 24, 2002, C4.

46. U.S. Patent No. 6,469,012.

47. Hensley, S. Pfizer wins Viagra patent, moves to block competition. *Wall Street Journal*, October 23, 2002, 3.

48. Caulfield, T., E. R. Gold, and M. K. Cho. Patenting human genetic material: Refocusing the debate. *Nature Reviews Genetics* 1 (2000): 227–231.

49. *Madey v. Duke University*, 307 F.3d 1351 (Fed. Cir. 2002) (emphasis added).

50. Ducor, P. Personal communication, 2003.

51. U.S. Patent and Trademark Office. Patent full-text and full-page image databases. Available at <http://www.uspto.gov/patft/index.html> (updated April 5, 2005; search performed March 30, 2005).

52. Henry, M. R., M. K. Cho, M. A. Weaver, and J. F. Merz. DNA patenting and licensing. *Science* 297 (2002): 1279.

53. Henry, M. R., M. K. Cho, M. A. Weaver, and J. F. Merz. A pilot survey on the licensing of DNA inventions. *Journal of Law, Medicine, and Ethics* 31 (2003): 442–449.

54. Walsh, J. P., W. M. Cohen, and A. Arora. Science and the law: Working through the patent problem. *Science* 299 (2003): 102.

Further Reading

Magnus, D., A. Caplan, and G. McGee, eds. *Who Owns Life?* Amherst, NY: Prometheus, 2002.

• A thoughtful exploration of the ethical issues raised by patents on life and genes

National Research Council, National Academy of Sciences. *A patent system for the 21st century*. Washington, DC: National Academies Press, 2004.

• An excellent overview of empirical research on the functioning of the patent system

U.S. National Academy of Sciences, Board on Science, Technology, and Economic Policy. *Reaping the benefits of genomic and proteomic research*. Washington, DC: National Academies Press, 2006.

• A thorough examination of the issue of whether patents are facilitating or inhibiting research leading to new therapies and diagnostics

Who Owns Life? Mr. Moore's Spleen

Before gene patenting became an issue, policymakers grappled with the task of determining whether living things might be patentable. As noted in chapter 3, the *Chakrabarty* Supreme Court decision in 1980 opened the door for patenting living organisms. Another case, described here, raised the issue of the ownership of cell lines derived from an individual.

Shortly after finding out that he suffered from hairy cell leukemia, a cancer of white blood cells, in 1976, John Moore came under the care of David Golde at the University of California at Los Angeles (UCLA) Medical Center. In October of that year, Golde removed Moore's spleen. Several times over the next six years, Moore returned to UCLA from his home in Seattle, and had blood and other tissue samples taken. During that time, Golde and researcher Shirley Quan successfully established and patented a cell line (Mo cells) from Moore's T-lymphocytes (cells of the immune system) and the products generated thereby [1]. The Moore cell line was unique in that it overproduces certain lymphokines (cell signaling chemicals) caused by infection with Human T-cell Leukemia virus Type II. Golde and Quan as well as other researchers have since found that such infection will cause normal T-lymphocytes to overproduce [2]. The cell line has commercial potential because the particular lymphokines produced by it have therapeutic value in stimulating immune system function. Golde, Quan, and UCLA entered into commercialization agreements with Genetics Institute, Inc. and Sandoz Pharmaceuticals Corporation, which gave the researchers salary support and interests in future earnings from the commercialization of the Mo cell line and the products derived from it.

Moore sued to assert a continuing property interest—a bundle of legal rights akin to ownership—in his cells. He argued that his cells were misappropriated, or in legal jargon, converted for their own use and benefit, by the above individuals, the regents of the University of California, and the two companies. He also alleged a dozen other bases for liability, importantly including a claim that he had not given an informed consent for the use of his cells. The trial court dismissed the case, deciding that Moore's complaint did not state a valid claim for which relief (i.e., monetary damages) could be granted by the court.

The Court of Appeal reversed, finding no good reason to refuse extension of common law and statutory definitions of property to include an interest one has in their bodily tissues and cells, and concluding that interest is not cut off once those bodily substances have been removed from the body [3]. In effect, this decision would require that any postremoval use of such bodily tissues be fully approved by the patient, and that the patient knowingly waive his or her rights in that property. This, further, violates federal policy for research, which holds that it is unethical to ask human subjects to waive any rights they may have in order to participate in research [4].

On appeal, the Supreme Court of California reversed, finding that the appellate court unwisely extended concepts of property and the law of conversion beyond that necessary to protect patients' interests [5]. The court felt that there are strong policy implications of any such recognition, and that the proper forum for extension of the law lies with the legislature [6].

The court nonetheless held that physicians should disclose personal interests unrelated to their patient's health, if potential conflicts of interest arising from financial or research interests may affect physician judgment (see chapter 8). The potential bias of physician's judgment may be "material" to a patient, to help the patient make an informed choice regarding the course of medical care. As the court stated: "The possibility that an interest extraneous to the patient's health has affected the physician's judgment is something that a reasonable patient would want to know in deciding whether to consent to a proposed course of treatment. It is material to the patient's decision and, thus, a prerequisite to informed consent" [5].

As the potential benefits of patenting cell lines become clearer, the issues surrounding ownership become more complex (see also the "Brave New World" section in chapter 4). What form of informed consent will best serve donors of their tissues? If individuals refuse to consent to the use of their tissues for research, researchers may find that their ability to gain insight into the nature of many diseases may be hampered. Should patients be able to refuse such uses? What protections are in place to assure that secondary uses of patients' tissues are used appropriately (see the "Seminal Events" section in chapter 5)?

Should patients be allowed to retain a share of profits derived from patents on their own cells? The potential impact on physician-patient relationships is considerable—if patients fear that they are merely tools to generate income for researchers. Do we want to create a market in human tissues? What are the potential problems with this? Do we allow markets for organ donation? Is the donation of blood (or other replenishable human tissue) different from the donation of solid organs (see chapter 12)?

References

1. U.S. Patent No. 4,438,032 (issued March 20, 1984).

2. Chen, T. S., S. W. Quan, and D. W. Golde. Human t-cell leukemia virus type II transforms normal human lymphocytes. *Proceedings of the National Academy of Science* 80 (1983): 7006–7009.

3. *Moore v. Regents of the University of California*, 215 Cal. App. 3d 709, 202 Cal. App. 3d 1230, 249 Cal. Rptr. 494 (1988).

4. 45 CFR 46.116(a) (2005).

5. *Moore v. Regents of the University of California*, 51 Cal. 3d 120, 793 P.2d 479, 271 Cal. Rptr. 146 (1990).

6. Parker, P. M. Recognizing property interests in bodily tissues: A need for legislative guidance. *Journal of Legal Medicine* 10 (1989): 357–375.

The Canavan Disease Patent Case

This section examines the use of a patent covering the molecular diagnosis of a specific disease, Canavan. This case has come to symbolize some of the negative effects gene patents may have on the provision of clinical medical services. The case has also caused many to consider the need for benefit sharing as one way of recognizing the contribution of patient groups and families to research and discovery.

The Research

In the early 1980s, Dan and Debbie Greenberg had two children—a son, Jonathan, and a daughter, Amy, both born with Canavan disease. Canavan is a recessive degenerative brain disease that irreversibly leads to the loss of body control and death, usually before the teen years [1]. There is no cure. Like the better-known Tay-Sachs disease, Canavan is more common in individuals of Ashkenazic Jewish descent. In 1987, Dan Greenberg approached Reuben Matalon and convinced him to study Canavan disease. Matalon ran a laboratory performing clinical testing and research of phenylketonuria and other familial disorders at the University of Illinois in Chicago. With blood, urine, and other tissue samples provided by the Greenbergs and another family affected by the disease, and "seed money" provided by the Greenbergs' Chicago chapter of the National Tay-Sachs and Allied Diseases Association (NTSAD), within a year Matalon identified the deficiency of an enzyme, aspartoacylase, as the cause of Canavan disease. This was great news because it offered the possibility of a prenatal screening test [2].

In 1988, the Greenbergs became the first couple to commence a pregnancy based on knowledge of the availability of the testing. They underwent prenatal testing with Matalon's enzymatic assay and gave birth to a healthy child. This was repeated for at least nineteen other couples over the next two years [3]. During this time,

Matalon, joined by his colleague Rajinder Kaul, moved to the Miami Children's Hospital (MCH) Research Institute. It was at the MCH, in the early 1990s, that Matalon's laboratory misdiagnosed four pregnancies that resulted in the birth of children with Canavan disease. The misdiagnoses also resulted in at least two lawsuits, which were settled [4]. Matalon and Kaul began looking for the gene, which offered the only reliable method for prenatal testing as well as carrier screening.

Guangping Gao, then a graduate student at Florida International University working in the lab and under the tutelage of Kaul, succeeded in cloning the Canavan gene by early 1993 [5]. The research drew on tissue samples provided to Matalon by the Greenbergs and over a hundred other families from around the world who had been stricken by the disease as well as blood samples provided by Josef Ekstein, executive director of Dor Yeshorim, Committee for Prevention of Jewish Genetic Diseases, in Brooklyn, New York. Rabbi Ekstein provided about six thousand stored blood samples that were used by the researchers to rapidly identify several mutations in Ashkenazi Jewish families and estimate population frequencies [6].

The Patent and Licensing

Unbeknownst to the families involved, a patent application was filed in September 1994, and U.S. Patent No. 5,679,635 was issued to the MCH Research Institute in October 1997. The MCH began to develop a marketing plan for its patent. At the same time, advocates from the Canavan Foundation in New York and the NTSAD in Boston were working with local and national groups to promote Canavan disease testing, and were successful in convincing the American College of Obstetricians and Gynecologists to issue guidelines recommending carrier screening of Ashkenazi couples [6]. Within two weeks after those guidelines were released, the MCH began sending letters to clinical laboratories informing them of the patent and the hospital's plans for commercializing the test. Ekstein received such a letter at Dor Yeshorim [7].

The marketing plan consisted of two stages of licensing. In the first one, a limited number of academic laboratories (likely to be a subset of the many already performing the testing) would be granted nonexclusive licenses to perform a limited number of tests per year. A fixed, $12.50 per test royalty was demanded, and a number of laboratories signed license agreements [8]. In the second stage, a large commercial laboratory would be licensed as a "market leader" with what would be in effect an exclusive license to the remainder of the testing volume. The justification for this plan was that a large reference laboratory would be able to spend the resources for outreach and education needed to ensure the screening and testing of all couples at risk. What this plan ignored was the role of community organizations—such as the NTSAD, local temple and consumer groups, and Dor Yeshorim—that were instrumental in making Tay-Sachs carrier screening widely available and used, and the United Leukodystrophy Foundation, which had developed a registry and screening program for Canavan disease promptly after the gene was discovered. Tay-Sachs testing methods have never been restricted, and the NTSAD and Dor Yeshorim stand as testaments to the ability of community-based organizations to develop and carry out population education and screening. In lieu of engaging these groups, the MCH sent letters indicating its intent to aggressively enforce the patent.

In 1999, the Canavan Foundation, the NTSAD, the National Foundation for Jewish Genetic Diseases, and the Canavan Research Fund created the Canavan Disease Screening Consortium. The Consortium started to bring pressure on the MCH to alter its licensing strategy, running advertisements and securing press coverage in the *Miami Herald* and several other newspapers [9, 10, 11]. The groups met with the MCH management in early 2000 (joined by two experts, including this author (Merz)), but were unable to secure any commitments to change the licensing plans. Shortly thereafter, in early April 2000, the MCH gave up its attempt to find a market leader laboratory. On April 3, 2000, the MCH mailed a letter to the consortium offering about $20,000 per year of an estimated $375,000 in royalties, to be used to increase public awareness and help provide testing to those unable to afford it. The consortium rightly believed that if it was unable to dissuade the MCH from collecting royalties on the test, then the hospital should dedicate some of the revenue to increasing awareness and access by the at-risk population. The offer, however, carried the condition that consortium members no longer be publicly critical of the MCH regarding the Canavan disease gene patent. In response, the consortium welcomed the financial help with its outreach programs, but refused to agree to restrictions on its right to free speech.

On October 31, 2000, Dan Greenberg, Dor Yeshorim, the NTSAD, and several other families who had participated in the research sued the MCH in federal district court in Illinois [12, 13, 14]. The case was moved to the Southern District of Florida in summer 2002. The trial court dismissed almost all of the plaintiffs' claims except one based on the equitable doctrine of unjust enrichment—the claim that the MCH profited unfairly at the plaintiff's expense [15]. Perhaps because of the difficulty of winning on this claim and the mounting costs of battling the hospital, the plaintiffs settled in August 2003 without any apparent concession on the MCH's part [16].

Ethical Issues

There are numerous ethical issues raised by this case. For example, one major concern is that any restriction on the delivery of clinical testing services may hamper medical practice as well as advances in medical training and knowledge (see chapter 3). Indeed, many new mutations have been discovered since the original article was published and the patent filed [17, 18, 19, 20]. This case raises other, salient public policy issues.

Business Secrecy

One issue that jumps off the page is the secrecy sought by the MCH. Secrecy is a fundamental value in business, but it strikes against the principles of public health and medicine [21]. The MCH sought to hide everything, from the meeting they agreed to hold with Canavan's activist families to the terms of the licenses with academic laboratories. They have even kept secret the identities of the laboratories they have licensed from those who most need the information.

Secrecy seems to have taken on some basic value, rather than being merely instrumental in the protection of trade information that could be competitively harmful to the organization. Given this, there simply is little known about how the licensing of genetic technology may be influencing the dissemination and use of tests. As the Canavan disease case exposes, licensing practices may impose serious costs and limitations on the availability of laboratory services.

Ethics of Research

According to some of the involved families, no consent form was used for much of the early research on Canavan disease, in violation of ethical norms (see chapter 5) [22]. Matalon simply collected blood, names, and social security numbers from affected children and parents. The families most directly involved helped by identifying, contacting, informing, and soliciting the participation of other affected families. It was these families that in 1994, suggested to Matalon that a consent form should be used, and helped generate a form that was used thereafter. According to Matalon, all his research was approved by the MCH's Institutional Review Board (IRB).

Furthermore, no participants ever suspected that the discovery of "their" gene could or would result in a patent [23]. These families—who had actively participated in Matalon's research enterprise in hopes of helping families like them avoid the ravages of Canavan disease—were thus dumbfounded when the MCH was issued a U.S. patent covering the genetic test for the disease and began to enforce it. Not only were the families uninformed about the scope of their research participation, they were betrayed by the ultimate commercialization and profiteering of the institution they believed was motivated, as were they, by altruism and the desire to help prevent this terrible disease. Participants should have a say about the nature of the benefits that may result from the research [24], and this becomes particularly acute when a community is involved closely in the performance of the research. In this case, there clearly was a mismatch between what participants and the community expected, and what the researcher and his institution sought and achieved.

Ownership of Genetic Invention

There were three necessary ingredients in the discovery of the Canavan gene. First was the active pursuit of research and willing study participation by families stricken by the disease. These parents were highly motivated by their desire to avoid births of children with Canavan disease in order to minimize the pain to families and their communities from this disease. They were willing to be research subjects so that medical science could be advanced, and their participation was a commonly accepted altruistic gift that would hopefully benefit them as well as others. Second was the effort by the researchers who worked on this particular research to build on the great advances in genetic technologies over the last two decades. These scientific advances were used effectively by Matalon, Kaul, and Gao in working from an identified enzyme to cloning the target gene. These individuals trained themselves to perform genetics research, and they performed wonderfully. Third was the financial support that was provided to these scientists to perform the research. Research funding totaling an estimated $5 million over several years was provided by numerous donors to the MCH Research Institute, including the NTSAD, the Canavan Foundation, various local Jewish organizations, and the United Leukodystrophy Foundation.

Interestingly, the patent only rewards those who have made financial or intellectual investments, not those who provided what Matalon (and other researchers) called "human resources" [25]. The families nonetheless, played what they believed to be significant roles in the research leading to the discovery of the Canavan disease gene. Dan Greenberg stated that the gene discovery was the result of an "extraordinarily productive partnership between lay parents and a medical researcher," and he felt a paternal role in the research that he described as being the "grandfather" of the discovery [22]. Reflecting a starkly contrasting view, Matalon stated that Greenberg played "no role" in the discovery of the gene, despite contributing samples, helping to identify and solicit families to participate in the research, and providing the aforementioned seed money for the early work [26]. Matalon's perspective apparently has changed over time because he was quoted in 1993 as saying, "This is a disease where a partnership between researchers and the families of affected children is critical for advancing knowledge for prevention, and, hopefully, for helping affected children" [27].

Some critics suggest that researchers may be characterized as being akin to predators in their approach to the persons who provide "resources" for their studies [28, 29]. If subjects are actively misled or even passively permitted by researchers and IRBs to proceed under false impressions and beliefs about issues in the research that could be material to their willingness to participate, then their consent is absolutely meaningless. As Fima Lifshitz, chief of the MCH's medical staff, was quoted in a *Miami Herald* article, "You voluntarily submitted a blood sample to be tested. As a result, I discover a gene that's patentable. What's wrong with that? This is done all the time. The issue should be quenched at once because these people are going to derive a great deal of benefit from this. They shouldn't be complaining" [9]. The problem with Lifshitz's analysis in part is that subjects were not submitting samples to be tested but were participants in an elaborate research scheme. More problematic is the assertion that the donation of blood samples was voluntary, when there was no informed consent, and as asserted above, there was such a complete mismatch between the expectations and goals of the subjects and the researchers.

Other researchers might have had equal success in identifying the Canavan gene; the only irreplaceable, critical resource—the sine qua non—in the discovery of the gene was the participation of the affected families. Ekstein's help in providing thousands of samples from his clinical collections was, simply, irreplaceable. Thus, it is extremely ironic that the system, in the end, fails to acknowledge their status and contribution. The ambiguous status of genes is perhaps best reflected in the quote of Gao, who performed the bulk of the laboratory research leading to the gene's discovery. At a press conference announcing the discovery, on being introduced to

Jacob Eisen, one of the subjects in the research, Gao said to Jacob's mother: "I cloned his gene. I held his gene in my hand. It's nice to meet him" [30].

References

1. Matalon, R., K. Michals, and R. Kaul. Canavan disease: From spongy degeneration to molecular analysis. *Journal of Pediatrics* 127 (1995): 511–517.

2. Matalon, R., K. Michals, D. Sebesta, M. Deanching, P. Gashkoff, and J. Casanova. Aspartoacylase deficiency and n-acetylaspartic aciduria in patients with Canavan disease. *American Journal of Medical Genetics* 29 (1988): 463–471.

3. Matalon, R., K. Michals, P. Gashkoff, and R. Kaul. Prenatal diagnosis of Canavan disease. *Journal of Inherited Metaboli Disease* 15 (1992): 392–394.

4. Winerip, M. Fighting for Jacob. *New York Times Magazine*, December 6, 1998, 56–63, 78–82, 112.

5. Kaul, R., G. P. Gao, K. Balamurugan, and R. Matalon. Cloning of the human aspartoacylase cDNA and a common missense mutation in Canavan disease. *Native Genetics* 5 (1993): 118–123.

6. American College of Obstetricians and Gynecologists, Committee on Genetics. ACOG committee opinion: Screening for Canavan disease. *International Journal of Gynaecology and Obstetrics* 65 (1999): 91–92.

7. Interview with Josef Ekstein, April 2000. The licensing arrangement was in part responsible for the breakdown of the testing arrangement that Dor Yeshorim had with Baylor University. Interview with Benjamin Roa, May 2000. Ekstein stated that there is "no question [the MCH's licensing program] is going to bring more Canavan's children into the world."

8. Rafinski, K. Hospital's patent stokes debate on human genes. *Miami Herald*, November 14, 1999, 1.

9. Peres, J. Genetic tests reduce neighborhood's grief; screening stops unwise matches. *Chicago Tribune*, September 12, 1999, 16.

10. Flap erupting over royalty for Canavan: Miami Children's Hospital exercises patent for test. *Forward* (New York), August 20, 1999, 15.

11. Saltus, R. Critics claim patents stifle gene testing. *Boston Globe*, December 20, 1999, C1.

12. *Greenberg v. Miami Children's Hospital*, Case No. 00C-6779 (N.D. Ill., 2000).

13. Gorner, P. Parents suing over patenting of genetic test. *Chicago Tribune*, November 18, 2000, 1.

14. Marshall, E. Families sue hospital, scientist for control of Canavan gene. *Science* 290 (2000): 1062.

15. Gitter, D. M. Ownership of human tissue: A proposal for federal recognition of human research participants' property rights in their biological material. *Washington and Lee Law Review* 61 (2004): 257–346.

16. See <http://www.canavanfoundation.org/news/09-03_miami.php> (accessed April 13, 2005).

17. Rady, P. L., J. M. Penzien, T. Vargas, S. K. Tyring, and R. Matalon. Novel splice site mutation of aspartoacylase gene in a Turkish patient with Canavan disease. *European Journal of Paediatric Neurology* 4 (2000): 27–30.

18. Rady, P. L., T. Vargas, S. K. Tyring, R. Matalon, and U. Langenbeck. Novel missense mutation (Y231C) in a Turkish patient with Canavan disease. *American Journal of Medical Genetics* 87 (1999): 273–275.

19. Kaul, R., G. P. Gao, R. Matalon, M. Aloya, Q. Su, M. Jin, A. B. Johnson, R. B. Schutgens, and J. T. Clarke. Identification and expression of eight novel mutations among non-Jewish patients with Canavan disease. *American Journal of Human Genetics* 59 (1996): 95–102.

20. Kaul, R., G. P. Gao, K. Michals, D. T. Whelan, S. Levin, and R. Matalon. Novel (cys152 > arg) missense mutation in an Arab patient with Canavan disease. *Human Mutation* 5 (1995): 269–271.

21. Rosenberg, S. A. Secrecy in medical research. *New England Journal of Medical* 334 (1996): 392–394.

22. Interviews with Judith Tsipis, Orren Alperstein Gelblum, and Daniel Greenberg, January 2000.

23. Indeed, in an interview, Matalon stated that he did not contemplate patent protection until he spoke with a colleague from the Scripps Research Institute, Ernest Beutler, who had several patents on mutations associated with Gaucher disease. See U.S. Patent Nos. 5,234,811 and 5,266,459.

24. Karlawish, J. H. T., and J. Lantos. Community equipoise and the architecture of clinical research. *Cambridge Quarterly Healthcare Ethics* 6 (1997): 385–396.

25. This is from Matalon's acknowledgment in the original enzyme article; see *American Journal of Medical Genetics* 29 (1988): 469–470.

26. Interviews with Reuben Matalon, Guangping Gao, and Rajinder Kaul, April 2000.

27. United Leukodystrophy Foundation. Press release: Canavan disease carrier screening program established by the United Leukodystrophy Foundation. Sycamore, IL: *United Leukodystrophy Foundation*, September 28, 1993.

28. Fox, R., and J. Swazey. *Spare Parts*. New York: Oxford University Press, 1992.

29. Andrews, L. and D. Nelkin. Whose body is it anyway? Disputes over body tissue in a biotechnology age. *Lancet* 351 (1998): 53–57.

30. Brecher, E. J. S. Florida research offers new insight into rare disease. *Miami Herald*, September 29, 1993, 1A, 21A.

4

Manufacturing Children: Assisted Reproductive Technologies and Self-Regulation by Scientists and Clinicians

As new discoveries are made in the biological sciences, policymakers can respond in several different ways. New regulations may be developed quickly to control the applications of a new technology. In many cases, however, the pace of scientific discovery is too rapid for policymakers to respond in a timely manner. In addition, since most policymakers are not trained scientists, their understanding of the science may limit their ability to judge what responses are appropriate. Finally, new developments may raise difficult ethical questions that complicate the development of new policy. This chapter will explore examples of biological developments for which regulatory policy has been developed within different timescales. These examples allow the examination of whether scientists are capable of "self-regulation"— whether they can judge the risks of new technologies and respond in ways to protect the public.

The Development of In Vitro Fertilization (IVF)

The first "test-tube baby," named Louise Brown, was born in Great Britain in 1978 and ushered in a new era of reproductive medicine. Two British researchers, Robert Edwards, a reproductive physiologist, and Patrick Steptoe, an obstetrician and gynecologist, had been conducting research on IVF since the 1960s.

Edwards began working with animal models, and joined with Steptoe in 1968 to study human applications. In 1969, Edwards and Steptoe published an article in *Nature* reporting that human oocytes had been successfully fertilized in vitro [1]. Their work was greeted with skepticism [2]. Most of their work was conducted with little funding from Great Britain's Medical Research Council (MRC), which expressed reluctance to fund research on human embryos. Edwards and Steptoe also skirted many of the requirements regarding the informed consent of human donors of tissues (see also chapter 5) [2]. Louise Brown's birth in 1978 was a media event, especially once it was ascertained that the baby was healthy; during the pregnancy, the health of the fetus was undetermined. Later that year, IVF babies were born in Australia and India; the first IVF baby was born in the United States in 1981. Since then, some hundreds of thousands of children around the world have been conceived via IVF.

The Biology of Assisted Reproductive Technologies (ARTs)

ARTs are methods to facilitate the joining of eggs and sperm necessary for human reproduction other than the natural means of sexual intercourse. The intent of ARTs is to permit individuals who are unable to reproduce through sexual intercourse (who may be infertile) to have their own children. The technology of ARTs arose as researchers and physicians gained knowledge about human reproductive physiology and embryology.

Infertility, defined as the inability to conceive after one year of trying, can arise from a variety of causes, and about equal numbers of men and women suffer from it [3]. The most common reasons for infertility in women are blocked fallopian tubes as a result of inflammatory disease or other causes, and failure to produce mature ova (anovulation). In men, infertility arises from low sperm counts, often caused by an abnormally high temperature in the scrotum as a result of abnormal blood flow. A second common cause is the poor motility of sperm because of malformation. As a result, the fertilization of eggs via sexual intercourse is not possible.

ARTs encompass a range of interventions based on the cause of infertility. They include any technological intervention to assist with reproduction, including medications that spur ovulation. In artificial insemination (AI), sperm is introduced into a woman's cervix or uterus using a catheter. The introduction is timed to match the woman's menstrual cycle, to increase the likelihood that a mature egg is present in her reproductive tract. AI may be employed if a woman's male partner is unable to carry out sexual intercourse, or stores sperm prior to chemotherapy or other treatment that may render him sterile. Donor sperm may be used if the male partner is already sterile or carries a genetic disorder. When donor sperm is used, the procedure is termed donor insemination (DI; earlier called "artificial insemination by donor"). Women who wish to become pregnant without a male partner may also use DI, although not all ARTs physicians will perform this because of their moral qualms about single, lesbian, or transgender parents. Donor sperm may be obtained from "sperm banks" that maintain collections frozen in liquid nitrogen. Sperm is kept frozen for six months to kill undetected viruses. Donors are repeatedly screened for HIV and other viruses, genetic diseases, and other illnesses. Recipients may select for the general physical characteristics of a donor (race, hair color, etc.), although the identity of the sperm donor usually remains anonymous. AI is a simple, noninvasive procedure. With good-quality sperm, success rates of around 75 to 80 percent are typical after a few months of treatment.

IVF is a procedure in which mature eggs (oocytes) are collected from a woman's ovaries and fertilized by sperm outside her body. IVF is often necessitated because the woman's fallopian tubes are blocked as a result of disease; eggs from her ovaries cannot reach the uterus. Newly fertilized eggs, or zygotes, are maintained for two to five days in a culture until they reach between the four to eight cell and blastocyst stages of development. Healthy embryos are then transferred into the uterus. In the best case, the embryos implant in the wall of the uterus, resulting in pregnancy and, hopefully, birth.

In contrast to AI, IVF requires both hormonal and surgical intervention for the woman providing the eggs. A woman undergoing IVF will be treated with hormones to stimulate her ovaries to produce more than the one egg typical in a given month (superovulation). Superovulation allows the collection of several oocytes at once, via laparoscopy, the insertion of a collection catheter with an attached fiber-optic camera through an incision in the abdominal wall. Alternatively, ova may be retrieved by threading a catheter through the vagina and uterus to the ovaries. If many eggs are collected and fertilized, extra embryos may also be frozen in liquid nitrogen (cryopreservation) to be used for later rounds of IVF. This avoids the need to collect more eggs if the first attempt at IVF is unsuccessful. Until recently, researchers have been unable to successfully maintain frozen unfertilized eggs, in contrast to sperm. Experimental technology is in development to permit the banking of unfertilized ova [4].

Two variations of IVF may also be employed. These approaches may improve the chances of a successful pregnancy by more closely mimicking the conditions occurring in natural reproduction. Zygote intrafallopian transfer involves the transfer of younger embryos into the fallopian tube of the woman, below the area of blockage. The embryo then migrates into the uterus in a manner similar to what occurs after natural fertilization. In gamete intrafallopian transfer, eggs and sperm are reintroduced into the fallopian tube where fertilization normally occurs. If fertilization takes place, the resulting embryo migrates into the uterus ready for implantation.

IVF is not without medical complications. Treatment with fertility drugs (combinations of gonad-stimulating hormones such as Pergonal) to increase ova production carries significant risk to the woman: if her ovaries are overstimulated, a serious or life-threatening medical condition, ovarian hyperstimulation syndrome, may result. The long-term effects of fertility drugs are not known; it has been suggested that treatment might increase the risk of cancer. In addition, harvesting ova subjects the woman to the risks of anesthesia and infection.

IVF may increase the chances of successful fertilization in cases of low sperm number by increasing the odds that a single sperm will find the ovum. If the sperm is unable to fertilize an egg because of defects, another technique may be employed. An individual sperm may be injected directly into an ovum's outer membrane; this triggers the normal response to fertilization. This process, known as intracytoplasmic sperm injection (ICSI), is a last resort for couples that wish to have their own biological children.

The source of eggs and sperm may not be restricted to a woman and her male partner; donor eggs and sperm, or donor embryos, may be used for IVF procedures. The use of donor eggs is common for older women, since the risk of genetic problems rises with the age of a woman. Some infertility problems may arise from genetic defects in mitochondria or the mis-expression of mRNA in the egg cytoplasm. The newest addition to the IVF armamentarium is oocyte cytoplasmic transfer: cytoplasm from a donor egg is injected into another egg prior to its fertilization. The goal is to replace missing cytoplasmic factors contributing to successful embryonic development. This technique is now used on an experimental basis.

In some cases, women may have medical conditions that preclude them from successful pregnancy. In these cases, couples (or in rare instances, individuals) may opt for a surrogate mother—a woman who acts as the gestational mother of a fetus. The baby is then given up to the couple after it is born. The surrogate may be inseminated by sperm or receive embryos created through IVF; she may or may not be a genetic parent of the fetus. Surrogacy can complicate the definition of parenthood: a baby may have genetic parents (those who provided the gametes), a gestational mother (the surrogate), and adoptive parents (the individual or couple who received the baby).

Because the embryos are typically maintained outside the body for a few days, embryos may be screened for a variety of genetic disorders by removing one cell from the early embryo (preimplantation genetic diagnosis [PGD]). Thus, IVF may be used to permit selection of embryos free of genetic disorders for introduction into the uterus. Sex and other selections may also be carried out using this analysis.

Despite twenty-five years of experience, the success rate of IVF techniques remains frustratingly low. In 2003, the national live birth delivery rate per initiated IVF procedure was 28 percent. Almost 122,800 IVF procedures were performed and over 48,700 babies were born [5]. About 87 percent of IVF cycles lead to the transfer of embryos to the uterus; the most common reason for failure is the inability to harvest viable eggs. If success rates are based on only those cycles that lead to transfer, then the success rate for IVF is 39 percent. Success rates are highest for women under thirty-five, and drop sharply for older women. The reasons for the low success rate are unclear, and may reflect the lack of information about human reproductive processes and embryonic development. In addition, most infertility is likely the result of a complex series of problems, not a single one. Sadly, many couples abandon IVF attempts at pregnancy only after spending large sums of money (see below for the economics of IVF).

One strategy to increase the IVF success rate is to increase the number of embryos transferred into a woman's uterus. Even if most embryos do not successfully implant, one or more may do so. The transfer of multiple embryos, however, increases the likelihood of twins, triplets, or higher multiple births. In 2003, over 56 percent of ARTs procedures in the United States involved the transfer of three or more embryos. In these cases, 31 percent of the births were twins, and 3.2 percent were triplet or higher-order births [5]. The overall multiple birthrate was 34.2 percent.

Multiple fetuses pose serious medical risks to the fetuses and their mother. The incidence of miscarriage, stillbirth, premature birth, birth defects, and long-term disability are all significantly higher as the number of fetuses increases. The medical costs of complications resulting from multiple-order pregnancies can range up to millions of dollars. Multiple births increase the risk of hypertension, hemorrhage, and other complications of pregnancy for women. If a woman finds she is carrying multiple fetuses, the number may be reduced by selectively aborting one or more fetuses. The remaining fetus(es) may have a better chance for developing normally. Nevertheless, fetal reduction carries considerable ethical challenges for parents and policymakers.

Is IVF Safe?

There is general agreement that multiple pregnancies pose a considerable risk to both fetuses and their mothers. Recent evidence suggests that singleton babies conceived via ARTs also may be at greater risk for low birth weight and birth defects [6, 7]. Yet the risks may be more related to infertility-associated issues, such as maternal age, and not to the procedures themselves. Data are lacking to determine whether children born using ICSI, in which defective sperm are used to fertilize eggs, will have developmental or fertility problems; the technique is simply too new to yield data.

The Regulation of ARTs in Great Britain

The regulation of IVF in Great Britain arose because of widespread public concern about embryo research and its applications, as IVF clinics began to expand after the birth of Louise Brown. In 1979, the MRC released a series of recommendations that, while endorsing the idea that there was no inherent moral problem with embryo research, tightened the requirements for IVF research with human subjects, put strict limits on how long embryos might be maintained in vitro, and indicated that if evidence accumulated that IVF posed a risk of abnormal offspring, the public would be advised. The MRC also limited access to IVF to married couples [2, 8]. Similar actions were also taken in Australia.

In 1984, the *Warnock Report* recommended that IVF clinics in Great Britain be subject to strict licensing and regulatory control [2]. An interim licensing authority was established in 1986 that recommended a series of guidelines for clinic and staff certification, reporting requirements, and the storage of embryos, limits on the number of embryos transferred to the uterus, and access to IVF. The board had no statutory authority to enforce these guidelines. The profession nonetheless responded well to the call for voluntary self-regulation until the Human Fertilization and Embryology Authority (HFEA) was established in 1990. The HFEA provides for the licensing and regulation of IVF clinics, treatment involving the use of donated gametes, storage of gametes and embryos, and limits on the uterine transfer of embryos (in general, two, though in some cases, three). It permits research on human embryos after preapproval by an ethics committee. The HFEA also mandates patient counseling prior to treatment and sets rules of disclosure of patient information. Preimplantation genetic testing remains tightly regulated. The HFEA has been criticized in recent years because it still limits access to IVF to married couples, even though the technology may benefit other people, such as lesbian couples [9].

The Regulation of IVF in the United States

The history of ARTs regulation in the United States is in marked contrast to Great Britain. After Edwards's and Steptoe's publications on IVF in 1969, public concern about the potential misuse of the method led to congressional hearings in 1971. Leading life scientists were asked by Congress to express opinions on the new technology. James Watson, Nobel laureate, acknowledged the likelihood that IVF would be a clinical reality within ten years, but offered strong moral reservations about leaving control of human embryo manipulation in the hands of scientists [10, 11]. Another critic was Leon Kass, a physician and bioethicist who served as a member of the Commission on Life Sciences at the NRC. Kass was a vocal opponent of IVF on two grounds: first, since the safety of the procedure was unknown, it was immoral to risk the health of future children; and second, that it distorted the natural concept of parenthood. He suggested that the profession self-regulate and impose a moratorium on research to allow for further discussion on the procedure's morality [12]. Kass's involvement in the early controversy over IVF presaged his policy engagement some thirty years later as chair of President George W. Bush's Council on Bioethics (see below).

In 1973, the U.S. Supreme Court ruled on *Roe v. Wade* (410 U.S. 133), which legalized abortion throughout the United States. The ruling energized a number of right-to-life groups, which focused their efforts on protecting fetuses. In their view, a fertilized human egg was equivalent to a human life. Also in 1973, the U.S. Department of Health, Education, and Welfare (DHEW, now DHHS) drafted guidelines that prohibited the federal funding of research involving human embryos unless each proposal was reviewed and approved by an Ethics Advisory Board (EAB) [13]. No EAB was created, so no research was funded. In 1977, however, Pierre Soupart, a reproductive endocrinologist at Vanderbilt University, submitted a proposal to the NIH to study the safety of IVF by fertilizing human eggs in vitro and looking for chromosomal abnormalities. After a delay, an EAB (with a mandate to expire in 1980) was created to review the proposal. Before the panel could complete its review, Louise Brown was born in Great Britain in July 1978. In September 1978,

Joseph Califano, the DHEW secretary, sent a letter to the EAB chair instructing the panel to consider the ethical and moral implications as well as the medical aspects of IVF.

The EAB released its report in May 1979 [14]. The board found no ethical objection to the federal funding of research on human embryos, although it did not specify what priority such research should have for funding. The EAB recommended funding of Soupart's proposal and future research, provided that it complied with provisions regulating human subjects research, informed individuals would voluntarily donate gametes used in the research, embryos would not be maintained longer that fourteen days in vitro, and if evidence of risk of abnormality appeared, the public would be informed. The board recommended that any embryo transfers be limited to married women, but argued against broad government regulation of reproduction. It also recommended that government agencies work with private organizations to assure that data would be available to all interested parties.

To this point, the history of regulation closely parallels events in Great Britain. What happened after the release of the EAB report is quite different. After receiving thousands of letters opposed to embryo research, the DHEW tabled Soupart's grant application without ever formally denying funding. The EAB's charter was allowed to lapse in 1980, with no action taken to reconstitute it [13]. The federal government effectively blocked federal funding of embryo research by the absence of the mandated EAB to review applications. It also failed to act on the EAB's oversight recommendations, leaving the growing ARTs industry unregulated. In 1993, in a different climate, Congress passed the NIH Revitalization Act; the legislation lifted the requirement for EAB approval for federal funding of human embryo research. The NIH then established a Human Embryo Research Panel (HERP) to make recommendations about what types of research might be acceptable for funding. Congressional opposition to research on human embryos shifted when the Republican Party took control of the House of Representatives, though, and in 1996, Congress passed legislation prohibiting the use of federal funds for research on human embryos as an attachment to the appropriations bill for the NIH (H.R. 3061, Sec. 510). The clause was added annually to the appropriations legislation until 2001, when President Bush issued an executive order permitting limited research on stem cells derived from human embryos (see "Brave New World" section in this chapter) [15].

In the meantime, clinics offering IVF and other ARTs increased rapidly in response to the growing demand by infertile couples. The first center in the United States,

Norfolk General Hospital in Virginia, began offering IVF in January 1980. By 1990, over three hundred clinics offered IVF services; this number increased to over four hundred in 2002 [5]. In 2002, about 1 percent of all births in the United States made use of ARTs procedures [5].

A single piece of federal legislation related to ARTs, the Fertility Clinic Success Rate and Certification Act, was enacted in 1992 [16]. It mandates that each clinic conducting IVF and related procedures file an annual report with the Centers for Disease Control (CDC) containing data on the number and type of procedures conducted, pregnancy and birth rates, and information concerning the cause of infertility and the age of patients. The legislation also mandated that the secretary of the DHHS develop a program for the certification of ARTs programs. In addition, the Clinical Laboratory Improvement Amendments of 1988 regulate clinical testing facilities [17]; fertility clinics may fall under its jurisdiction if certain kinds of testing (for HIV, say) are conducted.

The CDC released guidelines for ARTs clinics in 1999; they are intended to assist states in developing their own certification programs. No federal oversight is mandated. Also in 1999, the FDA proposed a rule for "suitability determination for donors of human cellular and tissue-based products" [18], a part of a comprehensive program to regulate human cellular and tissue-based products [19]. The rules are intended to reduce the risk of HIV transmission or other viruses through tissue transfers, although both egg and sperm donation would fall under their auspices. Both industry and private groups have expressed opposition to portions of the proposed rules, and the enactment of the rules was delayed. The FDA has also claimed statutory jurisdiction over oocyte cytoplasm transfer under regulations involving investigational new drugs (21 CFR Parts 50, 56, and 312).

Since the federal government did not establish regulatory oversight of IVF and other ARTs, it was left to the profession and individual states to develop their own regulations. The American Society for Reproductive Medicine (ASRM) and the Society for Assisted Reproductive Technology (SART) took the lead in developing recommendations for the appropriate practice of ARTs. The ASRM established guidelines in the 1990s on clinic standards and accreditation, the number of embryos that may be transferred, and other issues such as the use of preimplantation genetic diagnosis. The organizations also work with the CDC in gathering data on IVF on an annual basis. Most fertility clinics are accredited by either the College of American Pathologists Reproductive Laboratory Accreditation Program or the Joint Commission on Accreditation of Healthcare Organizations [5]. Best practices guidelines are only voluntary, however, and concern remains that abuses occur. For example, the ASRM recommends that no more than two embryos be transferred simultaneously into the uterus (depending on the quality of the embryos and the age of the recipient); statistics show that over 62 percent of IVF procedures involved the transfer of three or more embryos [5]. Clinics can boost their success rates by denying services to women over thirty-five, or to those whose infertility results from complex causes. Some clinics report pregnancy rates, not birthrates, thereby inflating their statistics [20].

At the state level, individual states have passed legislation requiring reporting by fertility clinics, mandating insurance coverage for certain infertility services, providing coverage under state Medicaid programs for limited services, and either supporting or banning research on human embryos. Because DI has been practiced since the 1950s, most states have statutes concerning the legitimacy of offspring. Reporting requirements vary from state to state. A small number of states, including California and New York, require the licensing of sperm banks.

About twenty-five states have passed legislation regarding surrogacy. While most states do not directly ban surrogacy arrangements, some ban compensation and others view written contracts as illegal [21, 22]. The current state legislation is conflicting and inconsistent. Existing laws against baby selling further complicate surrogacy agreements. As discussed below, states have also weighed in on parenthood issues.

Why did the federal government fail to establish a regulatory framework for ARTs? It appears likely that legislators were uncomfortable grappling with an issue entangled with the debate about abortion. Proposing legislation regulating IVF and other ARTs might be construed as tacit support for abortion. Second, Congress has tended to leave the regulation of medical practices to the medical profession and the states; while the FDA regulates drugs and medical devices, it does not regulate physicians [23]. Finally, reproduction is viewed as a private matter, although as demonstrated below, ARTs have brought reproductive issues into the public eye.

The Economics of ARTs

IVF and other ARTs are big business. Infertility affects an estimated 6.1 million women and their partners in the United States [24, 25], and about one in every six couples seeks fertility advice. Yet fewer than 5 percent make use of IVF and other more invasive procedures. In 2002, over 115,000 IVF procedures were reported to the CDC by 391 of the over 400 infertility centers [5]. These numbers do not include DI and other insemination procedures or hormonal treatments that did not involve embryo transfers. Since there are no reporting requirements for these procedures, the number conducted cannot be determined on a national basis.

Over one hundred sperm banks operate in the United States [26]. From twenty to thirty thousand babies are born each year conceived via DI. Typical charges for DI range from $200 to $400 per vial of sperm, not including shipping, insemination procedures, and other medical evaluations. Since several cycles are usually required for success, costs may reach several thousand dollars per live birth.

In 1988, the Office of Technology Assessment (OTA) estimated that $1 billion was spent on infertility-related health care, of which $66 million was spent on IVF procedures [27, 28]. The number of IVF procedures rose dramatically in fifteen years, while costs did not decline. Recent estimates for the cost of IVF in the United States range from $9,550 to $12,400 per cycle [24, 29]. This suggests that upward of $1 billion was spent on fertility treatments reported to the CDC alone. These costs include initial medical consultations, prescribed drugs, laboratory charges, ultrasound procedures, payments to "donors," IVF procedures (egg retrieval and embryo transfer), hospital charges, and other administrative and medical costs. The estimated cost per live birth is over $58,000 [29]. Complications related to multiple births can raise the cost of twins or higher-order multiples severalfold.

The brunt of these charges are borne by those seeking ARTs services, since most health insurance policies do not cover the cost of fertility treatment; an estimated 85 percent of IVF costs are paid by the patient [28]. Only fifteen states have passed legislation requiring insurance companies that cover pregnancy-related expenses to cover treatments for infertility [30]. But some of these states allow insurance companies to exclude coverage for IVF, and others only require that states offer policies that cover infertility. Legislation mandating insurance coverage for infertility has been repeatedly introduced in Congress, but no laws have been enacted.

The U.S. situation is in marked contrast to most European countries, where IVF is generally covered by insurance and costs are capped. The utilization rates of ARTs are higher in many European countries, while the costs are much lower [29, 31, 32]. For example, the utilization rate of IVF per million people in Great Britain was four times higher than in the United States, while the projected average cost per IVF cycle was only $2,955 [29].

Who Has Access to ARTs?

Although infertility is broadly distributed across the U.S. population, the vast majority of patients at fertility clinics are white, over thirty, educated, and middle class [28]. Poor women, especially minorities, have limited access to infertility services because of their inability to pay. States that offer insurance coverage for IVF have higher utilization rates than those that do not, but lower success rates [33]. The latter may be the result of a higher number of older women attempting IVF because it is covered by insurance. Notably, states in which insurance coverage was mandated had lower numbers of embryos transferred per cycle and lower rates of higher-order multiple births [33, 34]. For patients without insurance coverage, the economics of IVF may drive the demand for the transfer of a higher number of embryos, in hopes of a successful outcome. Since multiple births substantially increase the costs of IVF, these results suggest that insurance coverage might decrease the overall cost of IVF. Nonetheless, there are justice issues raised about society paying the costs for infertile individuals who wish to have their own children, but cannot afford the expensive medical care associated with ARTs, particularly when an overwhelming number of children in foster care go unadopted.

The popularity of donor gametes provides other economic opportunities. Sperm donors are typically paid around $75 to $100 per donation. The situation for egg donors is quite different. Over twelve thousand (11 percent) of the IVF procedures in 2002 involved the use of donor eggs [5]. It is not unusual to see advertisements in both college and leading newspapers asking for healthy young women (typically tall, white, athletic, and with high SAT scores) to serve as egg donors. Egg donors are typically paid from $3,000 to $5,000 per cycle, but couples seeking donors with particular characteristics may offer as much as $20,000 to $100,000. The evidence indicates that potential donors may be misled or misinformed about the medical risks associated with egg donation [22].

The commercial surrogacy industry began in California in the 1970s. About sixty centers nationwide offer surrogacy services [22]. Centers serve as brokers for interested couples and women interested in serving as surrogates. Surrogates may be paid from $10,000 to $20,000 and have their medical expenses covered. Additional fees are paid to the center for medical and administrative costs. Generally, agreements concerning parenthood are signed prior to initiating a pregnancy. These agreements often impose conditions on the surrogate regarding personal behaviors such as diet and drug use (tobacco, alcohol, and illegal drugs). Surrogacy arrangements may also be made between family members; in a notable case in 1991, Arlette Schweitzer gave birth to twins after being implanted with embryos generated from her daughter's ova and her son-in-law's sperm [22].

Ethical and Policy Challenges

ARTs raise a number of challenging questions regarding definitions of parenthood and the "rights" of embryos. There is considerable tension between those who believe that embryos are equivalent to any other human life and those who do not. Current policy is inadequate to guide decisions when conflicts arise.

Defining Parenthood

The use of ARTs has changed the definition of parenthood [23]. Babies can now have multiple parents: donors of ova and sperm are genetic parents, women who carry the fetuses to term are gestational mothers, and in some cases, babies are given up to adoptive (or intended) parents (babies created using oocyte cytoplasmic transfer may have two genetic mothers). Few laws have been passed regarding issues of parenthood. States tend to enact unique policies, leading to inconsistencies. One attempt to standardize law is the drafting of model uniform legislation to be adopted by each state. In the Uniform Parentage Act (UPA) of 1973, prepared by the National Conference of Commissioners on Uniform State Laws (NCCUSL), issues of the legitimacy of children born via DI were spelled out, and the husband of the wife receiving insemination was presumed to be the child's parent, assuming the husband had agreed to the procedure. Nineteen states adopted the proposed law; other states made their own laws.

The expansion of ARTs left the 1973 UPA hopelessly out of date. Issues of parentage were now left to the courts for resolution. A series of highly publicized court cases (see below) demonstrated the need for new laws defining parenthood. In 2000, the NCCUSL proposed the UPA of 2000, which contains detailed language regarding definitions of parenthood for children born of ARTs [35]. Central to the argument are the intentions of involved parties: donors of eggs and sperm are presumed to relinquish parental rights unless they specify otherwise. The husband of a woman undergoing IVF procedures using donated gametes is presumed to be the father of any children, if he consents to the procedure. The 2000 UPA also recognizes that there may be conditions under which children have no legal father—for example, a single or lesbian woman who undergoes DI or IVF with donated sperm. In cases of surrogacy, the child's parents are presumed to be those who *intended* to be parents—the individuals who contracted with a woman to serve as a gestational mother for the child. UPA was approved by the American Bar Association in 2002, and went to the individual states for possible adoption. As of July 2004, only four states had adopted the 2000 UPA [36].

The Courts Step In

Until the 2000 UPA or other bills become law, conflict resolution regarding parenthood takes place in the courts. Cases about parenthood have revolved around two areas: surrogacy and rights to frozen embryos. In 1988, the New Jersey Supreme Court decided the Baby M case [37], which reached national prominence. William Stern and his wife contracted with Marybeth Whitehead to serve as a surrogate. Whitehead was inseminated with Stern's sperm, and thus was both the genetic and gestational mother of the child. When the baby girl was born, Whitehead was unwilling to give her up; she took the child and fled to Florida. The trial court awarded custody to the Sterns and terminated Whitehead's parental rights. On appeal, however, the Supreme Court voided the lower court decision because the contract between the Sterns and Whitehead violated state laws prohibiting baby selling and prebirth adoption agreements. Whitehead was found to be the legal mother and granted visitation rights; the baby remained with the Sterns. Stern's wife has no legal relationship with the child.

In *Johnson v. Calvert* [38], Anna Johnson served as the gestational mother for the Calverts, who provided both the egg and the sperm. The California Supreme Court ruled that although a gestational mother might be defined as a parent, the intent was clear—that the Calverts had intended to become parents and initiated steps to achieve this. In this case, the genetic relationship with the baby was deemed more important than the gestational one. An even more complicated case involving gestational surrogacy was *In re Marriage of Buzzanca* [39]. The Buzzancas sought to have a child using a surrogate and *donated* gametes. Several weeks prior to the birth of the child, John Buzzanca filed for divorce and claimed he was not the child's father to avoid paying child support. Mrs. Buzzanca sued for custody and sought child support from her ex-husband. The trial court ruled that the child had *no* legal parents, since neither the Buzzancas nor the surrogate had a genetic relationship to the child, the surrogate had disavowed parenthood by signing the surrogacy arrangement, and the identities of the gamete donors were unknown. On appeal, the California Court of Appeals ruled that the Buzzancas were the child's legal parents since they initiated a medical procedure with the intention of becoming parents. The court rejected the more traditional arguments that linked maternity to either genetics or gestation. These cases demonstrate the need for consistent law regarding parenthood, such as that proposed by the 2000 UPA.

The Right Not to Procreate

In many cases, all embryos are not used in IVF procedures. If a couple divorces, should one member of the pair be permitted to use the embryos to produce more children? At issue is whether there is an inherent right to procreate (or not to procreate). Several cases at the state level make clear that the courts determined that an individual may not be forced to become a genetic parent against his or her wishes. Implicit in this reasoning is the presumption that an embryo itself is not a living human with "parents." For example, in *Davis v. Davis* [40], a divorced couple disagreed on the fate of seven frozen embryos. Mrs. Davis wished to have the embryos donated to another couple; Mr. Davis wished to have the embryos destroyed. He claimed that he had a right *not* to become a parent. The Tennessee Supreme Court ruled that Mr. Davis could not be forced to become a parent and that the embryos could not be transferred.

Inheritance Issues

It is possible for women to conceive children from the sperm of their deceased husbands if the sperm is stored in a sperm bank or is collected immediately after the man's death. Might such children be heirs? Colorado, Louisiana, North Dakota, and Virginia approved statutes clarifying the legal parentage in cases of posthumous conception [36]. In *Woodward v. Commissioner of Social Security* [41], a case in Massachusetts, the court ruled that the marriage ended with the death of the husband and that any children conceived posthumously using stored sperm were born outside the marriage. The state was not obligated to provide Social Security survivor benefits to such children. The mother has to provide unequivocal evidence for both the paternity of the child and the prior consent of the sperm donor (in this case, her deceased husband) in order for the child to inherit [42].

In 1984, a wealthy American couple, Mario and Elsa Rios, died in a plane crash, leaving two "orphaned" embryos frozen at a fertility clinic in Melbourne, Australia. Speculation arose as to whether these embryos, if implanted in a surrogate, might inherit the Rios's fortune. The court in Victoria (Australia) ruled that the embryos might be donated to another couple, but any resulting children would not inherit. This ruling was affirmed by a California court, which awarded the Rios estate to surviving relatives [43]. The Rios case raised complex issues about the fate of frozen embryos, as discussed below.

The Fate of Frozen Embryos

As of 2003, there are an estimated four hundred thousand embryos in frozen storage in the United States [44]. As the cases above demonstrate, the existence of frozen embryos maintained in fertility clinics creates a number of policy challenges [43, 45]. Who controls the fate of embryos? Are embryos "property" or are they potential humans with "rights"? What may be done with unwanted embryos (donated to others, destroyed, used in research)? Case law provides limited guidance on these issues.

A couple undergoing IVF procedures typically completes an agreement prior to initiating procedures that direct how their unused embryos should be used: embryos may be stored indefinitely for the couple's use, donated to other couples, donated for research purposes, or destroyed. Couples who successfully have children using IVF may effectively "abandon" embryos that remain frozen, though. As the number of frozen embryos increases, fertility clinics are under pressure to make determinations as to their fate. The vast majority (88 percent) of frozen embryos are designated for parent use [44].

Should clinics be allowed to dispose of unclaimed embryos after perhaps ten years? Great Britain and several other countries have passed controversial laws permitting the disposal of embryos after waiting periods of five or more years. In the United States and elsewhere, some organizations strongly oppose the disposal of embryos; they argue that embryos are "prehumans" and should be transferred to permit them to fully develop. This contention flies in the face of the general perception (as supported by the courts) that forcing gamete donors to become genetic parents against their will is unethical. The proposal to transfer unwanted frozen embryos also has a technical problem. Although recent improvements in technology have enhanced embryo viability after cryopreservation, embryos maintained in long-term storage were frozen using less effective methods. Many frozen embryos will most likely not thaw out to viable, healthy embryos for transfer [44].

A criminal case in California provides another example of the complex emotional and ethical elements that influence the development of policy. In 1994, whistleblower employees of a fertility clinic associated with the University of California at Irvine (UCI) revealed that eggs and embryos had been taken from couples without their consent and used for IVF procedures in other women. The recipients were unaware that the embryos or eggs had not been provided with the consent of the donors. As many as a hundred babies may have been born using "stolen" embryos. The physicians directing the clinic were charged with insurance fraud and other crimes, but not with embryo theft. There was no law against the misappropriation of embryos. In 1996, the state of California passed a law making embryo theft a felony. UCI ultimately settled close to a hundred lawsuits filed by individuals whose eggs or embryos had been used without their permission [46].

The issue of what to do with frozen embryos has languished, and only the new potential for using embryos to generate stem cells (see the "Brave New World" section) has stimulated minimal state legislation. Only California and New Jersey currently explicitly allow research on embryos. In contrast, North Dakota bans all research on embryos. Other states have passed a hodgepodge of laws that limit research in different ways.

Preimplantation Genetic Diagnosis
PGD has been available since 1990; its use increased in recent years. Prenatal screening via amniocentesis has been available for many years, and is strongly recommended for women over the age of thirty-five or for couples at risk for known genetic diseases. The essential difference between amniocentesis and PGD is that in the latter, a positive form of selection occurs; only "healthy" embryos are selected for transfer [47]. By choosing embryos free of genetic flaws, the odds of having a healthy child increase, and the moral dilemma and health risks of abortion are minimized.

PGD may be useful for couples who do not suffer from infertility. For example, couples with a known risk for single gene disorders, such as Tay-Sachs disease or sickle cell anemia, may also opt for PGD to avoid transferring an embryo carrying a defective gene. A more controversial application was the case of John and Lisa Nash, who used PGD not only to select for healthy embryos but also for an immunologic match so that the resulting child could serve as a bone marrow donor for an older sibling suffering from a genetic disease [22]. PGD may also be used to select for a particular trait, such as deafness or dwarfism, which may be viewed by some as a disability but by those in the community of people possessing such traits as normal and desirable.

An ethical challenge of PGD is its potential for screening for nonmedical traits. As information about more genes becomes available, it will be possible to screen for a variety of risk factors for diseases or nonmedical physical characteristics. Both physicians and ethicists express increasing unease about where to draw the line between acceptable and unacceptable applications of PGD as well as other screening techniques. Most problematic is the use of PGD for nonmedical sex selection. When using amniocentesis for sex determination, parents wishing a child of a particular gender might choose to terminate a pregnancy if the fetus was of the undesired sex. The same procedure can be carried out prior to implantation using PGD; only embryos of the desired sex would be transferred into the uterus. The motivations surrounding sex selection are complex. While some reasons, such as "balancing" one's family, may seem justified, others may appear sexist and biased. If cultural belief systems value one gender over the other, it is possible that significant skewing of natural sex ratios may occur [47]. In some regions of China and India, sex selection has led to wildly unbalanced population sex ratios, with males comprising more than 60 percent of the babies born [48].

Should PGD be regulated? The HFEA in Great Britain has legal authority over PGD practices and rules on potential new applications. No similar authority exists in the United States, and PGD use is left to the discretion of patients and physicians. As of 2003, about ten thousand children have been born in the United States after using PGD [47]. Because of the expense of IVF, it seems unlikely that PGD use will be widespread. Nevertheless, in 2004, the President's Council on Bioethics called for the regulation of PGD as part of a broad package of proposals concerning ARTs [49].

Lessons from ARTs

The ARTs industry in the United States has developed almost entirely unregulated. Despite attempts by professional organizations to encourage best practices, the potential for abuse remains high. The population that seeks treatment for infertility is a highly vulnerable one; individuals may not be in a position to rationally search out treatment options and select the best ones. The costs of IVF procedures are high, and show no sign of declining, despite improved technology. Access to IVF is inequitable; the absence of broadly mandated insurance coverage limits access to only those who can afford it. The high frequency of multiple births that result from IVF raises the cost of medical care for everyone, since complications resulting from pregnancies are generally covered by insurance. Many clinics limit access to IVF to married couples, leaving individuals and unmarried or nonheterosexual couples who wish to share in procreation barred from the experience.

The existence of ARTs has complicated legal definitions of parenthood, and the absence of consistent state laws has led to resolution by litigation. This is an expensive and inefficient approach to answering fundamental policy questions. The legal and moral status of frozen embryos also remains unresolved, even as their numbers keep growing. Surrogacy is particularly problematic for feminists, since the potential for the commodification of women for reproductive purposes is evident.

Have scientists and physicians done a good job of self-regulation? Depending on one's viewpoint, ARTs are either a boon to the infertile or an example of technology run amok. Many express concern that infertility treatment has become a "business"—best medical practice may be neglected in search of profit. IVF has created a market for human gametes with escalating prices and increasing selection criteria. It is this creation of a reproductive industry that most strongly suggests that self-regulation has not been successful, even if evidence of direct harm is lacking.

Despite these problems, many happy couples have been able to become parents using ARTs. For them, the hazards and costs of the procedures are not an issue; their children are the rewards. Between 1981 and 2000, over two hundred thousand babies were born in the United States via IVF [24]. As success rates improve, more infertile persons may be able to have children of their own.

Should ARTs be regulated? The Supreme Court cases *Roe. v. Wade* (1973), which legalized abortion, and *Griswold v. Connecticut* (1965), which legalized access by married couples to contraception, attest to the view that the Bill of Rights protects "private matters" such as reproduction. Nevertheless, the highly publicized and dramatic abuses of ARTs suggest that some measure of industry regulation may be needed. The President's Council on Bioethics gingerly proposed some regulatory steps in 2004: requiring better reporting and enforcement of current guidelines, placing moratoriums on activities such as the creation of human-animal hybrid embryos (see also "Brave New World" section), and prohibiting the sale of embryos [49]. These proposals reflect the conservative viewpoints of a majority of members of the council; whether Congress chooses to enact them is uncertain.

References

1. Edwards, R. G., P. C. Steptoe, and J. M. Purdy. Fertilization and cleavage in vitro of preovulator human oocytes. *Nature* 227 (1970): 1307–1309.

2. Gunning, J., and V. English. *Human in vitro fertilization: A case study in the regulation of medical innovation.* Aldershot, UK: Dartmouth Publishing Company Ltd, 1993.

3. laVelle, L. B. Starting human life: The new reproductive technologies. In *Bioethics for scientists*, ed. J. Bryant, L. B. laVelle, and J. Searle, 201–231. Chichester, UK: John Wiley and Sons, Ltd., 2002.

4. McGee, G. *The perfect baby: Parenthood in the new world of cloning and genetics.* 2nd ed. Lanham, MD: Rowan and Littlefield Publishers, Inc., 2000.

5. Centers for Disease Control. 2003 Assisted reproductive technology success rates: National summary and fertility clinic reports. *CDC.* December 2005. Available at <http://www.cdc.gov/ART/index.htm> (accessed December 23, 2005).

6. Kovalevsky, G., P. Rinauldo, and C. Coutifaris. Do assisted reproductive technologies cause adverse fetal outcomes? *Fertility and Sterility* 79, no. 6 (2003): 1270–1272.

7. Lambert, R. D. Safety issues in assisted reproductive technology: Aetiology of health problems in singleton ART babies. *Human Reproduction* 18, no. 10 (2003): 1987–1991.

8. Waller, L. Development of regulation of assisted reproduction: A world view of early days. In *The regulation of assisted reproductive technology*, ed. J. Gunning and H. Szoke, 67–74. Aldershot, UK: Ashgate, 2003.

9. Deech, R. The HFEA: 10 years on. In *The regulation of assisted reproductive technology*, ed. J. Gunning and H. Szoke, 21–38. Aldershot, UK: Ashgate, 2003.

10. Gunning, J. Regulating ART in the USA: A mixed approach. In *The regulation of assisted reproductive technology*, ed. J. Gunning and H. Szoke, 55–66. Aldershot, UK: Ashgate, 2003.

11. Grobstein, C. *From chance to purpose: An appraisal of external human fertilization.* London: Addison-Welsey Publishing Co., 1981.

12. Kass, L. R. *A conversation with Dr. Leon Kass.* Washington, DC: American Enterprise Institute for Public Policy Research, 1979.

13. Bonnicksen, A. *In vitro fertilization: Building policy from laboratories to legislatures.* New York: Columbia University Press, 1989.

14. Ethics Advisory Board. *Report and conclusions: HEW support of research involving human in vitro fertilization and embryo transfer.* FR Doc. 79–18925. Washington, DC: Department of Health, Education, and Welfare, May 4, 1979.

15. Center for Science TaC. AAAS policy brief: Stem cell research. *AAAS.* August 26, 2004. Available at <http://www.aaas.org/spp/cstc/briefs/stemcells/index.shtml> (accessed February 20, 2005).

16. The Fertility Clinic Success Rate and Certification Act of 1992, Pub. L. 102–493, 106 Stat. 3146 (1992) (codified at 42 U.S.C. §§ 263a-1 et seq.).

17. Clinical Laboratory Improvement Act of 1988, Pub. L. 100–578, section 2, 102 Stat. 2903 (1998) (codified at 42 U.S.C. 263a et seq.).

18. Department of Health and Human Services, Food and Drug Administration. Suitability determination for donors of human cellular and tissue-based products. *Federal Register* 64, no. 189 (September 30, 1999): 52696–52723.

19. Human cells, tissues, and cellular and tissue-based products: Establishment registration and listing. *Food and Drug Administration* 21 (2001): CFR parts 207, 807, and 1271.

20. Caplan, A. Ethics of in vitro fertilization. *Primary Care* 13 (1986): 241–254.

21. The American Surrogacy Center. I. Legal overview of surrogacy laws by state. *TASC, Inc.* March 1997. Available at <http://www.surrogacy.com/legals/map.html> (accessed October 17, 2003).

22. Merrick, J. C., and R. H. Blank. *Reproductive issues in America: A reference handbook.* Santa Barbara, CA: ABC CLIO, 2003.

23. Blank, R., and J. Merrick. *Human reproduction, emerging technologies, and conflicting rights.* Washington, DC: CQ Press, 1995.

24. American Society for Reproductive Medicine. Fact sheet: In vitro fertilization (IVF). *ASRM.* Available at <http://www.asrm.org/Patients/FactSheets/invitro.html> (accessed October 10, 2003).

25. American Society for Reproductive Medicine. Frequently asked questions about fertility. *ASRM.* Available at <http://www.asrm.org/Patients/faqs.html#Q6> (accessed October 13, 2003).

26. Schuster T. G., K. Hickner-Cruz, D. A. Ohl, E. Goldman, and G. D. Smith. Legal consideration for cryopreservation of sperm and embryos. *Fertility and Sterility* 80, no. 1 (2003): 61–66.

27. U.S. Congress Office of Technology Assessment. *Infertility: Medical and social choices.* OTA–BA–358. Washington, DC: U.S. Congress, 1988.

28. Ryan, M. A. *The ethics and economics of assisted reproduction: The cost of longing.* Washington, DC: Georgetown University Press, 2001.

29. Collins, J. A. An international survey of the health economics of IVF and ICSI. *Human Reproduction Update* 8, no. 3 (2002): 265–277.

30. National Conference of State Legislatures. Fifty states summary of legislation related to insurance coverage for infertility treatment. *NCSL.* August 2003. Available at <http://www.ncsl.org/programs/health/50infert.htm> (accessed October 13, 2003).

31. Garceau, L., J. Henderson, L. H. Davis, et al. Economic implications of assisted reproductive techniques: A systematic review. *Human Reproduction* 17, no. 12 (2002): 3090–3109.

32. Katz, P., R. Nachtigall, and J. Showsack. The economic impact of the assisted reproductive technologies. *Nature Cell Biology and Nature Medicine* (October 2002): fertility supplement, s29–s32.

33. Jain, T., B. L. Harlow, and M. D. Hornstein. Insurance coverage and outcomes of in vitro fertilization. *New England Journal of Medicine* 347, no. 9 (2002): 661–666.

34. Reynolds, M. A., L. A. Schieve, G. Jeng, and H. B. Peterson. Does insurance coverage decrease the risk for multiple births associated with assisted reproductive technology? *Fertility and Sterility* 80, no. 1 (2003): 16–23.

35. National Conference of Commissioners on Uniform State Laws. Uniform Parentage Act (last amended or revised in 2002). *NCCUSL.* 2002. Available at <http://www.law.upenn.edu/bll/ulc/ulc_frame.htm> (accessed October 13, 2003).

36. National Conference of State Legislatures. State laws and legislation: Use, storage, and disposal of frozen embryos. *NCSL.* May 2003. Available at <http://www.ncsl.org/programs/health/embryodisposition.htm> (accessed October 14, 2003).

37. *In the Matter of Baby M*, 537 A.2d 1227 N.J., 1988.

38. *Johnson v. Calvert*, 5 Cal. 4th 84, 1993.

39. *In re Marriage of Buzzanca*, 61 Cal. App. 4th 1410, 1418, 1998.

40. *Davis v. Davis*, 842 S.W. 2d 588 Tenn., 1992.

41. *Woodward v. Commissioner of Social Security*, 760 N.E. 2d 257 Mass., 2002.

42. Komoroski, A. L. After *Woodward v. Commissioner of Social Services*: Where do posthumously conceived children stand in the line of descent? *Boston University Public Interest Law Journal* (Spring–Summer 2002): 297–316.

43. Forster, H. The legal and ethical debate surrounding the storage and destruction of frozen human embryos: A reaction to the mass disposal in Britain and the lack of law in the United States. *Washington University Law Quarterly* 76 (1998): 759–780.

44. Hoffman, D. I., G. L. Zellman, C. C. Fair, et al. Cryopreserved embyos in the United Stated and their availability for research. *Fertility and Sterility* 79, no. 5 (2003): 1063–1069.

45. Fischer, J. D. Misappropriation of human eggs and embryos and the tort of conversion: A relational view. *Loyola of Los Angeles Law Review* 32 (1999): 381–431.

46. Norton, F. Assisted reproduction and the frustration of genetic affinity: Interest, injury, and damages. *New York University Law Review* 74 (1999): 793–843.

47. The President's Council on Bioethics. Beyond therapy: Biotechnology and the pursuit of happiness. Available at <http://bioethics.gov/reports/beyondtherapy/index.html> (accessed November 3, 2003).

48. French, H. W. As girls "vanish," Chinese city battles tide of abortions. *New York Times*, February 17, 2005, A3.

49. President's Council on Bioethics. *Reproduction and responsibility: Regulation of new biotechnologies*. Washington, DC: President's Council on Bioethics, March 2004.

Further Reading

Blank, R. and J. C. Merrick. *Human reproduction, emerging technologies, and conflicting rights*. Washington, DC: CQ Press, 1995.

• A thorough analysis of issues surrounding reproduction

Bonnicksen, A. I. *In vitro fertilization: Building policy from laboratories to legislatures*. New York: Columbia University Press, 1989.

• A thoughtful exploration of IVF development in the United States

Gunning, J., and H. Szoke, eds. *The regulation of assisted reproduction technology*. Hampshire, UK: Ashgate, 2003.

• An international perspective on ARTs and their regulation

McGee, G. *The perfect baby: Parenthood in the new world of cloning and genetics.* 2nd ed. Lanham, MD: Rowan and Littlefield Publishers, Inc., 2000.

• An exploration of the ethics of ARTs

Merrick, J. C., and R. H. Blank. *Reproductive issues in America: A reference handbook.* Santa Barbara, CA: ABC-CLIO, 2003.

• A source book with clear descriptions of ARTs and other issues, plus a chronology and excerpted documents

Brave New World Revisited: Human Cloning and Stem Cells

In February 1997, the world was electrified by the birth announcement of the first mammal cloned from an adult cell, a sheep named Dolly [1]. A clone shares genetic information with the nucleus donor. In many ways, it is a "twin" born at a different time, but it contains different mitochondrial DNA (this form of "cloning" is different from the production of single genes or chromosome fragments; see chapter 2). Cloning is asexual; it does not involve the joining of gametes characterized by other forms of natural or assisted reproduction, and therefore does not generate genetic diversity. Ian Wilmut and his colleagues at the Roslin Institute in Scotland took the nucleus from a somatic cell from an adult sheep and transplanted it into an enucleated egg from another. The resulting embryo was transferred to the uterus of a surrogate sheep, leading to Dolly's birth. Of 277 attempts, only Dolly was produced. Before Dolly, scientists had been unable to clone mammals, although they had success with other species. After Dolly's birth, researchers successfully cloned other mammals, among them mice, cattle, goats, pigs, a dog, and cats, but notably, no primates [2].

Although most discussion of cloning has centered on the nuclear transfer method described above, cloning is also possible by mechanically dividing a two-, four-, or eight-cell embryo into individual cells (blastomeres); this "twinning" process occurs naturally in the creation of identical twins. In 2000, a healthy rhesus monkey was born after an eight-cell embryo was split into four embryos and transferred to surrogates [3].

Is Cloning Safe?

The mechanics of cloning are simple in principle, but difficult in practice. The challenge of cloning is to "reprogram" the nucleus from the donor cell to recapitulate the developmental program of a fertilized egg; genes that may have been turned off during the process of development must be turned on again. Wilmut used shocks not unlike Frankenstein's method of reawakening his creature. Scientists have only limited knowledge of embryogenesis and the regulation of development. This undoubtedly contributes to the extremely low success rate of cloning.

A subset of offspring produced by cloning has exhibited a variety of developmental abnormalities; the most common are placental defects involving both maternal and fetal tissue that may lead to miscarriage, increased size at birth, and kidney and lung abnormalities [2]. Some problems are similar to those observed in cattle produced using IVF; "large offspring syndrome" may result from the suboptimal culture conditions for embryos prior to transfer into the uterus.

A second concern centers on the reprogramming of the donor nucleus. Under normal conditions, a subset of mammalian genes is "imprinted." The copies of these genes from the mother and the father behave differently in offspring. During gamete production, a chemical mark is placed on these imprinted genes that affects the pattern of expression. Defects in imprinting in humans can lead to fetal death or give rise to syndromes associated with severe mental retardation. Cloned embryos may not be subject to the mechanisms that allow imprinting to occur, since normally the process occurs only in gametes.

Other safety concerns include possible incompatibilities between a donor nucleus and host egg mitochondria or cytoplasm, and whether the life span of the offspring might be shortened as a result of shortened telomeres at the ends of donor chromosomes. Dolly was euthanized in 2003 at age six, after developing a serious illness; the normal life span of a sheep is twelve years [4]. Conclusions cannot be drawn from a single example, however.

Initial Policy Responses

Dolly's birth raised the real possibility of cloning humans. The public concern was immediate and intense. President Clinton instructed the National Bioethics Advisory Committee (NBAC) to immediately review the ethical and legal issues associated with cloning, and report back within ninety days. The NBAC built on the analysis conducted by the HERP that was established by the NIH in 1993. In 1994, the HERP recommended a moratorium on any attempt to clone humans in its report on acceptable applications of human embryo research [5, 6]. The NBAC's report recommended continuing the moratorium and that legislation be enacted to prohibit attempts to create a child by nuclear transfer. The committee remained concerned that legislation banning cloning should not interfere with research on other aspects of reproduction. The NBAC also discussed genetic determinism—the misperception that individuals are solely the product of their genes—and was careful to distinguish between possible scenarios and far-fetched ones. Its report attempted to skirt the issue of the moral status of preimplantation embryos—an issue also avoided in IVF because of its complicated relationship to abortion politics. The president proposed the Cloning Prohibition Act in 1997 in an attempt to act on the NBAC's recommendations, but the legislation was not passed [7].

In 1997, the state of California enacted legislation banning human cloning for reproductive purposes [8]. Other states soon followed suit, and by 2005, fourteen states had adopted laws banning human cloning. Several bills were introduced in Congress, but initially did not move out of committee. This changed dramatically in 2001.

That year, Advanced Cell Technology (ACT), a Massachusetts-based biotechnology company, announced it had successfully cloned a human embryo for the stated purpose of generating human embryonic stem (hES) cells [9]. Whether ACT was actually successful in cloning a human embryo is questionable, but the attempt was widely criticized. A bill criminalizing all forms of human cloning was passed in the House of Representatives (Human Cloning Prohibition Act of 2001 [10]) and introduced in the Senate, where the bill failed to come up for a vote. In late December 2002, Clonaid, a company founded by a fringe group called the Raelians, announced that the first cloned baby had been born. No proof was ever made available that the child existed. The hoax nevertheless led to renewed congressional efforts to ban all forms of human cloning. The House once again passed legislation to ban cloning [11], and debates took place in the Senate.

In 2002, the NRC weighed in to the controversy with a report that argued there was currently no justification for cloning for reproductive purposes, and that it should be banned. The safety of the procedure for future children was questionable [2]. Such a recommendation is unprecedented—no panel had ever proposed that a line of research be banned. The panel drew a sharp distinction between reproductive cloning and so-called therapeutic cloning (or cloning for biomedical research), the creation of cloned embryos as a source of hES cells or for research on human diseases. The NRC recommended that research into cloning for the generation of stem cells should continue.

Stem Cells Enter the Picture

Although federal policy prevented the federal funding of research on human embryos, it did not bar research in the private sector. Several private companies continued to conduct research on human embryos and techniques for human cloning. The two lines of research became inextricably tangled in the process.

Stem cells are cells that have the potential to give rise to many kinds of differentiated tissues (pluripotent), depending on the signals they receive. Although stem cells may be isolated from both embryos and adult tissues, adult stem cells appear to be much more limited in their potential to become different kinds of cells. Stem cells are touted as possible sources of treatments for a variety of human degenerative diseases. The idea is to replace damaged cells with healthy ones derived from the pluripotent stem cells. Stem cells from *cloned* embryos may be useful in the study of human genetic diseases by providing a reliable source of cells carrying the genetic defect. In 1998, John Gearhart and James Thomson of the University of Wisconsin reported the derivation of pluripotent human stem cells from embryos [12], and argued that they had the potential for treatment of a variety of degenerative diseases, including Alzheimer's and Parkinson's.

In March 2004, a group of South Korean researchers led by Woo Suk Hwang reported the derivation of hES cells from cloned blastocysts, indicating that stem cells could successfully be made from cloned embryos [13]. The resulting stem cells would be genetically identical to the nucleus donor's cells, and any tissues or organs generated from them would not be rejected by their immune system. The same group reported in June 2005 the derivation of eleven lines from different individuals using the same cloning technique [14]. The improved success rate suggested that the production of patient-specific hES might be both feasible and practical. But allegations of misconduct and data fabrication led to the retraction of a second article in December 2005, calling into question the earlier piece as well [15]. In January 2006, investigators in Korea reported that Hwang also fabricated the data from his 2004 article [16]. The successful creation of cloned human stem cells now awaits further experimentation.

Human ES cells are controversial because their best source is from embryos at the blastocyst stage, when the embryo consists of a hollow ball. Stem cells are collected from the interior of the embryo, and the embryo is destroyed in the process. Should embryos be destroyed to generate stem cells that might be used to treat illnesses in people? Should embryos left over from IVF procedures be used to generate stem cells? An alternative approach would be for donors to designate their gametes for the generation of embryos to be used in stem cell research. Should embryos be deliberately created to provide a source of stem cells? For those who believe that preimplantation embryos are "human lives," the answer is no. For those who view embryos as tissue, the potential for saving lives is a strong argument for generating stem cells.

In August 2001, President Bush, in a compromise move, announced that federal funds might be used to study stem cells, but only using cell lines that had been derived from embryos prior to the date of his proclamation. No new stem cell lines would be created from embryos using federal funds [17]. The federal government announced that over sixty cell lines were available for study. Unfortunately, most of the existing cell lines were insufficiently characterized and turned out not to be useful; others were owned by private companies and therefore not available for research by others. Stem cell researchers complain that their research is stymied by the federal regulations.

In November 2004, California voters approved Proposition 71, a constitutional amendment to create the California Institute of Regenerative Medicine (CIRM) to support stem cell research [18]. The bill authorized the state government to issue bonds to finance up to $3 billion toward grants and loans to support research and research facilities. California hopes to become the leader in stem cell research by attracting researchers both nationally and internationally. Since the bill's passage, the CIRM named Zach Hall, a prominent neuroscientist, to head the institute, established a twenty-nine-member advisory panel, and developed procedures for the awarding of grants. The CIRM, however, is mired in lawsuits that challenge the use of state funds without adequate state control, limiting its ability to raise funds [19].

California became the second state to explicitly support stem cell research; New Jersey passed legislation enacted in January 2004 to support stem cell research at the more modest level of $6.5 million [20]. In 2005, Illinois and Massachusetts also authorized the use of state monies for stem cell research.

The Ethics of Human Cloning and Policy Recommendations

There is general agreement that there is little justification for reproductive cloning or cloning to produce babies. Most concur that safety issues for future offspring make such research unacceptable. The President's Council on Bioethics recommended unanimously that a permanent ban on cloning to produce children should be instituted [21, 22]. The majority also recommended that a five-year moratorium on cloning for biomedical research be instituted. A minority recommended that research continue with regulation [21].

The council was split on whether cloning to produce stem cells was acceptable. Leon Kass, the council chair, and other council members argue that creating embryos only to destroy them violates the sanctity of human life, exploits humans, and does moral harm to society. These assertions closely mirror Kass's opposition to IVF over twenty-five years ago [23, 24]. Council members who support further research on stem cells contend that while embryos may have an intermediate (or no particular) moral status, the potential benefits of research outweigh these concerns. In 2004, two supporters of continued research were not renewed on the council and were replaced by more conservative members—a move criticized by many scientists as religiously and politically motivated [25]. The reconstituted council issued a report expressing stronger concerns, but stopped just short of proposing an outright research ban [22]. The death of former president Reagan from Alzheimer's disease in June 2004 led to public calls for increased stem cell research [26, 27], even from staunch antiabortion politicians like Senator Orrin Hatch of Utah.

Concerned by the lack of federal oversight, the NRC issued a report in April 2005 providing detailed guidelines for the ethical conduct of research on hES cells [28]. The proposed guidelines set limits on how long embryos may be maintained in vitro, place controls on human/animal chimera studies, ban any payment for embryos, and require strict informed consent for donors of gametes or embryos. Institutional committees modeled after those that approve human or animal research (see chapter 5) should be established to oversee all research on hES cells. Researchers must also assure that proper medical care is provided to all potential donors.

In 2004, the President's Council on Bioethics embraced the possibility of developing hES using alternative methods [29]. Suggested approaches include stem cells derived from adults or umbilical cord blood, or deriving cell lines from "dead" embryos, disabled embryos, or by removal of single cells from embryos, leaving them still viable (akin to PGD). The approaches were theoretical, though, and their potential was undetermined. In October 2005, two groups reported new methods for developing stem cells from mice that may not pose the same degree of ethical difficulty if applied to humans. One method removes single cells (blastomeres) from early embryos from which to derive stem cells; the embryos are not destroyed in the process [30]. The second method creates cloned embryos with a disabled gene that are unable to develop to a stage that would permit implantation; such embryos have no potential for full development [31]. These studies in animals provide a "proof of concept" for these approaches, but it remains to be determined whether they are feasible in humans. Still, there have been calls from Congress to place a moratorium on "traditional" methods until the utility of these approaches is determined [32]. In addition, with the controversy surrounding the South Korean publications on cloned hES cells, whether it is actually possible to clone human embryos is once again uncertain.

The issues surrounding human cloning and embryo research, and the rhetoric they inspire, are almost identical to those voiced when IVF was under development. Will Congress act to ban human cloning, or will the government adopt the same laissez-faire approach it took for ARTs? In the United States, the states have taken the lead in developing regulatory policy in these areas. Congress has given some indication of a movement to support human embryo research—for instance, majority leader Bill Frist announced his support of embryo research in fall 2005. While controversy continues in the United States, similar debates are in progress around the world, and thus whether stem cell research and human cloning flourish in the future is an open question.

References

1. Wilmut, I., A. E. Schnieke, J. McWhir, A. J. Kind, and K. H. S. Campbell. Viable offspring derived from fetal and adult mammalian cells. *Nature* 385 (1997): 810–813.

2. Committee on Science Engineering and Public Policy, National Academy of Sciences, National Academy of Engineering, Institute of Medicine. *Scientific and medical aspects of human reproductive cloning*. Washington, DC: National Academy Press, 2002.

3. Chan, A. W. S., T. Domino, C. Leutjens et al. Clonal propagation of primate offspring by embryo splitting. *Science* 287 (January 14, 2000): 317–319.

4. Merrick, J. C., and R. H. Blank. *Reproductive issues in America: A reference handbook*. Santa Barbara, CA: ABC CLIO, 2003.

5. Lauritzen, P., ed. *Cloning and the future of human embryo research*. Oxford: Oxford University Press, 2001.

6. Green, R. M. *The human embryo research debates: Bioethics in the vortex of controversy*. Oxford: Oxford University Press, 2001.

7. Forster, H., and E. Ramsey. The law meets reproductive technology: The prospect of human cloning. In *Cloning and the future of human embryo research*, ed. P. Lauritzen, 201–221. Oxford: Oxford University Press, 2001.

8. National Conference of State Legislatures. State human cloning laws. June 21, 2005. Available at <http://www.ncsl.org/programs/health/genetics/rt-shcl.htm> (accessed December 24, 2005).

9. Cibelli, J. B., A. A. Kiessling, K. Cunniff, C. Richards, R. P. Lanza, and W. D. West. Somatic cell nuclear transfer in humans: Pronuclear and early embryonic development. *e-biomed: The Journal of Regenerative Medicine* 2 (November 26, 2001): 25–31.

10. H.R. 2505: To amend Title 18, United States Code, to prohibit human cloning. 107th Cong., 1st sess., 2001.

11. H.R. 534: To amend Title 18, United States Code, to prohibit human cloning. 108th Cong., 1st sess., 2003.

12. Thomson, J. A., J. Itskovitz-Elder, S. S. Shapiro et al. Embryonic stem cell lines derived from human blastocysts. *Science* 282, no. 5391 (1998): 1145–1147.

13. Hwang, W. S., Y. J. Ryu, J. H. Park et al. Evidence of a pluripotent human embryonic stem cell lined derived from a cloned blastocyst. *Science* 303 (March 12, 2004): 1669–1674 (published online February 12, 2004).

14. Hwang, W. S., S. I. Roh, B. C. Lee et al. Patient-specific embryonic stem cells derived from human SCNT blastocysts. *Science* 308 (June 17, 2005): 1777–1783 (published online May 19, 2005).

15. Kennedy, D. Editorial expression of concern. *Sciencexpress*. December 23, 2005. Available at <http://www.sciencexpress.org/22December2005/Page1/10.1126/science.1124185> (accessed December 24, 2005).

16. Wade, N., and C. Sang-Hun. Researcher faked evidence of human cloning, Koreans report. *New York Times*, January 10, 2006, A1.

17. Bush, G. W. Remarks by the president on stem cell research. Available at <http://www.whitehouse.gov/news/releases/2001/08/print/20010809-2.html>   (accessed November 3, 2003).

18. League of Women Voters of California. Proposition 71. December 15, 2004. Available at <http://www.smartvoter.org/ca/state> (accessed January 28, 2005).

19. Pollack, A. California's stem cell program is hobbled but staying the course. *New York Times*, December 10, 2005, C1.

20. N.J. Stat. § 26:2Z-1 (2006), Title 26. Health and Vital Statistics, Chapter 2Z. Human Cell Research, § 26:2Z-1. Findings, declarations relative to human stem cell research, LexisNexis (TM) New Jersey Annotated Statutes.

21. The President's Council on Bioethics. Human cloning and human dignity: An ethical inquiry. July 2002. Available at <http://bioethics.gov/reports/cloningreport/cloning.pdf> (accessed November 5, 2003).

22. President's Council on Bioethics. Reproduction and responsibility: Regulation of new biotechnologies. March 2004. Available at <http://www.bioethics.gov> (accessed July 15, 2004).

23. Kass, L. R. *A conversation with Dr. Leon Kass*. Washington, DC: American Enterprise Institute for Public Policy Research, 1979.

24. Kass, L. R., and J. Q. Wilson. *The ethics of human cloning*. Washington, DC: AEI Press, 1998.

25. Holden, C. Researchers blast U.S. bioethics panel shuffle. *Science* 303 (March 5, 2004): 1447.

26. Coalition for the Advancement of Medical Research. *Letter to President George W. Bush.* Washington, DC: Coalition for the Advancement of Medical Research, June 23, 2004.

27. Kirkpatrick, D. D. Bush defends stem-cell limit, despite pressure since Reagan death. *New York Times*, June 16, 2004, A18.

28. Committee on Guidelines for Human Embryonic Stem Cell Research, National Research Council. *Guidelines for human embryonic stem cell research.* Washington, DC: National Academy Press, 2005.

29. Holden, C., and G. Vogel. A technical fix for an ethical bind? *Science* 306 (December 24, 2004): 2174–2176.

30. Chung, Y., I. Klimanskaya, S. Becker et al. Embryonic and extraembryonic stem cell lines derived from single mouse blastomeres. *Nature* 439, no. 7073 (January 12, 2006): 216–219 (published online October 16, 2005).

31. Meissner, A., and R. Jaenisch. Generation of nuclear transfer-derived pluripotent ES cells from cloned CDX2-deficient blastocysts. *Nature* 439, no. 7073 (January 12, 2006): 212–215 (published online October 16, 2005).

32. Weissman, I. L. Politic stem cells. *Nature* 439, no. 7073 (January 12, 2006): 145–147 (published online October 16, 2005).

Further Reading

Committee on Science, Engineering, and Public Policy, National Academy of Sciences. Scientific and medical aspects of human reproductive cloning. Washington, DC: National Academy Press, 2002.

• The NRC report that reviews the science of cloning, with an excellent bibliography

Lauritzen, P., ed. Cloning and the future of human embryo research. Oxford: Oxford University Press, 2001.

• An excellent collection of essays on the implications of cloning and stem cell research

The Asilomar Conference on Recombinant DNA: A Model for Self-Regulation?

The development of recombinant DNA technology (or "gene splicing") is marked by an event unprecedented in the history of science: a deliberate moratorium on scientific research. This moratorium was not called by policymakers outside the scientific community but rather by leading scientists themselves.

In 1973, Stanley Cohen and Herbert Boyer reported the successful recombination of DNA molecules in vitro using newly characterized restriction enzymes [1]. The scientific community immediately recognized the potential of the new technology for studying molecular genetics and fundamental biochemical processes in cells. Some scientists also expressed concern that recombinant DNA research might lead to the unintended release of dangerous new pathogens into the environment [2, 3]. At a 1973 conference, researchers discussed the issues surrounding certain kinds of experiments involving recombinant DNA [4]. Among the concerns were experiments using a particular strain of *E. coli* bacteria, K-12, that was capable of colonizing the human gut. Letters from leading scientists were published calling for a moratorium on research and asking the NRC to establish a panel to address areas of concern. In April 1974, the NRC committee published a letter in *Science* urging researchers not to conduct experiments in some areas of recombinant DNA research [5].

In February 1975, 140 participants—scientists, lawyers, and journalists—met at the Asilomar Conference Center in Pacific Grove, California, to discuss scientific and policy matters regarding recombinant DNA. There was considerable disagreement about the risks of recombinant DNA. Some scientists felt that the concerns were overblown, while other believed that there was simply not enough information available to make a judgment. Nevertheless, those who thought that regulation was necessary prevailed [3, 4]. A strong motivation was concern that congressional action might place severe limits on the conduct of recombinant DNA research. By developing their own regulations, scientists hoped to avoid legislation that might have a devastating effect on their research. The conferees agreed that research on recombinant DNA should continue, using biological containment facilities of increasing stringency (designated P1 to P4) for experiments viewed as potentially more hazardous [6]. For example, research that involved splicing DNA from eukaryotes into prokaryotes would be subject to the strictest containment because of the possibility of transferring viral genomes hidden in the DNA. The conference also recommended that certain kinds of research should be deferred indefinitely: those that involved DNA from dangerous pathogens or toxins, and large-scale experiments that might make potentially harmful products (see also chapter 9). Furthermore, the conference called on governments to establish codes of conduct for recombinant research.

The guidelines proposed at the Asilomar Conference became the framework for the NIH's Recombinant DNA Advisory Committee (RAC) and regulations that continue to provide research oversight in this area. The first RAC regulations were published in 1976 [7, 8]. As it became clear that the risks of recombinant DNA were not as great as initially feared, many of the restrictions were lifted. Nevertheless, a RAC review of grant proposals involving some types of recombinant DNA techniques still is mandated today. Research funded by private sources is not subject to the regulations imposed by the NIH.

After the conference recommendations were published, the public became aware of the new technology for the first time. Most members of the public did not understand the science behind the new technology and were fearful of it. The possibility of "genetic engineering" of humans raised concerns about the ethical and social legitimacy of the research [9]. In Cambridge, Massachusetts, home of both Harvard University and MIT, the city council voted in 1976 to ask both institutions to observe a three-month moratorium on research while a citizens' board reviewed whether such research should be permitted [4, 8]. The board ultimately ruled that research was permissible under the just-released NIH guidelines. States and other municipalities took similar action to regulate such research.

The federal government also participated in the debate. Joseph Califano, the DHEW secretary, convened a Federal Interagency Committee on Recombinant DNA Research in November 1976 to assure that all federal agencies would adopt the guidelines being developed at the NIH. Califano testified before Congress that regulation was necessary and outlined legislation that might serve this purpose [4, 8].

Although numerous bills were introduced in Congress in 1977–78, no laws were passed. Beginning in 1977, it became increasingly clear that recombinant DNA research was not as risky as initially believed. Scientists actively lobbied Congress not to impose stringent regulations. After 1978, recombinant research was more generally accepted, and calls for tight regulations diminished.

These events were remarkable for several reasons. First, scientists voluntarily halted lines of research in order to consider the potential risks of moving ahead. Second, they suggested that there were certain kinds of experiments that should be forbidden. Finally, scientists themselves developed regulations of their activities using the NIH to provide oversight.

The outcomes of the Asilomar Conference are not without their critics. The focus on biohazards led to a culture of "worst-case scenarios" that continues to pervade research discussion today [10]. For some researchers, the expense of building high-level containment facilities discouraged them from continuing certain lines of research. Others suggest that regulations slowed progress in areas of molecular biology. Of course, science has no formal organization, and the organizers and participants in Asilomar had only the power of persuasion, peer influence, and effects on journal editors, government, and university administrators to alter scientific practices.

Is It Time for Asilomar Two?

As the twenty-fifth anniversary of the Asilomar Conference approached, there were calls for a reexamination of the issues surrounding many areas of biotechnology [11, 12, 13]. Paul Berg and Maxine Singer, organizers of the original conference, wrote in 1995 that few attendees could have anticipated the impact of applications of recombinant DNA technology on society, or the profound ethical, legal, and societal debates that progress engendered [14].

The first Asilomar Conference took place in a unique environment: the technology was brand-new, and the issues were not yet public [15]. The conference could thus operate as "an early warning system within the scientific community" without pressures from a public debate. The absence of public involvement permitted a scientific approach to the issues and led to the first voluntary moratorium on research. At the present time, debates on biotechnological issues take place in the public arena, with no similar opportunity for a purely scientific discussion. Hence, the Asilomar Conference may not serve as a good model for future debates.

References

1. Cohen, S. N., A. A. Y. Chang, H. W. Boyer, and R. B. Helling. Construction of biologically functional plasmids in vitro. *Proceedings of the National Academy of Sciences U.S.A.* 70 (November 1973): 3240–3244.

2. Krimsky, S. Regulating recombinant DNA research and its applications. In *Controversy: Politics of technical decisions*, ed. D. Nelkin, 219–248. 3rd ed. Thousand Oaks, CA: Sage Publications, 1992.

3. Holton, G., and R. Morison, eds. *Limits of scientific inquiry*. New York: W. W. Norton and Co., 1979.

4. Grobstein, C. *A double image of the double helix: The recombinant DNA debate*. San Francisco: W. H. Freeman and Company, 1979.

5. National Academy of Sciences. Report of the committee on recombinant DNA molecules: Potential biohazards of recombinant DNA molecules. *Proceedings of the National Academy of Sciences U.S.A.* 71 (1974): 2593–2594.

6. Berg, P., D. Baltimore, S. Brenner, R. O. Roblin III, and M. F. Singer. Asilomar conference on recombinant DNA molecules. *Science* 188 (1975): 991–994.

7. National Institutes of Health. Guidelines for research involving recombinant DNA molecules. *Federal Register* 41, no. 131 (1976): 27911–27922.

8. Wright, S. *Molecular politics: Developing American and British regulatory policy for genetic engineering, 1972–1982*. Chicago: University of Chicago Press, 1994.

9. Callahan, D. The involvement of the public. In *Research with recombinant DNA: An academy forum*, ed. National Academy of Sciences, 31–44. Washington, DC: National Academy of Sciences, 1977.

10. McClean, P. Historical events in the rDNA debate. Available at <http://www.ndsu.nodak.edu/instruct/mcclean/plsc431/debate/debate3.htm> (accessed October 24, 2003).

11. Davatelis, G. N. Commentary: Asilomar 2: An idea whose time has come. *Scientist* 13, no. 8 (1999): 12.

12. Russo, E. Another Asilomar? Preliminary plans underway. *Scientist* 13, no. 8 (1999): 11.

13. Special issue: Symposium on science, ethics, and society: The 25th anniversary of the Asilomar Conference. *Perspectives in Biology and Medicine* 44, no. 2 (2001).

14. Berg, P., and M. F. Singer. The recombinant DNA controversy: Twenty years later. *Proceedings of the National Academy of Sciences U.S.A.* 92 (1995): 9011–9013.

15. Singer, M. F. Commentary. In *AAAS science and technology policy yearbook 2001*, 161–164. Washington, DC: American Association for the Advancement of Science, 2001.

Further Reading

Berg, P., D. Baltimore, S. Brenner, R. O. Roblin III, and M. F. Singer. Asilomar conference on recombinant DNA molecules. *Science* 188 (1975): 991–993.

• A summary of the guidelines proposed at the Asilomar Conference

Grobstein, C. *A double image of the double helix: The recombinant DNA debate.* San Francisco: W. H. Freeman and Co., 1979.

• A clear discussion of the history and issues surrounding recombinant DNA; also contains reprints of important documents

Holton, G., and R. S. Morison, eds. *Limits of scientific inquiry.* New York: W. W. Norton and Company, 1979.

• An edited volume generated in the aftermath of Asilomar discussing the concept of social, political, and scientific constraints on science

Special issue: Symposium on science, ethics, and society: The 25th anniversary of the Asilomar conference. *Perspectives in Biology and Medicine* 44, no. 2 (2001).

• A collection of papers from the twenty-fifth anniversary meeting

Wright, S. *Molecular politics: Developing American and British regulatory policy for genetic engineering, 1972–1982.* Chicago: University of Chicago Press, 1994.

• A thorough account of the development of regulatory policy involving molecular biology

5

Protecting the Public: The FDA and New AIDS Drugs

Americans have a general expectation that the food they consume, the drugs they take, and the products they use are safe. Many look to the federal government to assure that industry is regulated appropriately to achieve this goal. Yet regulation is also unpopular: it increases costs for industry (and therefore for consumers), and may slow the pace of introduction of new drugs and consumer products. Some also view regulation as intrusive and paternalistic, denying individuals the freedom to make decisions about their own lives. This part explores these issues by looking at the FDA, the agency with primary responsibility for both the safety and efficacy of drugs as well as the safety of food. The development of drugs to treat AIDS was a defining issue for the FDA in recent years, and hence will be a focus here.

A Short History of the FDA

Prior to the beginning of the twentieth century, the U.S. federal government played a small role in regulating industry. Laissez-faire capitalism was the accepted practice, allowing low pay and long hours for workers, no workplace or product safety, few social support services, and highly inequitable tax structures [1]. The government did little to regulate drug marketing and food safety. Anyone could market "patent" medicines without testing—all one needed to do was claim someone had been cured. A single piece of legislation, the Drug Importation Act of 1848, banned the entry of adulterated drugs from abroad [2], but little enforcement took place. Manufacturers commonly adulterated food with toxic additives (claimed to be preservatives), added chemicals to disguise rotten food, or failed to remove dirt, insects, or other contaminants.

By the early 1900s, President Theodore Roosevelt ushered in a new spirit of reform. He was aided by a doctor, Harvey Wiley, who gathered evidence of the dangers of food adulteration and poisonings by patent medicine, and presented them to Congress. Investigative reporters published articles in leading magazines about the dangers of patent medicines. Congress, however, expressed little will to enact legislation until the publication of Upton Sinclair's muckraking novel, *The Jungle*, in early 1906, which described in graphic detail the practices of the meatpacking industry. Within months, Congress passed laws regulating the meatpacking industry and also the Pure Food and Drug Act (PFDA) of 1906 [2]. The responsibility for enforcing the new law went to Wiley's Bureau of Chemistry in the U.S. Department of Agriculture (USDA), which became the precursor of the FDA, which was created in 1927.

The PFDA contained provisions to protect the public against consumer fraud; it required drugs to be pure and clearly labeled. But it contained no provisions that required drugs to be effective or, for that matter, safe to consume. A producer need only claim a belief that the medicine was effective to escape prosecution for fraud [3]. A tragedy stimulated new law. In 1937, the Massengil Company developed and marketed a liquid formulation of the antibiotic, sulfanilamide, using the highly toxic chemical diluent, diethylene glycol [2, 4]. Over one hundred people, primarily children, died after taking the "Elixir Sulfanilamide." The fledgling FDA had authority to intervene and pull the product from stores solely because the manufacturer had mislabeled the product as an elixir, which is an alcohol-based mixture, and not because the product was poisonous. The resulting public outcry led to the passage of the Federal Food, Drug, and Cosmetic Act (FFDCA, Ch. 675, 52 Stat. 1040) in 1938.

For the first time, the federal government granted the FDA genuine jurisdiction to ensure that drugs were safe. Drug manufacturers had to demonstrate that new products were safe before marketing. False claims about drug effectiveness became illegal; however, no direct demonstration of effectiveness was required. The addition of toxic substances to food was prohibited except when unavoidable, and limits were set in such cases [3]. The FDA was given authority to conduct inspections of the food industry. In 1951, the Durham-Humphrey Amendments placed controls on drugs to be available only by prescription [2]. An additional series of amendments in the 1950s further limited food additives. The Delaney Clause, added in 1958, prohibited any substance found to cause cancer to be used as a food additive. This zero-tolerance clause proved problematic in later years.

After passage of the FFDCA in 1938, no major changes to FDA law occurred until 1962, in the aftermath of yet another tragedy. Thalidomide was marketed in

Europe between 1959 and 1962 as a sedative and antinausea drug for pregnant women. Yet it caused severe birth defects in the babies born to pregnant women taking the drug, including children born with flipperlike limbs (phocomelia). An estimated ten thousand babies were born in Europe with birth defects before the drug was withdrawn in 1962 [1, 2].

The United States largely escaped the tragedy because thalidomide was not sold in the States, having been delayed in the FDA's drug approval process (the mothers of the small number of babies born in the United States with phocomelia had obtained thalidomide while in Europe). Frances Kelsey, an FDA reviewer, is credited with blocking the drug's approval by demanding that adequate safety and effectiveness data be provided by the manufacturer [1, 5]. The 1962 amendments (Public Law No. 87–781, 76 Stat 780) gave the FDA authority to fully approve new drugs prior to their marketing. A critical element of the approval was a demonstration that the new drug was effective as well as safe. The 1962 amendments established the current system of FDA drug evaluation and approval [2, 4]. It is somewhat paradoxical that the changes to the law were unrelated to the problem of unsafe drugs exemplified by thalidomide, as thalidomide might in fact have been (and has since been shown to be) an effective, if not a "wonder," drug, albeit not for pregnant women. Nonetheless, this pattern of regulation following seminal, even tragic, events and public outcries typifies the evolution of federal regulations of drugs and devices as well as the use of humans and animals in research (see the "Seminal Events" section in this chapter).

The FDA Drug Approval Process

The new drug-screening process begins with the submission by a drug developer of an investigational new drug (IND) application. The testing of medical devices is similarly permitted under investigational device exemptions. The IND must include data on animal and other preclinical studies demonstrating the experimental drug's safety and chemical action. Once the FDA approves the IND, the manufacturer may conduct research on human beings, known as clinical trials.

There are three stages to the clinical testing of a new drug [4, 6]. Phase I trials examine whether the drug has significant toxic side effects by administering it to a small group of twenty to thirty volunteers. Normally, healthy individuals are paid to be in these studies, which involve increasing doses of drug to subsequent cohorts of subjects and assessing toxicity. For especially dangerous drugs, like cancer chemotherapies, sick patients who have failed conventional treatments are typically used. As a general rule, phase I trials do not examine whether the drug is effective against disease. The purpose of the phase I trial is to identify the maximum tolerated dose, which dose is then used in a phase II study; if no dose presumed to be effective is acceptably tolerated, the development of the drug may be halted.

Phase II trials are pharmacokinetic and pharmacodynamic studies that address how the body metabolizes the drug, in which the drug is given at a single dose level to larger numbers of subjects (often ranging from several dozen to more than a hundred). These trials typically have no controls, and involve looking primarily at how the human body processes and excretes the drug. Data are also collected on safety as well as effectiveness. Drugs that are shown to be biologically active and reasonably safe after testing in this larger group of subjects may then go into phase III trials. Phase II trials may reveal that a new drug appears to be effective, but do not adequately assure that the drug is efficacious.

In phase III trials, a large number of patient-subjects, as many as several thousand, are given the drug. Phase III trials are randomized, controlled trials, meaning that subjects who meet entry criteria are randomly assigned to either receive the experimental drug or be a control (for comparison purposes). Controls may be placebo (receiving a biologically inactive agent) or active (receiving a standard drug provided to patients for the disease under study). Trial designs vary widely, with some including both active and placebo control arms, and in cases where there are effective drugs and it would be unacceptable to withhold treatment, the experimental drug and placebo may be given in addition to standard therapy. Phase III trials are designed to determine whether an experimental drug works—that it is efficacious and reasonably safe.

Substantial evidence of the drug's efficacy must be obtained if the drug is to be approved by the FDA [2]. The final step in the process is the filing of a new drug application (NDA) requesting approval to market the new drug. The FDA reviews the data generated in the clinical trials, and determines whether the drug is both safe and effective. The entire process typically takes years, and costs pharmaceutical companies many millions of dollars. Note that a drug will be licensed if it is at all effective; it is not necessary for licensure that a drug company establishes that a new drug works better than other available drugs. Thus, drug companies always wish to test their experimental drugs against placebos to establish efficacy, but placebo control trials are not always ethical to perform (see "Placebo Controls" section below).

Criticism of the slow pace of new drug approval began in the 1970s, when studies by William Wardell suggested that new drugs were approved much more rapidly in the United Kingdom than in the United States [2]. Political conservatives, who opposed regulation, claimed that the slow pace of new drug approval did not protect consumers from adverse drug effects, since well-documented cases such as thalidomide were purely fortuitous [7]. Instead, they argued that Americans were deprived of important drugs because of the "drug lag" between development and approval.

In response to rising criticism, the FDA developed a "compassionate use" provision that allowed drug companies to make unapproved experimental drugs available to patients on a case-by-case basis with FDA approval [4]. The FDA also decided that patients might be given permission to import small quantities of unapproved drugs for personal use only; the drugs were not to be sold to others. Neither move silenced the FDA's critics. For example, in the 1970s, the FDA refused to approve a popular alternative anticancer drug, laetrile, because there was no clinical evidence suggesting it was effective (see part 7); desperate patients went to other countries to obtain it. A group of cancer patients filed suit against the FDA for its policy on laetrile, contending that regulation interfered with interstate commerce. The U.S. Supreme Court upheld the right of the FDA to regulate drugs in *United States v. Rutherford* (442 U.S. 544) in 1979 [4]. By the 1980s, AIDS and related activism ushered in a new era of pressure to change drug policies.

The Biology of HIV/AIDS

AIDS is caused by infection with a retrovirus called the human immunodeficiency virus (HIV). Viruses contain genetic material surrounded by a protein package, and operate by infecting cells and forcing the cell's synthetic machinery to make more virus particles. The genetic material in retroviruses is in the form of RNA, rather than DNA. When HIV infects cells, it causes cells to translate the genes encoded by the RNA. Among the genes is one that encodes the enzyme, reverse transcriptase, which makes DNA from the RNA template. More RNA can then be made, along with the proteins needed to package the RNA. When new virus particles escape from the cell, it is killed because holes are punched into its membrane. The new virus particles then go on to infect other cells.

HIV selectively attacks a population of cells in the immune system called helper T-cells. These cells assist in immune responses involving both B-cells, which make antibodies, and T-cells, which directly attack foreign cells (see chapter 12). HIV recognizes a protein on the surface of helper T-cells called CD4, and uses it to enter the cells.

Initial infection with HIV produces only mild flu-like symptoms, including nausea, chills, and night sweats. The virus may then enter a dormant phase, when it "hides" in the body and remains inactive. Over time, however, the viral infection of helper T-cells causes the population of CD4+ cells to drop lower and lower. Normal levels of CD4+ cells range from eight hundred to twenty-two hundred per cubic millimeter per mm of blood [3]; once levels drop below two hundred, AIDS symptoms are likely to appear [8]. When helper T-cells are absent, the immune system cannot mount an adequate response to infectious agents. Affected individuals become vulnerable to a variety of pathogens and develop symptoms of other diseases that are characteristic of HIV/AIDS. Patients may develop a rare form of pneumonia (PCP) produced by infection with *Pneumoncystis carinii*, a parasite. They may develop a form of malignant skin cancer called Kaposi's sarcoma usually seen only in elderly or immunosuppressed patients. Others may develop severe fungal or viral infections, leading to blindness and dementia. These illnesses result from so-called opportunistic infections, as the immune system function declines. Without aggressive treatment, the mortality rate of HIV/AIDS is close to 100 percent.

When the first cases of HIV/AIDS were described in the early 1980s, most patients were homosexual men or intravenous drug users. AIDS affects both men and women, though, and can be transmitted from infected women to their fetuses in utero. HIV is commonly spread by sexual contact or other exchange of body fluids, such as blood transfusions or the sharing of needles by drug users. HIV/AIDS is a global problem, with the highest infection rates in sub-Saharan Africa. In 2003, an estimated forty million people were infected worldwide [9], with close to four hundred thousand cases in the United States in 2002 [10].

Current Treatments for HIV/AIDS

The treatments for HIV/AIDS are aimed at killing the virus or preventing it from infecting other cells. The genome of HIV is now known, and researchers work to develop drugs targeted at different stages of the virus's life cycle. For example, some drugs, such as azidothymidine (AZT), are artificial nucleosides (DNA building blocks) and interfere with the activity of the reverse transcriptase enzyme, preventing the synthesis of DNA from the viral RNA. Others attack specific viral proteins, making it difficult for the virus to either infect cells (fusion inhibitors) or make new virus particles (protease inhibitors). Researchers are attempting to develop a vaccine against the virus; unfortunately, this has met with little success so far.

A second approach to treating HIV/AIDS is to develop drugs that effectively combat the serious opportunistic infections that appear in AIDS. Such drugs may prolong the life of patients with AIDS, but ultimately are less successful than drugs that directly attack the virus itself.

The challenge in treating HIV/AIDS comes from the high rate of virus mutation. The virus can become resistant to antiviral drugs, as mutated forms proliferate. Current treatments use "cocktails" of several anti-HIV drugs to limit the development of drug-resistant strains. Since the mid-1990s, the death rate from HIV/AIDS has declined markedly in the United States; HIV/AIDS begins to appear more like a chronic illness than a "death sentence." Nevertheless, treatment remains expensive, and the public health consequences of a growing HIV/AIDS population, even a healthier one, are considerable. There is concern, too, that changing AIDS from a deadly disease to a chronic condition will lead to increased rates of infection as people who are at risk stop taking precautions against infection (e.g., condom use).

The AIDS Era and Drug Development

The first cases of the syndrome to become known as AIDS appeared in 1981; the CDC reported the unusual occurrence of PCP in five homosexual men in Los Angeles [8]. The number of cases grew rapidly, reaching fifteen hundred by January 1983 and over ten thousand by January 1985 [8]. The politically and socially conservative Reagan administration, however, failed to respond to the growing epidemic for several years; the outrage this inaction generated in the gay community contributed to the rapid rise of AIDS activism [11]. The Gay Men's Health Collective was formed in 1981 to disseminate information about HIV/AIDS to the gay community and lobby the government for support. The more militant AIDS Coalition to Unleash Power (ACT-UP) was founded in San Francisco in 1987, and became a major influence in AIDS drug policy. The antiregulatory stance of the Reagan administration also affected the FDA; its budget was slashed and its staff reduced [1].

Researchers in France and the United States isolated the HIV virus in 1984, allowing for the development of screening tests for HIV infection, but no FDA-approved drugs were available to treat HIV/AIDS. Beginning in 1984, the FDA allowed drug companies to provide available drugs to treat AIDS-associated infections such as PCP and cytomegalovirus on a compassionate use basis. Yet the increasing frustration of activists over the limited government response led to independent measures.

Black-market buying clubs, such as the People with AIDS Health Group, began to import unapproved purported anti-AIDS drugs from abroad; most proved ineffective, if not dangerous. Buying groups made use of the personal use import exemption, in a quasi-legal manner [4].

The pharmaceutical industry was not interested in developing anti-HIV drugs initially because it assumed that the number of affected individuals would be small and the disease would "burn out" in a short time. The FDA granted the Burroughs-Wellcome Company (now GlaxoSmithKline) an "orphan drug" exemption for AZT, the first promising AIDS drug. The Orphan Drug Act was passed in 1983 to encourage companies to develop treatments for rare diseases [8]. Drug developers received tax credits and exclusive marketing rights to make up for limited sales. Burroughs-Wellcome earned a windfall given the extent of the HIV/AIDS epidemic.

AZT entered clinical trials in July 1985, after FDA approval of an IND filed by Burroughs-Wellcome. Almost immediately, AIDS activists demanded that the drug be made available to patients prior to approval. In 1986, early phase II trials indicated that AZT was effective in reducing the progression of AIDS; the trial was halted, and all patients were given AZT. The FDA sped through the evaluation process and approved AZT in 1987, only two years after it approved the IND. While the FDA viewed the speed of the approval as a testament to its ability to respond to community concerns [12], the AIDS community complained that patients were denied the drug during the approval process [8]. In addition, the initial price of $10,000 per year was viewed as beyond the reach of most AIDS patients; Burroughs-Wellcome dropped the price to $8,000 in the face of intense public and congressional pressure [8].

In 1987, the FDA reorganized and separated the division that evaluated new drugs and biologics (such as vaccines) into two separate entities. The Center for Drug Evaluation and Research had authority for new drugs. A separate authority was responsible for medical devices such as heart valves or breast implants (see chapter 6). One effect of the split was to reduce the caseload for evaluation. Given the FDA's limited staff as a result of budget cuts, the reorganization was helpful in speeding the evaluation process. The FDA also announced an expansion of the compassionate use protocol. It authorized the use of investigational new drugs to treat patients, or "treatment INDs" (21 C.F.R., Sec. 312.34). Drug companies could sell drugs to patients suffering from serious or life-threatening illnesses while clinical trials were underway [2, 4, 12]. A notable policy change allowed companies to charge for investigational drugs; under compassionate use, they usually were provided for free.

While the FDA argued on ethical grounds that treatment INDs allowed seriously ill and dying patients access to drugs, critics charged that the approach seriously compromised the clinical trials themselves. Why would patients agree to be part of a clinical trial, during which they might be given a placebo compound, when they might get the drug outside the trial? In addition, since the safety and efficacy of the drug was not known, drug companies were in a position to exploit patients by selling them unapproved drugs. On the other hand, AIDS activists criticized the new program for limiting patient access to experimental drugs by requiring FDA approval of the treatment IND [4]. The FDA could delay approval until adequate initial clinical data were available to suggest that the drug was promising; activists contended that patients died in the interim. In response, the FDA expanded the early access program by developing a "parallel track" that permitted HIV/AIDS patients who did not qualify for clinical trials access to investigational drugs [12, 13].

A third component of the new FDA procedures was codified in subpart E of the FDA's regulations (21 C.F.R., Sec. 312.80). These rules were intended to speed drugs aimed at serious or life-threatening illnesses through the approval process. They called for close collaboration between the FDA and drug developers. Promising drugs might be approved prior to stage III clinical trials; clinical trials might continue after drug approval (so-called stage IV postapproval trials) [13]. The policy change led to a new line of criticism: that the FDA was a tool of the pharmaceutical industry [1].

The pace of new AIDS drug development remained frustratingly slow. No new anti-HIV drugs were released until 1991, when dideoxyinosine (ddI), another nucleoside analogue, was approved [14]. Both AZT and ddI had serious side effects, and both lost effectiveness as the virus mutated. ACT-UP conducted a series of militant acts publicizing the limited number of anti-AIDS drugs, including occupying the offices of Burroughs-Wellcome to protest the high cost of AZT, and infiltrating the New York Stock Exchange to unfurl a banner telling investors to sell Burroughs-Wellcome stock [8]. Activists continued to vocally criticize the FDA for the slow rate of new drugs approval and the rigidity of its review guidelines. Another group, Project Inform, sponsored "underground" clinical trials within the gay community [2, 8]. It then provided information to the FDA in hopes of speeding new drug approval. In most cases, the FDA did not deem the data as adequate.

In the early 1990s, the FDA made further changes to expedite the review process. It announced the establishment of an "accelerated approval" mechanism for new drugs directed against life-threatening illnesses. Drugs must offer a significant benefit over existing treatments. Accelerated approval may be granted based on the drug's effect on a "surrogate endpoint marker" rather than direct evaluation of a clinical benefit to patients [2, 4, 12, 13]. In the case of new AIDS drugs, the surrogate marker might be the number of CD4+ cells; an increase in cell counts would suggest improved survival for the patients. Given the uncertainty of this approach, final FDA approval was contingent on demonstrating that the surrogate marker did reflect clinical efficacy using postapproval clinical trials (stage IV).

The FDA also instituted a parallel track program targeted directly at HIV/AIDS [4]. Patients were given access to investigational drugs as in treatment INDs, but the amount of evidence required for access was considerably less. In other words, the FDA responded to pressure from activists by allowing HIV/AIDS patients access to investigational drugs using criteria other than those for other diseases. Only one drug, stavudine, was distributed to patients under the parallel track program, however. [13]

The FDA approved more anti-HIV drugs in the early 1990s, with shorter times between the filing of INDs and the final approval of the NDAs. In December 1995, the first protease inhibitor, saquinavir, was approved [15]. Protease inhibitors prevent new virus particles from escaping from infected cells by blocking the attack of the cell membranes. Other protease inhibitors followed. The FDA also approved the use of drug cocktails, combinations of drugs that proved more effective than a single drug. Finally, in 2003, the first of a new class of drugs was approved, Fuzeon, which blocks the fusion of HIV with T-helper cells [16]. Fuzeon, made by Roche, is extremely expensive at $20,000 per year, and like other anti-AIDS drugs, has serious side effects.

Although the death rate from HIV/AIDS in the United States has declined by about 70 percent since the mid-1990s, the pandemic continues to spread in the developing world [17]. Because of the high cost of HIV/AIDS medications, most are not available to patients worldwide. The search for low-cost strategies for controlling the spread of the disease continues. Research aimed at reducing the transmission of HIV from a mother to a fetus is fraught with ethical problems (see the "Placebo Controls" section below).

User Fees and New Law

Throughout the 1980s and early 1990s, the FDA remained short of staff and funds to conduct NDA reviews. In 1992, Congress authorized the FDA to collect user fees from pharmaceutical companies to help pay for the costs of new drug reviews [2,

18]. The Prescription Drug User Fee Act (PDUFA) provided revenue to permit the hiring of an additional six hundred reviewers. Congress also linked the improvement in the speed of review to future appropriations for the FDA [2]. Recent evidence suggests that the PDUFA did indeed reduce the time between an NDA and approval. The PDUFA was reauthorized in 1997, and again in 2002.

The reauthorization of the PDUFA was part of the FDA Modernization Act of 1997 (FDAMA) [19]. The FDAMA provided, for the first time, a clear mission statement for the FDA that called for the prompt and efficient review of clinical research with the aim of protecting public health. It also allowed the FDA to weaken its own standards of evidence for approval of new drugs; a fast-track approval process was part of the new law. This provision came in response to continuing criticism, particularly from conservatives, that the FDA regulatory review process denied access to new drugs [1].

The Dark Side of Accelerated Approval

Does accelerated approval represent a risk to Americans? A cautious approach may increase the likelihood that new drugs are safe, but it also delays the marketing of potentially lifesaving new medications. Accelerated approval mechanisms may shorten the time taken to review clinical data, but they may also increase the risk that a new drug carries significant side effects not detected during testing.

Before the 1990s, the number of NDAs approved in less than a year was under 20 percent [2]. By the decade's end, the percentage increased to over sixty. In 1990, the FDA approved around 75 new drugs; the number increased rapidly to around 150 per year by the late 1990s [20]. The FDA, however, withdrew 12 drugs between 1997 and 2001, 9 of which had been approved under the new accelerated procedures. In the previous eight-year period, the FDA withdrew only 6 drugs for safety reasons. Among the drugs withdrawn in the latter half of the 1990s were Rezulin [21], a diabetes drug that caused liver failure, and Baycol [22], a cholesterol-lowering drug that caused organ failure. A popular off-label combination of two diet drugs, "fen-phen," also was withdrawn because of damage to heart valves [23]. Both diet drugs were approved initially only for use separately, but physicians and patients found they were much more effective for weight loss when taken together. The FDA withdrew both drugs to prevent further misuse.

Several drug withdrawals received heavy media coverage, leading to criticism of the FDA for lax oversight [1]. Others countered that the risk of promising drugs was outweighed by their potential benefits at the time of approval, and it sometimes takes more data to change that perception [24]. In September 2004, Merck and Company withdrew its blockbuster painkiller, Vioxx, after accumulating evidence that the drug was linked to increased risk of heart attack. Vioxx was one of a new generation of painkillers called Cox-2 inhibitors, which reduce pain and inflammation, but without causing the stomach upset associated with popular drugs such as ibuprofen. Critics charged that both Merck and the FDA had been aware of the risks, but chose to keep the drug on the market [25, 26, 27]. Thousands of lawsuits have been filed; the first two cases resulted in decisions split with one for the plaintiff and one for the defendant, Merck (a third case ended in a mistrial) [28]. In February 2005, the FDA announced the withdrawal of a drug to treat multiple sclerosis, Tysabri, because of links to fatal brain infection. Tysabri had been approved only three months earlier under the FDA's accelerated approval process [29]. In June 2006, the FDA approved renewed marketing to MS patients within a strict risk monitoring program [30].

Public concern over the withdrawal of Vioxx and other drugs stimulated calls for better oversight of new drug approval by the FDA, as well as calls for congressional investigations on the possible collusion between drug manufacturers and the FDA [31, 32, 33]. Charges of intimidation of FDA examiners who questioned drug safety were reported in congressional testimony. The stock values of Merck and other pharmaceutical companies selling Cox-2 inhibitors dropped dramatically. In February 2005, the FDA announced the establishment of a new drug oversight board to keep tabs on drug safety issues and provide more timely information to consumers [34]. The FDA's ability to order postmarketing safety reports is limited, however, since pharmaceutical companies are not required to report evidence of side effects [35]. If drug companies conceal negative studies (see chapter 8), the FDA has limited resources to uncover the action.

Off-Label Drug Applications
Off-label drug use is the prescribing of drugs approved for one condition for a different one [2]. It is not unusual for cancer chemotherapy agents to be used off-label to treat other cancers; frequently, they are then approved for use for additional forms of cancer [36]. Yet some off-label applications, such as fen-phen, can have disastrous results. Physicians and hospitals are encouraged to open INDs whenever trying drugs for new indications, to allow for the collection of good evidence of safety and efficacy. In 1994, new legislation limited the FDA's ability to regulate off-

label drug use (see below). Then, in 1998, the U.S. District Court for the District of Columbia ruled that the FDA's guidelines for drug company distribution of published material and their influence on continuing medical education activities regarding off-label uses of drugs violated the First Amendment guarantee of free speech (*Washington Legal Foundation v. Friedman*) [37]. This ruling meant that a pharmaceutical firm's promotion of drugs for off-label indications is permitted, but it is limited to the distribution of "bona fide peer-reviewed professional journal" articles or "reference textbook (including any medical textbook or compendium) or any portion thereof published by a bona fide independent publisher," and they may suggest "content or speakers to an independent program provider in connection with a continuing medical education seminar program". [37]

The Regulation of Dietary Supplements and Herbal Medicines

The FDA has broad jurisdiction over the safety of prescription drugs and food additives, but its regulatory capability is limited when dealing with over-the-counter dietary supplements or herbal medicines. In the early 1990s, the FDA proposed rules to more tightly regulate the burgeoning dietary supplement industry [1]. The impetus for the proposed rules involved an epidemic in 1989 of cases of eosinophilia-myalgia syndrome, a serious disabling condition that causes severe muscle pain and scarring, and is associated with elevated white blood cell counts. The outbreak was linked to the dietary supplement L-tryptophan (an amino acid), which is marketed as an aid to relieving depression. All cases were associated with L-tryptophan from a single Japanese manufacturer; it remains uncertain whether a contaminant caused the illnesses. The CDC estimates that thirty-seven people died and at least fifteen hundred were injured [38]. The FDA pulled L-tryptophan off the market and pressed for tighter regulation. Its ban was overturned in the 1990s, and L-tryptophan remains on the market despite continuing FDA concerns [38].

The manufacturers of vitamins, herbal medicines, and other supplements argued that the proposed FDA regulations would drive them out of business. Their lobbying proved successful. Congress passed the Dietary Supplement Health and Education Act (DSHEA) in 1994 [38]. The statute established new definitions for dietary supplements, separating them from food additives and drugs, and placing them outside the FDA's regulatory jurisdiction. The DSHEA also set up an Office of Dietary Supplements within the NIH, with a mandate to explore the role of supplements in promoting health and preventing chronic diseases. The DSHEA asserted that the benefits and safety of dietary supplements were clear, and placed the burden of proof on the FDA to demonstrate that a given supplement was hazardous. Should the FDA believe a supplement was dangerous, it had to apply to the DHHS secretary for permission to ban the supplement. The DSHEA also freed manufacturers from many of the labeling restrictions dating back to 1938. Manufacturers are free to claim health benefits for supplements, as long as they don't claim to cure a disease.

Problems with supplements continue to surface, because no premarket testing is required. Manufacturers are expected to report problems to the FDA's voluntary Medical Products Reporting Program [40], but they are not required to do so. For example, St. John's Wort, a herbal supplement touted as a treatment for depression, can block the action of other medications, including anti-HIV/AIDS drugs [1]. Supplements containing kava-kava, intended to relieve stress and promote relaxation, are linked to a number of cases of liver failure requiring transplants [41]. In response, the FDA issued a warning to consumers, but it lacks sufficient evidence to ban the supplements [42].

In 2001, the Office of the Inspector General of DHHS issued a scathingly critical report of procedures for reporting the adverse effects of dietary supplements [43]. It found that only a small number of problems are reported, the FDA receives only limited information, and the manufacturers rarely report adverse events. As a result, the FDA is unable to take actions to protect the public. The report recommended requiring manufacturers to report problems, requiring the registration of dietary products and their manufacturers, and setting guidelines for the types of information to be reported. The report also urged the FDA to seek further oversight authority for dietary supplements. While supported by consumer groups, most manufacturers opposed the recommendations.

In December 2003, the FDA announced its intention to ban the dietary supplement ephedra, a weight-loss and energy supplement [44]. Ephedra is popular with athletes for enhancing performance. It and other ephedrine-containing products are linked to an increased risk for psychiatric, gastric, and cardiovascular problems. The highly publicized death of a young professional baseball player in spring 2003 brought the risk into the public eye. The risks of ephedra were not new, though. The FDA issued its first warning on the hazards of ephedra in 1997, and required a warning statement to be included with the product [44]. A detailed analysis of a number of studies of ephedra confirmed a significant increased risk for serious complications while demonstrating only modest benefit for short-term weight loss [45]. Armed with these data, the FDA had the ammunition to ban the supplement.

An editorial in the *Journal of the American Medical Association* used ephedra to argue for the tighter regulation of dietary supplements by the FDA [46]. Asserting that ephedra is not an isolated case, the authors called for new legislation to define and regulate dietary supplements. They also called for tighter regulations on product advertising.

Drug Marketing and Public Education

In 1997, the FDA altered a long-standing provision that barred drug manufacturers from marketing new drugs direct-to-consumer (DTC). The FDAMA partially mandated this change, allowing pharmaceutical companies to promote approved drugs for off-label applications [2]. The impetus for these changes was a desire to better inform the public about new treatments for diseases and make the operations of the FDA more transparent. The development of the Internet provides additional sources of information about prescription drugs for consumers; advertising also appears in the print and broadcast media.

Pharmaceutical companies argue that advertising is needed to allow them to recoup the considerable costs associated with new drug development; the industry estimates it spent $30.3 billion on research and development in 2001 [47]. It spent $19.1 billion on promotional activities in the same year.

The FDA's Division of Drug Marketing, Advertising, and Communications regulates the content of advertisements. Ads cannot provide false or misleading statements, and must include a balance of information concerning the risks and benefits of the drug. Ads must include descriptions of side effects. Advertisers also must provide sources of additional information for consumers. Several kinds of ads may be run: the most common are "product-claim" ads that link a named drug to a specific medical condition; "reminder" ads mention a drug by name, but do not define its use; and "help-seeking" ads mention a disease, but not a drug [2].

The impact of DTC is considerable. Although 80 percent of marketing is still directed at physicians, the amount spent on DTC increased from $1.1 billion in 1997 to $2.7 billion in 2001 [47]. DTC advertising also appears to increase both the utilization and sales of prescription drugs, most notably on those that are heavily advertised. A GAO report in 2002 indicated that an estimated 8.5 million consumers annually request and receive prescriptions for drugs they saw advertised [47].

The GAO study also found that the FDA's ability to enforce its regulations concerning false advertising is limited, largely because the DHHS has responsibility for reviewing FDA letters of complaint prior to their being sent to pharmaceutical companies [47]. Some drug companies repeatedly violate FDA regulations, despite warnings.

The increased spending on DTC may be interpreted in two ways. Proponents argue that advertising provides a means to educate the public about medical conditions and possible treatments. This allows individuals to be more proactive and engaged in their own health. On the other hand, advertising may lead patients to demand new, more expensive medicines over cheaper, equally effective ones, raising the cost of medical care. Physicians may be pressured into providing drugs for relatively trivial medical conditions.

Lessons from the FDA

The FDA has been a political punching bag since its formation, subject to criticism from both the political left and right wings, consumers, and companies. Its effectiveness in protecting the health and safety of American citizens depends strongly on one's viewpoint: the FDA can deny access to drugs desired by the public while at the same time approving inadequately tested drugs. Recent criticism charges that the FDA is too strongly influenced by pharmaceutical companies to exert appropriate oversight of drug safety. The pharmaceutical and dietary supplement industries, on the other hand, maintain that FDA regulation interferes with free enterprise and damages the U.S. economy. Industry spokespersons suggest that the business is capable of self-regulation, despite one hundred years of history that indicates the opposite [1].

Support for the FDA rises and falls with the politics of the party in power. Policymakers do not agree on whether citizens should have the right to access unapproved drugs regardless of whether there is evidence that they are effective and safe. Policymakers do not agree on whether the regulation of new drug development and drug usage represent a benefit to society, or rather reflect inappropriate government intrusion into individual freedom. The FDA provides a clear example of how politicians use scientific evidence to bolster their arguments. What is remarkable is that the same data are subject to entirely different interpretations.

Nevertheless, the FDA's response to public pressure during the AIDS era demonstrates that the regulatory agency maintains some degree of policymaking flexibility. Whether the changes made to accelerate the approval of new drugs and enhance access to unapproved medications prove to be in the best interests of the public remains to be determined.

References

1. Hilts, P. J. *Protecting America's health: The FDA, business, and one hundred years of regulation.* New York: Alfred A. Knopf, 2003.

2. Ceccoli, S. J. *Pill politics: Drugs and the FDA.* Boulder, CO: Lynne Renner Publishers, 2004.

3. Janssen, W. F. The story of the laws behind the labels, part II: 1938—The federal food, drug, and cosmetic act. *FDA Consumer.* June 1981. Available at <http://vm.cfsan.fda.gov/~lrd/histor1a.html> (accessed December 17, 2003).

4. Greenberg, M. D. AIDS, experimental drug approval and the FDA new drug screening process. *New York University Journal of Legislation and Public Policy* 3, no. 2 (2002): 295–350.

5. Janssen, W. F. The story of the laws behind the labels, part III: 1962 drug amendments. *FDA Consumer.* June 1981. Available at <http://vm.cfsan.fda.gov/~lrd/histor1b.html> (accessed January 5, 2004).

6. Farley, D. Benefit vs. risk: How FDA approves new drugs. *FDA Consumer* (January 1995): 23–29.

7. Gieringer, D. H. The safety and efficacy of new drug approval. *Cato Journal* 5, no. 1 (1985): 177–201.

8. Arno, P. S., and K. L. Feiden. *Against the odds: The story of AIDS drug development, politics, and profits.* New York: HarperCollins Publishers, 1992.

9. World Health Organization. AIDS epidemic update: December 2003. *WHO.* Available at <http://www.who.int/hiv/pub/epidemiology/epi2003/> (accessed January 12, 2004).

10. Centers for Disease Control. Cases of HIV infection and AIDS in the United States, 2002. *HIV/AIDS surveillance report,* October 17, 2003, 14.

11. Shilts, R. *And the band played on: Politics, people, and the AIDS epidemic.* New York: St. Martin's Press, 1987.

12. Flieger, K. FDA finds new ways to speed treatments to patients. *FDA Consumer* (January 1995): 19–22.

13. Shulman, S. R., and J. Brown. The Food and Drug Administration's early access and fast-track approval initiatives: How have they worked? *Food and Drug Law Journal* 50 (1995): 503–517.

14. U.S. Food and Drug Administration. FDA approved HIV AIDS drugs. Available at <http://www.fda.gov/oashi/aids/virals.html> (accessed January 5, 2004).

15. Schwartz, J. FDA approves first of new potent anti-AIDS drugs in just 97 days. *Washington Post,* December 8, 1995, 3.

16. New AIDS drug wins FDA approval. *CNN.* March 14, 2003. Available at <http://www.cnn.com> (accessed January 5, 2004).

17. Folkers, G. K., and A. S. Fauci. AIDS agenda still daunting. *Issues in Science and Technology* 19, no. 4 (Summer 2003): 37–39.

18. Henkel, J. User fees to fund faster reviews. *FDA Consumer* (January 1995): 50–52.

19. U.S. Food and Drug Administration Modernization Act. Vol. 21, U.S.C. 393. 105th Cong., 1997.

20. Center for Drug Evaluation and Research. Drugs @ FDA. *FDA.* February 26, 2005. Available at <http://www.accessdata.fda.gov/scripts/cder/drugsatfda> (accessed February 27, 2005).

21. FDA Press Office. Rezulin to be withdrawn from the market. *Department of Health and Human Services.* March 21, 2000. Available at <http://www.fda.gov/bbs/topics/ NEWS/NEW00721.html> (accessed January 19, 2004).

22. U.S. Food and Drug Administration. Baycol information. August 8, 2001. Available at <http://www.fda.gov/cder/drug/infopage/baycol/default.htm> (accessed January 19, 2004).

23. U.S. Food and Drug Administration. FDA announces withdrawal of fenfluramine and dexfenfuramine (Fen-Phen). September 15, 1997. Available at <http://www.fda.gov/ cder/news/fenphen81597.htm> (accessed January 19, 2004).

24. Ross, G. M. Rezulin proves the system works. *American Council on Science and Health.* March 29, 2000. Available at <http://www.acsh.org/press/editorial/rezulin032900.html> (accessed January 19, 2004).

25. Vioxx: An unequal partnership between safety and efficacy. *Lancet* 364 (October 9, 2004): 1287–1288.

26. Topol, E. J. Failing the public health: Rofecoxib, Merck, and the FDA. *New England Journal of Medicine* 351 (October 21, 2004): 1707–1709.

27. Maugh, T. H., II, and L. Girion. Journal slams Vioxx study; Questions about 3 heart attacks not included in data analysis could hurt Merck's legal defense. *Los Angeles Times,* December 9, 2005, C1.

28. Berenson, A. Merck is winner in Vioxx lawsuit on heart attack. *New York Times,* November 4, 2005, A1.

29. Pollack, A. F.D.A. panel weighs fate of a drug for cancer. *New York Times,* March 5, 2005, A8.

30. U.S. Food and Drug Administration. Natalizumab (marketed as Tysabri) Information. Available at <http://www.fda.gov/cder/drug/infopage/natalizumab/default.htm> (accessed June 14, 2006).

31. Harris, G. F.D.A.'s drug safety system will get outside review. *New York Times,* November 6, 2004, A11.

32. Horton, R. Vioxx, the implosion at Merck, and the aftershocks at the FDA. *Lancet* 364 (December 4, 2004): 1995–1996.

33. Wadman, M. The safety catch. *Nature* 434 (March 31, 2005): 554–556.

34. U.S. Food and Drug Administration. FDA improvements in drug safety monitoring. *FDA.* February 15, 2005. Available at <http://www.fda.gov/oc/factsheets/drugsafety.html> (accessed February 27, 2005).

35. Frantz, S. Chasing shadows. *Nature* 434 (March 31, 2005): 557–558.

36. Food and Drug Administration. Fast track, priority review, and accelerated approval. January 5, 2004. Available at <http://www.accessdata.fda.gov/scripts/cder/onctools/ Accel.cfm> (accessed January 5, 2004).

37. *Washington Legal Foundation v. Friedman*, 13 F. Supp. 2d 51 (Appeal dismissed 202 F.3d 331 [DC Cir 2000]). U.S. District Court for the District of Columbia, 1998.

38. U.S. Food and Drug Administration. Information paper on L-tryptophan and 5-hydroxy-L-tryptophan. April 4, 2001. Available at <http://vm.cfsan.fda.gov/~dms/ds-tryp1.html> (accessed January 21, 2004).

39. Public Law 103–417: Dietary Supplement Health and Education Act of 1994. 103rd Cong., 1994.

40. Ropp, K. L. MedWatch: On the lookout for medical product problems. *FDA Consumer* (January 1995): 42–45.

41. Centers for Disease Control. Hepatic toxicity possibly associated with kava-containing products: United States, Germany, and Switzerland, 1999–2002. *Morbidity and Mortality Weekly Report*. 51, no. 47 (November 29, 2002): 1065–1067.

42. U.S. Food and Drug Administration. Kava-containing dietary supplements may be associated with severe liver injury. *Center for Food Safety and Applied Nutrition*. Available at <http://www.cfsan.fda.gov/~dms/addskava.html> (accessed January 20, 2004).

43. Office of Inspector General. *Adverse event reporting for dietary supplements*. OEI–01–00–00180. Boston: U.S. Department of Health and Human Services, April 2001.

44. U.S. Food and Drug Administration Press Office. FDA announces plans to prohibit sales of dietary supplements containing ephedra. *U.S. Department of Health and Human Services*. December 30, 2003. Available at <http://www.hhs.gov/news/press/20031230.html> (accessed January 20, 2004).

45. Shekelle, P. G., M. L. Hardy, S. C. Morton, et al. Efficacy and safety of ephedra and ephedrine for weight loss and athletic performance: A meta-analysis. *JAMA* 289, no. 12 (March 26, 2003): 1537–1545.

46. Fontanarosa, P. B., D. Rennie, and C. D. DeAngelis. The need for regulation of dietary supplements: Lessons from ephedra. *JAMA* 289, no. 12 (2003): 1568–1570.

47. U.S. General Accounting Office. *Prescription drugs: FDA oversight of direct-to-consumer advertising has limitations*. GAO–03–177. Washington, DC: General Accounting Office, October 2002.

Further Reading

Ceccoli, S. J. *Pill politics: Drugs and the FDA*. Boulder, CO: Lynne Renner Publishers, 2004.

• A detailed and up-to-date study of the FDA

Hilts, P. J. *Protecting America's health: The FDA, business, and one hundred years of regulation*. New York: Alfred A. Knopf, 2003.

• A readable exposé of FDA politics.

Shilts, R. *And the band played on: Politics, people, and the AIDS epidemic*. New York: St. Martin's Press, 1987.

• A critique of the federal government's inaction during the emerging AIDS crisis

Seminal Events: The Evolution of Regulations for Research

In the United States, federal statutes and regulations were adopted throughout the twentieth century, placing licensing requirements on those seeking to sell their drugs and medical devices in interstate commerce, as well as requirements for those using humans or animals in research supported by the federal government. While small modifications to these regulations have occurred over time to correct or refine the rules, their initial adoption along with major extensions and changes were made in response to specific, identifiable problems that came to the public's attention.

Drugs and Medical Devices

As mentioned earlier, the FDA and the regulation of drugs initially resulted from public concerns about food and drug quality (Pure Food and Drug Act of 1906, Pub. L. No. 59–384, 34 Stat. 768). Following the elixir of sulfanilamide tragedy [1], the law was replaced by a new statute that required safety testing (Food, Drug, and Cosmetic Act, Ch. 627, 52 Stat. 1040–1059 [1938]), and after the thalidomide tragedy, this law was amended to further require proof that a drug was both safe and effective (Kefauver-Harris Amendment, Pub. L. No. 87–781, 76 Stat. 780 [1962]). This latter change yielded the current research practices consisting of preclinical studies (in vitro and animals), phase I, II, and III clinical trials in humans, and postmarketing phase IV surveillance studies (sometimes referred to as seeding studies because of their utilization by pharmaceutical firms to get doctors used to using their new drugs). In 1976, the Medical Device Amendments were enacted (Pub. L. 94–295, 90 Stat. 539 [1976]), following revelations about the harms caused by the Dalkon Shield, a type of intrauterine contraceptive device that caused pelvic inflammatory disease, a serious infection that results in scarring of the fallopian tubes, causing sterility. The law gave the FDA jurisdiction over medical devices, including those implanted in patients' bodies.

At the same time that requirements were developing for testing drugs in animals and humans, there were revelations in the mid-1960s of problems in the ways research was being carried out.

Animal Research

In February 1966, *Life* magazine published an article and photo exposé titled "Concentration Camp for Dogs" about the use of pound animals in research, and there was an outcry by a public concerned by the chance that their family pets might wind up in lab cages. This led Congress to enact the Laboratory Animal Welfare Act (Pub. L. No. 89–544, 2[h], 80 Stat. 350, 351 [1966]), which was amended and renamed the Animal Welfare Act (AWA) in 1970 (Pub. L. No. 91–579, 3[g], 84 Stat. 1560, 1561 [1970]). The act required the humane care of animals by dealers and researchers. Subsequently, regulations issued under the act require that researchers consider alternative designs, implementing the "three Rs" first proposed by William Russell and Rex Burch: replace (replace animals with lower-order creatures, computer models, or in vitro assays when feasible), reduce (lower the number of animals used by the tight control of experimental methods), and refine (minimize the pain, suffering, or stress experienced by the animals) [2].

In 1985, another public exposé occurred when the activist group People for the Ethical Treatment of Animals (PETA) broke in to a laboratory at the University of Pennsylvania and stole about sixty hours of videotape. The group extracted twenty-five minutes of video that graphically displayed primates undergoing head trauma in the lab and revealed several problems such as unsanitary surgical conditions [3]. The video was shown on television and in Congress, leading to the 1985 amendment of the AWA (Improved Standards for Laboratory Animals Act, Pub. L. No. 99–198, 1751–1759, 99 Stat. 1354, 1654). This amendment required exercise for certain animals, care for the psychological welfare of nonhuman primates, and the creation of Institutional Animal Care and Use Committees that have authority to approve and oversee research involving animals.

Interestingly, the regulations covering laboratory animal research excluded birds, rats, and mice. This was in large part a pragmatic decision by the USDA that it would be simply too expensive to apply the AWA to these animals, considering that rats and mice comprise some 90 percent of the estimated twenty million vertebrate animals used in research in the Untied States. A lawsuit was brought to compel the USDA to comply with the AWA by the Alternatives Research and Development Foundation and the American Anti-Vivisection Society, and in late 2000 the department agreed to promulgate regulations for birds, rats, and mice. Congress intervened, though, refusing to allocate funds for the development of the regulations [4]. In 2002, Congress enacted a farm bill that explicitly excludes birds, rats, and mice from protections under the AWA (Farm Security and Rural Investment Act of 2002, Pub. L. No. 107–171, 116 Stat. 134).

While heavily regulated in the United States, research with animals is nonetheless permissible and widely done with appropriate protections for the welfare of the animals used [5]. In contrast, many countries in Europe have taken more protective approaches to animal use, expressing a public concern for the rights of animals to not be used for research [6].

Human Subjects Research

Also in 1966, a highly influential article appeared in the *New England Journal of Medicine* authored by Henry Beecher, a Harvard professor of anesthesiology [7]. He identified fifty published research articles in leading biomedical journals, and described twenty-two of them in detail, discussing what he felt were serious ethical problems. The article led the Public Health Service and the FDA to require, for the first time, the approval of human research by an institutional review board (IRB). Then, in 1972, the national press exposed the Public Health Service's "Study of Untreated Syphilis in the Male Negro in Macon County, Alabama," which has come to be known as the "Tuskegee Syphilis Study." A large group of poor black men with syphilis (and later, a group of two hundred initially uninfected controls) were monitored for decades without receiving any treatment, long after effective antibiotics (specifically penicillin) were readily available; the men were told that the blood tests and other examinations were helpful for their "bad blood." The investigators, with the knowledge of Public Heath Service administrators, went to great lengths to block subjects from getting care—most notably by preventing them from joining the military, where their syphilis would be diagnosed and treated. The subsequent investigation concluded that the results of the study were without value [8]. Nonetheless, Art Caplan showed that publications generated throughout the study, as well as with the results of other unethical research, have been heavily cited in the scientific literature without mention of the ethical deficiencies [9].

In response, Congress enacted the National Research Act in 1974 (Pub. Law 93–348, 88 Stat. 342), which among other things, created the National Commission for the Protection of Human Subjects of Biomedical and Behavioral Research. The National Commission (as it has come to be known) issued its findings in the so-called Belmont Report in 1979, calling for the federal regulation of research practices [10]. Regulations were issued in their final form in early 1981, requiring that, in general, human subjects in research give informed consent to participation and that research be approved by an IRB [11]. In 1991, the regulations were formally adopted by fifteen federal agencies that sponsor or perform human research (note that the FDA has its own regulations, which are quite similar [12]), and thus became known as the "Common Rule." Since then, the Common Rule has become the de facto minimal standard for performing research on human beings in the United States.

Conclusion

The regulation of research has been highly reactive to revelations of problems in scientific practices, from drugs and medical devices placed on the market without an adequate testing of their safety and efficacy, to the general use of animals and human beings in studies. The unacceptability of certain scientific practices demanded social intervention and the imposition of standards of behavior. Some complain that regulations are too strict, and stifle or impede research [13]. Others assert that we really do not know how well these protections work, and worry that the protections are inadequate. It seems clear that scientists who want to skirt the rules can get away with it, and the ultimate safeguard of animal and human welfare in research is an ethical researcher. Indeed, the most critical risk of overregulation is that investigators may feel that they are not responsible for their own ethical conduct, and that as long as an IRB has approved their research, they are not morally responsible [14].

References

1. Wax, P. M. Elixirs, diluents, and the passage of the 1938 Federal Food, Drug, and Cosmetic Act. *Annals of Internal Medicine* 122 (1995): 456–461.
2. Russell, W. M. S., and R. L. Burch. *The principles of humane experimental technique.* London: Methuen, 1959.

3. Orlans, F. B., T. L. Beauchamp, R. Dresser, D. B. Morton, and J. P. Gluck. The Philadelphia head-injury studies on primates. In *The human use of animals: Case studies of ethical choice*, 167–183, Oxford: Oxford University Press, 1998.

4. Swanson, K. M. Carte blanche for cruelty: The non-enforcement of the Animal Welfare Act. *University of Michigan Journal of Legal References* 35 (2002): 937–968.

5. Office of Laboratory Animal Welfare. *Public health service policy on humane care and use of laboratory animals*. Washington, DC: U.S. Department of Health and Human Services, 2002.

6. Brody, B. A. The use of animals in research. In *The ethics of biomedical research: An international perspective*, 11–30. New York: Oxford University Press, 1998.

7. Beecher, H. K. Ethics and clinical research. *New England Journal of Medicine* 274 (1966): 1354–1360.

8. Brandt, A. M. Racism and research: The case of the Tuskegee Syphilis Study. *Hastings Center Report* 8, no. 6 (1978): 21–29.

9. Caplan, A. L. The intrusion of evil: The use of data from unethical medical experiments. In *Am I My Brother's Keeper? The Ethical Frontiers of Biomedicine* (Bloomington: Indiana University Press, 1988), 22–29.

10. The National Commission for the Protection of Human Subjects of Biomedical and Behavioral Research. *The Belmont report: Ethical principles and guidelines for the protection of human subjects of research*. Washington, DC: Department of Health and Human Services, 1979.

11. 45 Code of Federal Regulations, Part 46 (2004).

12. 20 Code of Federal Regulations, Part 50 (2004).

13. Snyderman, R., and E. W. Holmes. Oversight mechanisms for clinical research. *Science* 287 (2000): 595–596.

14. Merz, J. F. The hyperregulation of research. *Lancet* 363 (2004): 89.

Further Reading

Brody, B. A. *The ethics of biomedical research: An international perspective*. New York: Oxford University Press, 1998.

• A good overview and comparison of international norms for human subjects biomedical research

Moreno, J. D., and P. R. Josephson. *Undue risk: Secret state experiments on humans*. New York: W. H. Freeman and Co., 1999.

• A seminal history of the use of human beings in research by the U.S. government

Placebo Controls in Clinical Research

Placebos are inert, nonbiologically active agents used in the clinical trials of experimental drugs or other therapies [1]. The typical placebo is a pill made to look like an experimental drug, the purpose of which is to fool the research subject as well as the researchers about whether a subject is getting the drug or not. This is called "blinding," which is used to prevent bias from affecting a study's results. Placebos are not the same as no treatment at all, because placebos are known to elicit physiological effects. For example, a typical placebo trial may show that some 20 to 30 percent of subjects "benefit" or show improvement on the outcomes being studied. This provides a standard that must be exceeded by an experimental treatment.

Typically, clinical trials have two or more treatment arms, in which subjects are given different drugs. Experimental treatments are studied to determine if they are safe and effective, comparing them to either placebo or active controls. Active controls are standard therapies for the disease or condition being studied. From a scientific perspective, placebo controls are preferred over active ones for a number of reasons. First, federal law only requires applicants for FDA licenses to establish safety and efficacy (that is, that a drug works, not that it works better than alternative treatments) [1]. Second, placebo control trials generally require smaller sample sizes (leading to faster and cheaper trials), because the expected differences in outcomes between the experimental treatment that hopefully will work are greater when comparing to placebo than to active control. Third, the use of active controls may lead to ambiguous results, because the studies are typically designed to show noninferiority, not superiority; an experimental drug could be shown to be roughly as effective as the standard treatment in a study, but nonetheless inferences about whether the drugs worked (or had less severe side effects) could not be drawn [1, 2].

From an ethical perspective, however, there are serious reservations about placebo control trials. The primary concern stems from Immanuel Kant's "categorical imperative" that we should not use persons as mere means to ends but rather as ends in themselves [3]. Thus, if there are treatments available, we should not deny them to patients just for the social purpose of studying new potential ones [4]. This ideal is embodied in the World Medical Association's Declaration of Helsinki, an international code for clinical research. As article 29 of the declaration states, "The benefits, risks, burdens and effectiveness of a new method should be tested against those of the best current prophylactic, diagnostic, and therapeutic methods. This does not exclude the use of placebo, or no treatment, in studies where no proven prophylactic, diagnostic or therapeutic method exists" [5].

Resolution of the conflict takes various forms, including using placebos only when patient-subjects are not placed at risk of irreversible or serious harm or death from participation, and designing studies to minimize the time and risk from withholding treatment [6]. The ethical concerns raised by forgoing existing therapy are answered to the extent that subjects give a free and informed consent to participate. People can be altruists, and be willing to expose themselves to minor discomforts for the benefit of others.

In the early 1990s, trials were performed in the United States and elsewhere to develop prophylactic methods to prevent the vertical transmission of HIV, the cause of AIDS, from a mother to a child. In the United States, the critical multicenter trial was run by the AIDS Clinical Trials Group (ACTG 076) and established the efficacy of a course of AZT treatment in substantially reducing the transmission rate. The results were so strong that the study was stopped early and the results announced to the public [7]. The regimen required treatment of the pregnant woman during pregnancy, and at the time of birth, followed by treatment of the newborn for six weeks. New mothers were not permitted to breast-feed their babies. The treatment cost over $800, and the method was rapidly adopted as the standard of care for treating HIV+ pregnant women in the United States.

The method was not adopted in developing countries, because of a lack of the financial resources and related medical services to provide such intensive drug therapy. Nonetheless, faced with mounting HIV infection rates, many developing countries sought methods to arrest this pathway of infection. Toward this end, many trials were begun in an attempt to develop effective, practical, and affordable interventions, using lower and less frequent doses of anti-HIV drugs. Most of these trials were placebo controlled.

In 1997, an article by Peter Lurie and Sidney Wolfe [8], and an accompanying editorial by Marcia Angell [9], in the *New England Journal of Medicine* attacked the placebo control trials, asserting that the quoted language from the Declaration of Helsinki mandated that the studies be actively controlled, using the ACTG 076 protocol as the comparison group. They went so far as to draw an analogy of these trials to the Tuskegee Syphilis Study, in which over four hundred infected poor black sharecroppers in Alabama were actively denied access to penicillin after it became available, without their knowledge (see previous section). In response, Harold Varmus (then director of the NIH) and David Satcher (who became the U.S. surgeon general the following year) dismissed the analogy, largely based on the fact that the placebo control trials were all supported by local governments and, in stark contrast to the syphilis study, all had the express purpose of helping the local populations develop strategies for overcoming an extremely serious health problem [10].

The debate turns, in part, on the ambiguity of "best current" treatments in article 29 of the declaration. Lurie and Wolfe assert that this means the best standard of care anywhere in the world. In contrast, the researchers justified the use of a placebo because the local standards of care involved no preventive treatment at all. The placebo trial did not place subjects at any disadvantage because if they did not participate in the research, they would not have had any chance of receiving prophylactic care.

The debate also turns on the problems of active control trials. As described above, the ACTG 076 protocol was relatively intensive. Any less rigorous treatment regimen presumably would be less efficacious. Thus, in lieu of performing an equivalence trial, an active control trial would be designed to see if a cheaper regimen (or perhaps a range of cheaper ones) would be only an acceptably bit worse than the 076 protocol. For the study design, the difference in transmission rates between the two groups would be arbitrarily selected. Any cheaper regimen, involving a much shorter course of treatment of a mother and a newborn, would presumably not work as well as the 076 protocol. The study therefore could answer the question, how much worse than the 076 protocol is the cheaper protocol? But it could never reliably show that the new regimen actually works better than nothing.

Finally, the Lurie and Wolfe analysis criticizes the placebo control studies for their failure to satisfy the ethical requirement of clinical equipoise. Equipoise is a state of medical uncertainty about which of two or more treatments will work better. It is an ethical condition for justifying a comparison study in which subjects are randomly assigned to either of two or more treatment arms [11]. They assert that from the efficacy of the 076 regimen, it may be presumed that shorter-duration treatment with AZT will likewise reduce the maternal-child transmission rates. Hence, equipoise is not satisfied in a placebo control trial. Their analysis, however, did not address the parallel question about equipoise in the active control trials that they preferred. They believe trials should be designed to study cheaper treatment regimens that will hopefully be "approximately as effective" as the 076 protocol. Nonetheless, any such trial, as described above, would be based on the expectation that the cheaper treatment would not work as well as the 076 protocol, in clear violation of the requirement of equipoise. They cannot have their cake and eat it too.

Conclusion

This case highlights the ethical complexities of clinical trials, and the trade-offs of different values and goals that must sometimes be made. At the end of the day, effective treatments were developed that cost less than about $10, and that could practically be delivered to women who did not have good prenatal care and for whom breast-feeding was desirable to help infants fight other infections. While a placebo control trial likely led to some children being infected, whose infections could have been avoided by running an active control trial, the value of the trial to help determine how best to treat future mothers and children arguably makes the placebo control trials worthwhile.

References

1. Federal Food, Drug, and Cosmetic Act. U.S. Code Title 21, Ch. 9, § 355 (b)(1) (2004).

2. Temple, R., and S. S. Ellenberg. Placebo-controlled trials and active-control trials in the evaluation of new treatments. Part 1: Ethical and scientific issues. *Annals of Internal Medicine* 133 (2000): 455–463.

3. Kant, I. *Groundwork of the metaphysics of morals.* Trans. M. Gregor. Cambridge: Cambridge University Press, 1996.

4. Rothman, K. J., and K. B. Michels. The continuing unethical use of placebo controls. *New England Journal of Medicine* 331 (1994): 394–398.

5. World Medical Association. Declaration of Helsinki. September 10, 2004. Available at <http://www.wma.net/e/policy/b3.htm> (accessed April 16, 2005).

6. Ellenberg. S. S., and R. Temple. Placebo-controlled trials and active-control trials in the evaluation of new treatments. Part 2: Practical issues and specific cases. *Annals of Internal Medicine* 133 (2000): 464–470.

7. Connor, E. M., R. S. Sperling, and R. Gelber et al. Reduction of maternal-infant transmission of human immunodeficiency virus type 1 with zidovudine treatment. *New England Journal of Medicine* 331 (1994): 1173–1180.

8. Lurie P., and S. M. Wolfe. Unethical trials of interventions to reduce perinatal transmission of the human immunodeficiency virus in developing countries. *New England Journal of Medicine* 337 (1997): 853–856.

9. Angell, M. The ethics of clinical research in the third world. *New England Journal of Medicine* 337 (1997): 847–849.

10. Varmus, H., and D. Satcher. Ethical complexities of conducting research in developing countries. *New England Journal of Medicine* 337 (1997): 1003–1005.

11. Freedman, B. Equipoise and the ethics of clinical research. *New England Journal of Medicine* 317 (1987): 141–145.

Further Reading

Barnett, A., and A. Whiteside. *AIDS in the twenty-first century: Disease and globalization.* New York: Palgrave Macmillan, 2002.

• Provides a comprehensive assessment of the social and economic consequences of AIDS, with an emphasis on sub-Saharan Africa

6

Cosmetic Science: Breast Implants and the Courts

It is not unusual for scientific information to be used as evidence in legal proceedings. Who determines what information should be presented at trial, and what standards are applied to assure that the information is accurate and appropriate? The standards for what constitutes expert scientific testimony are remarkably flexible. This part explores the role of science in the courts by focusing on a particular kind of case: the product liability tort case. A variation on the product liability case, the "toxic tort case," is based on the assertion that exposure to chemicals leads to injury; litigation claiming illness from silicone breast implants is the example used here.

The FDA and the Regulation of Breast Implants

Prior to 1976, the FDA had no authority to regulate the safety of medical devices (see chapter 5). Breast implants had been on the market for fourteen years when Congress passed amendments (21 USC, Sec. 360[c]) to the Federal Food, Drug, and Cosmetic Act of 1938 that mandated FDA oversight of medical devices, including breast implants [1, 2]. The FDA was required to classify medical devices into three groups: class I refers to devices such as tongue depressors that require only general manufacturing oversight; class II devices include items such as hearing aids that are subject to performance standards, postmarket surveillance, and use guidelines; and class III devices, such as heart valves, require proof of their safety and effectiveness prior to sale [2].

Breast implants temporarily were assigned to class II, although the FDA proposed placing them into class III in 1982 [3]. Testing was not required, under a long-standing "grandfather clause" that permitted approval of devices that were already marketed and apparently safe at the time of the 1976 amendments. The classification was changed to class III in 1988, as concerns about medical complications of rupture and leakage of silicone arose. The FDA asked the manufacturers to provide safety information akin to premarket approval under the new standards, and gave the manufacturers thirty months to collect the needed information [1, 2]. Because breast implants were developed at a time when little testing was required, the manufacturers had only limited data on the implants themselves to present—and thirty months gave them little time to conduct additional tests. The reporting requirements were further tightened in 1990, when Congress passed new legislation regulating medical devices [4]. Not surprisingly, the FDA ruled in 1991 that the data provided by the manufacturers were inadequate, particularly because long-term studies were lacking. Citing inadequate information, David Kessler, the newly appointed FDA commissioner, asked for a voluntary moratorium on the use of silicone breast implants in January 1992. In April 1992, Kessler announced a ban on silicone implants for cosmetic purposes, but allowed the use of implants for breast reconstruction under restricted circumstances. In announcing the ban, Kessler correctly asserted that the burden of proof of safety lay with the manufacturers [5]. Saline-filled implants remained on the market with additional testing and FDA approval [3].

Given that breast implants had been on the market for thirty years, why did the FDA choose to ban them? Critics suggest that public pressure caused the FDA to overreact [1, 2, 6]. Kessler also suggested that because breast implants were used for cosmetic purposes, their relative benefit was minor [5]. Even a small potential and avoidable risk was justification for withdrawal.

The Public Furor over Breast Implants

The FDA decision to ban silicone breast implants—like many FDA decisions—was not made in a dispassionate vacuum. Public opinion and heavy media coverage, and ultimately politics, undoubtedly influenced the decision. As we'll see below, these forces are still at work. Lawsuits against implant manufacturers received considerable media coverage. In 1990, journalist Connie Chung covered the breast implant story in a sensational manner on her television show, *Face to Face with Connie Chung*. She claimed that there was no question that breast implants were dangerous and that the FDA was negligent in permitting their sale [1]. Chung accepted the anecdotal evidence of the dangers of breast implants without question. Advocacy groups also lobbied the FDA to ban the devices. After losing a major court case in December 1991, Dow Corning announced it was pulling out of the breast implant market in March 1992—a move that many viewed as an admission of guilt.

The ban on implants triggered a flood of lawsuits. Sixteen hundred suits were filed in federal and state courts on behalf of women with implants in the two years following the ban [1]. A class action suit that ultimately involved over 440,000 women was filed in federal court in 1992 [1, 2].

The Science of Silicone Breast Implants

Silicone-filled breast implants came on the market in 1962, when two plastic surgeons worked with Dow Corning to design an implant that felt similar to breast tissue [1]. Silicone is made from the elements silicon and oxygen, and can vary from a liquid to a firm solid. Implants represent a marked improvement over previous approaches to breast augmentation dating back to the late nineteenth century. A variety of substances, including paraffin, beeswax, various oils, and liquid silicone, were tried with poor cosmetic results. In many cases, women suffered from disfiguring and life-threatening infections and other complications [1, 2].

Breast implants are used to augment the size of normal breasts, and as prosthetic replacements for women who undergo breast removal because of cancer or other illness; about 60 to 80 percent of surgeries are done for augmentation [1, 2, 7]. The implants, silicone-based bags containing silicone gel or saline, are inserted through small incisions into a pocket behind the breast while the patient is under general anesthetic. Breast implantation carries the general surgical risks of anesthesia and infection. Nevertheless, between 1 and 1.5 million women had breast implant surgery by 1992, the year that silicone breast implants were withdrawn by the FDA. Saline-filled implants remain available; breast augmentation was second only to liposuction in plastic surgery procedures in 1999 [2], when close to 250,000 women had implant surgery for augmentation or reconstruction.

Complications

When any foreign material is introduced into the body, inflammation is triggered. An inflammatory response involves increased circulation to an injured area, causing swelling and redness, and the accumulation of white blood cells that engulf bacteria and other infectious agents. If the material remains in the body, scar tissue ultimately forms around it. Over time, scar tissue tends to contract and can produce hard lumps within the tissue.

Scar tissue normally forms a capsule around the implant after surgery [7], since the implant is treated as a foreign object. The contracting scar tissue can distort the implant, leading to pain from pressure on surrounding tissues. The implant may rupture, allowing the silicone gel to leak into the surrounding tissue. Small amounts of silicone may "bleed" through the membrane of the implant, even if it remains intact. Most of the gel remains within the capsule of scar tissue, but some molecules may escape into the bloodstream or lymphatic circulation. In cases of severe contracture or rupture, the implant may be removed; however, many women opt for a new implant at the time [1, 2, 7]. These local complications are fairly common; a study in 1997 reported that 24 percent of implant recipients required reoperation within an average of eight years after implantation [7, 8].

Is Silicone Toxic?

The central issue in lawsuits involving breast implants is whether silicone is toxic. Since silicone is used in a wide range of medical devices, including syringes, catheters, artificial heart valves, and shunts, extensive data are available. Studies in animals dating back to the 1940s indicate that various forms of silicone are well tolerated, with few signs of toxicity beyond some skin irritation [7]. Silicone exposure is not associated with an increased risk of cancer, nor with birth defects or other developmental problems. The extensive use of silicone in medical devices stems from studies indicating only a limited inflammatory response by the body.

In 1988, a case study published in *JAMA* reported that three women receiving breast implants had subsequently developed scleroderma, a serious autoimmune disease of connective tissue [1, 6]. (It is important to note that case studies are not subject to rigorous peer review; their methodology and conclusions are not closely examined. As such, they are viewed as "anecdotal reports.") Other anecdotal evidence also suggested a link between silicone breast implants and other autoimmune diseases; physicians reported a variety of symptoms including fatigue, joint pain, headaches, and difficulty swallowing. Physicians postulated that exposure to silicone triggered abnormal immune responses. Autoimmune diseases, among them rheumatoid arthritis and lupus, develop when the immune system begins to attack normal body cells as though they are foreign. In general, the reasons for this immune response are unknown. Were these cases of connective tissue disease and other illness associated with the implants, or was this merely coincidence?

Little epidemiological evidence about breast implants was available prior to 1994, when a series of retrospective studies of large cohorts of women with breast implants was published [9, 10, 11]. These studies found no clear link between silicone implants and connective tissue diseases. In 1999, a congressionally mandated study by the Institute of Medicine (IOM) concluded that there was no convincing evidence linking breast implants to immunologic diseases [7]. The report concluded that the implants were associated with significant local complications, however.

How did researchers conclude that breast implants do not cause connective tissue disease? Animal studies, although useful, cannot rule out the possibility that implants cause disease in humans. Epidemiology is the science that analyzes the association of disease prevalence with characteristics of a human population. In some cases, the link between disease and exposure to an agent is unmistakable, as is the case for the drug thalidomide and a specific type of birth defect, phocomelia (see chapter 5). In other cases, exposure to a toxic agent produces a characteristic disease not seen otherwise, as in the case of asbestos exposure and a particular lung cancer called mesothelioma. For many suspected toxic agents, though, the links are not as strong. This was the case for silicone breast implants.

Epidemiologists conducted retrospective studies in which breast implant recipients were matched with a similar group of women without implants. To carry out an epidemiological study, researchers hypothesize that breast implants do not increase the risk of connective tissue disease, and determine the frequency of disease in the two populations. If, for example, ten in a thousand women without breast implants develop a connective tissue disease, and eleven in a thousand implant recipients do, then the relative risk of developing the disease is 1.1. Given natural variation between populations, these risk ratios are unlikely to differ statistically. A 95 percent confidence level is generally accepted as statistically significant, which means the likelihood that the result occurred by chance is 5 percent. Yet an epidemiological study cannot absolutely rule out a small increase in risk, particularly when a disease is rare, because the number of persons in a study is limited. An epidemiological study can also never prove causation. Nevertheless, if several studies yield similar results, the argument becomes more compelling. A combined statistical analysis of numerous epidemiological studies, a method known as meta-analysis, indicated that added risk from breast implants is extremely small [7, 11]. Scientific studies can suggest that the probability that breast implants cause autoimmune disease is quite low, but cannot rule it out completely since nothing can be scientifically proven to be impossible.

Tort Law and Product Liability

The term tort is defined as a legal wrong [12]. If an act or failure to act results in injury or other harms, the injured party may seek monetary compensation—called damages—in a civil lawsuit. A product manufacturer may be held liable under two primary tort theories: negligence and product liability. Negligence generally requires that a plaintiff prove the four elements of negligence liability: a duty (a responsibility to provide a safe product); a breach of that duty; injury for which monetary damages can be awarded; and proximate cause, which is a causal link between the defendant's actions and the injury suffered by the plaintiff. There are numerous causes of every accident, so the causal link must be close enough that liability is perceived as being fair [13].

Manufacturers have a duty to make products that are not overly dangerous, and that obligation may be owed to both product users and other parties who could be harmed by the use of the product in planned as well as unplanned, but again foreseeable, ways. For example, an automobile with defective brakes poses a hazard not only to the driver but also to passengers and people in other vehicles. The primary difficulty for plaintiffs in product lawsuits involved establishing a breach of duty—that is, that the manufacturer didn't make the product safe enough or that the product as marketed was defective. There are three distinguishable types of product defects: manufacturing defects, design defects, and defective warnings. Because of the need for clear evidence that a manufacturing problem caused a defect, the courts fashioned a variant of strict liability that required proof only that a product left the manufacturer in a condition that did not meet the manufacturer's own quality standards. A manufacturer is liable for all injuries proximately caused by such products [14].

The remaining product defects are more complex to adjudicate because they require a judgment about how safe is safe enough. Design defects require a judgment that a product was designed in a way that it is unreasonably dangerous, focusing not on the product per se but on the reasonableness of the action taken by the company in marketing the product. Products arguably could always be made safer, or products that present any risk of harm could simply not be put into the stream of commerce. We know from looking around us—at guns that can kill, knives that can cut, chemotherapy drugs that have serious side effects, and automobiles that kill forty thousand Americans every year—that products that have risks are not necessarily *unreasonably* dangerous. Defective design will turn on whether the risks posed by a product are reasonable in light of the benefits derived from its use, whether a product could have been made safer, how much alternative safety measures would cost, and whether safety measures would reduce the utility of the product (see also chapter 11). Finally, for unavoidable (and not obvious) risks inherent in the use of a product, a manufacturer must provide information—typically warning labels, markings, or package inserts—to enable the user to control or avoid the foreseeable risks.

The second difficulty facing plaintiffs in product lawsuits is establishing that the negligent act or defective product caused the plaintiff's injury. This is particularly true for injuries allegedly caused by exposures to chemicals, drugs, or radioactive materials. These exposures may be in small doses over extended periods of time, the diseases or harms suffered may have multiple or unknown causes, and there may be no science establishing a causal link.

Injured parties may recover compensatory damages, the aim of which are to make a person whole again by paying for pain and suffering, lost wages, medical costs, and the like. In cases when defendants have acted wantonly and in blatant disregard of the rights and welfare of others, exemplary or punitive damages may be assessed. Punitive damages both punish a defendant for gross negligence and act as a general deterrent to socially undesirable behaviors.

The size of damage awards in product liability cases has increased steadily. Multimillion dollar awards are becoming more common, as sympathetic jurors seek to compensate a victim for his or her suffering [1, 2, 15]. Lawyers often agree to represent clients on contingency; their fee is waived unless the litigation is successful. The lawyers then collect a substantial fraction—typically a third—of any award as payment for the costs of trying the case. An inevitable consequence of a successful product liability case is to encourage additional lawsuits. Large multinational manufacturers that make and distribute many thousands to millions of units of a particular product may be sued if public perceptions of common injury come to be associated with the product. A manufacturer may choose to settle out of court rather than absorb the costs of additional litigation. Numerous cases may be combined into class action suits, where a small number of plaintiffs represent a much larger group of individuals seeking restitution. The costs of litigation are considerable for both plaintiffs and defendants, and damage awards, if large enough, may drive manufacturers into bankruptcy.

At trial, lawyers for both the plaintiffs and defendants seek to present the strongest possible case supporting their arguments. For the plaintiff, it is critical to demonstrate that the product caused the injury; the defendant needs to show that the product could not have caused the injury. Both sides call experts to present opinions, and those opinions must be narrowly focused on a factual issue such as causation, drawing on evidence that is offered by the parties and admitted by the court. After hearing the arguments presented on both sides, the jury is left to decide which side presented a more compelling case supported by a preponderance of the evidence [1].

Who are expert witnesses, and who determines whether the information they present reflects scientific knowledge? A typical expert witness has advanced training and experience in the matter at hand, considers the available evidence, and presents an opinion supporting the contention of the plaintiff or defendant. Acceptance of the expert's testimony is dependent on his/her credibility—credentials that suggest that this opinion is based on superior knowledge and experience. As demonstrated in the case study, however, standards for experts and expert testimony sometimes may not reach those familiar to scientists. Some expert testimony has been criticized as "junk science" that fails to achieve any reasonable standard of accuracy or relevance [1, 2, 15].

Breast Implant Litigation

Prior to 1984, a small number of lawsuits against breast implant manufacturers were focused on rupture of the implant [2, 16], and received little publicity. In 1984, a jury in California awarded Maria Stern over $200,000 in compensatory damages and $1.5 million in punitive damages; her lawyers claimed that she developed an autoimmune disorder as a result of silicone exposure (*Stern v. Dow Corning Corp.*, C–83–234–MHP [N. D. Cal. Nov. 5, 1984]) [2]. Lawyers for the plaintiffs presented proprietary Dow documents and claimed these suggested that the manufacturer had concealed evidence that implants were dangerous. The documents showed that Dow Corning researchers were aware that breast implants caused long-term inflammation and that silicone could bleed through the envelope of the implant. By agreement, the documents were sealed after the trial and thus not made public. This action proved damaging in later trials, in which Dow Corning again was accused of concealing evidence that it was aware that silicone breast implants were defective.

In December 1991, Mariann Hopkins was awarded $7.3 million (including $6.5 million in punitive damages) by a jury in California—an award that was later upheld by the federal Court of Appeals (*Hopkins v. Dow Corning Corp*) [17]. In 1979, Hopkins was diagnosed with mixed connective tissue disease (an extremely serious illness) three years after receiving implants for reconstruction. She did not blame her implants initially for causing her illness; nine years after diagnosis she saw Stern's attorney, Dan Bolton, on television and contacted him [1]. At trial, Bolton presented experts who concluded that the breast implants caused Hopkins's illness, based on their own theories about silicone and immune function. Lawyers for the plaintiff argued that Dow Corning had concealed evidence that the manufacturer was aware of the risks, using the same documents as in the *Stern* trial. The defense asserted that there was no causal link between implants and autoimmune disease, and the testimony of the plaintiff's experts should be inadmissible. Remarkably, Hopkins's own physician suggested that she exhibited some symptoms of connective tissue disease prior to receiving her implants [1]. Yet the appeals court found that the jury had made a reasonable decision based on the medical evidence, given that no epidemiological evidence was available [17]. The Court of Appeals also found that Dow Corning had "concealed evidence" and affirmed the punitive damage award. Dow Corning appealed the ruling to the U.S. Supreme Court, which declined to hear the case.

The *Hopkins* case demonstrated that juries might award huge damages in breast implant cases. This success led to thousands of other cases directed against Dow Corning and other breast implant manufacturers. A breast implant tort "industry" developed in the legal profession, with thousands of lawyers searching for injured clients [1, 6]. In 1992, a Texas jury awarded Pamela Jean Johnson $25 million in her lawsuit against Bristol-Myers Squibb, even though her flu-like symptoms were not diagnosed as a genuine connective tissue disease [1, 2]. Another Texas jury awarded $27.9 million to three plaintiffs in 1994.

As the number of cases grew, judges sought to consolidate them within a single court. Congress created the Judicial Panel on Multidistrict Litigation (MDL) in 1968 to oversee class action lawsuits (P.L. 90–296). Consolidation increases the efficiency of the legal process by collecting pretrial information and resolving common issues only once. Cases may then be tried separately, or a group settlement may be negotiated. In 1992, a class action suit on behalf of breast implant litigants was transferred to an Alabama federal judge, Sam Pointer. In September 1994, the judge certified a class settlement of $4.23 billion, of which 25 percent (over $1 billion) would go to the plaintiffs' lawyers [18]. Eight companies—representing the implant manufacturers Dow Corning, Bristol-Myers Squibb, Baxter Healthcare, and 3M—

made payments into the settlement fund, with Dow Corning contributing over $2 billion. The manufacturers did not admit responsibility for the plaintiffs' illnesses. A grid established a level of compensation depending on the type of disorder, age at onset, and severity [1]. Women who were not yet ill could also claim lesser awards. The settlement allowed women to opt out of the settlement and seek restitution individually. But the defendants could withdraw from the settlement if the number of opt out cases grew too large.

Two factors led to the collapse of the 1994 class action settlement. By spring 1995, 440,000 women submitted claims; almost 250,000 of them claimed to be sick already [1]. This represented close to 20 percent of all breast implant recipients [1]. The large number of class members meant that the level of compensation to individuals had to be reduced; women complained that the new levels were too low, and 15,000 withdrew from the settlement [2]. In May 1995, Dow Corning filed for Chapter 11 bankruptcy protection in the face of over nineteen thousand individual breast implant lawsuits and forty-five other class action suits [2]. By declaring bankruptcy, all debts are consolidated in a single bankruptcy court that determines payments to creditors. (Undeterred, lawyers filed suits against Dow Chemical Company, Dow Corning's parent company, with some success.) Dow Corning withdrew from the settlement, halving the available funds for payout.

A new settlement, amounting to about $3 billion with the remaining three manufacturers, was developed in fall 1995, limited to women who had implants made by these companies. Women who become ill have fifteen years to join the settlement. Checks were mailed to 110,000 women in 2000 [19].

Litigation versus Epidemiology
The first epidemiological study examining the relationship between breast implants and connective tissue disease was published in 1994, with additional studies in 1995. These studies found no evidence to suggest that breast implants caused autoimmune diseases. Juries continued to award large sums to plaintiffs, however. Attorneys simply discounted the evidence, using experts who claimed that silicone triggered "atypical" forms of illnesses that did not fit into the categories examined in the published studies [20]. Some experts suggested that a new autoimmune disease, found *only* in women who received breast implants, was responsible for their illness [2, 6, 7]. Lawyers also sought to discredit the research by claiming that any research funding by implant manufacturers automatically biased the study. Researchers publishing studies found themselves targets of harassment. Marcia

Angell, then editor of the *New England Journal of Medicine*, was served subpoenas asking for documents linking "payments" to the publication of such articles; the subpoenas were quashed as "fishing expeditions" [1].

The first case to specifically exclude expert testimony claiming that breast implants caused connective tissue disease was *Hall v. Baxter Healthcare* in 1996, a class action suit in Oregon [21]. Hall's lawyers claimed that silicone migrating from her breast implants led to the development of "atypical connective tissue disease." The defense filed a series of motions to exclude the plaintiff's expert witnesses, based on the admissibility of evidence. A 1993 Supreme Court decision, *Daubert v. Merrell Dow Pharmaceuticals, Inc.* (509 U.S. 579) influenced the case; this decision gives judges "gatekeeper" responsibility for determining the appropriateness of scientific evidence (see "Setting Limits on Expert Testimony" section below). The judge appointed a panel of independent technical advisers to review the plaintiff's evidence, which found that the material presented was of questionable scientific validity. The judge thereby excluded the evidence. The plaintiff could still sue for local injury, but could not claim any systemic injury or fear of systemic disease.

Judge Pointer, in charge of the MDL class action suit, also appointed a panel of scientific experts to review the growing body of evidence regarding breast implants and systemic disease. The National Science Panel (NSP) released its report in December 1998, concluding there was no association between breast implants and connective tissue or autoimmune disease [22]. The NSP report was followed by the IOM report in 1999, which again failed to find any link between breast implants and systemic disease.

By the late 1990s, an increasing proportion of breast implant cases resulted in verdicts for the defense. Some punitive damage awards were reversed on appeal (for example, *Meister v. Medical Engineering Corporation and Bristol-Myers Squibb* [23]), but other appeals are still pending.

Scientific Evidence in the Courtroom

What evidence convinced jurors that breast implants were responsible for the plaintiffs' illnesses? The first question is whether plaintiffs were suffering from connective tissue disorders. In some cases, there was no question that the plaintiff was ill, as in the case of Hopkins. In others, a constellation of vague symptoms was presented; these symptoms, such as fatigue and muscle aches, are nonspecific and common. Many of these symptoms might develop as a result of suggestion; certainly, few complaints of illnesses were reported prior to the FDA's withdrawal of silicone breast implants [1, 2, 6, 7]. The IOM report was highly critical of so-called atypical disease, and argued that the definition of a novel syndrome that has as a precondition the presence of silicone breast implants, cannot be studied as "an independent health problem" [7]. This conclusion differs from cases where the epidemiological link between exposure and disease is unquestionable, such as for asbestos and the otherwise-rare lung cancer, mesothelioma.

The second central issue in breast implant litigation is whether the breast implants were responsible for the plaintiffs' illness. The absence of clearly defined illness did not deter plaintiffs from exploring this issue. Epidemiologists frequently turn to the so-called Bradford Hill criteria for causality. In 1965, Austin Bradford Hill gave an address to the British Royal Society of Medicine in which he outlined nine "issues" relevant when distinguishing causal and noncausal associations between events. Bradford Hill proposed that the factors should inform determinations based on statistical evidence, but instead often they have been misused as a checklist for proving causality [24, 25]. Such rigidity creates problems for epidemiologists and the courts.

Proving causality involves several components: the insertions of breast implants must precede the illness; the time between surgery and the development of symptoms must be medically plausible; other primary factors potentially causing the illness must be ruled out; and the mechanism of injury must be biologically plausible and supported by scientific evidence [1, 15, 26]. A mechanism that ignores background information in basic biology ought to be thrown out by the courts.

Much of the expert testimony in breast cancer litigation came from a small group of self-proclaimed experts on silicone's effects on the immune system. Few published any peer-reviewed papers in reputable scientific journals, and instead claimed expertise based on unpublished work. They made heavy use of anecdotal reports linking breast implants and illness. Some witnesses were later discredited for manufacturing data or deliberately distorting their credentials [1, 6]. Nevertheless, some courts permitted their testimony, and juries believed their opinions.

The Conflict between Science and the Law

Science and the law operate by distinctly different mechanisms. In science, a hypothesis (a possible explanation for an observation) is subject to controlled experimentation, and based on the results, the hypothesis is either supported or rejected. Additional studies are then conducted to provide further support for the hypothesis, or new information may lead to its rejection. When alternative hypotheses exist, each is studied until the bulk of evidence tends to support one hypothesis; incorrect assumptions are thus weeded out as research continues. When there are cases of genuine disagreement, usually scientists agree to continue to examine the question until the conflict is resolved (or in some cases, simply "agree to disagree"). Some individuals will continue to maintain positions far outside the mainstream view—and continue to do so even in the face of overwhelming evidence to the contrary. In some cases, however, "heretical" views eventually become part of the mainstream as published evidence accumulates to support the hypotheses. Scientists are human, of course, and may cling to viewpoints for personal or political reasons, even when scientific data do not support their positions [27]. Yet in principle, there are no "sides" in science: either the evidence supports a contention or it does not.

Law, on the other hand, has an adversarial basis. Expert witnesses are intended to help jurors understand and interpret complex issues, and the law does place limits on what evidence may be admissible at trial (see the following two sections). Still, the courts allow considerable latitude in the interpretation of what constitutes acceptable evidence. The two sides in a court proceeding present alternate views of the scientific facts of the case, and may present research data that support their contentions. No steps are necessarily taken to demonstrate that the research has any scientific validity, although cross-examination by opposing attorneys may reveal weaknesses. Lawyers may deliberately omit scientific information that does not support their clients and question experts only on selected material [28]. Experts present their credentials, refer to their experience, and then offer their opinions. Lawyers seek experts who present the strongest views in support of their "side." As a result, sometimes fringe scientists with viewpoints that may differ significantly from the mainstream prove to be the most persuasive [15, 26]. Attorneys may select witnesses for their physical appearance and ability to present science in an effective manner to the jury, rather than for genuine expertise [26]. The opinions of expert witnesses then become legal evidence in the trial [1, 15]. This effectively turns science on its ear: the conclusion becomes the evidence.

The jury as fact finder in legal cases is given the primary responsibility to determine which evidence is most compelling. The ultimate questions of duty, breach, causation, liability, and damages must be left to the jury, and experts cannot opine on those questions left squarely in its province. Most jurors, however, have only limited exposure to science and are not able to distinguish real science from junk science [1, 15, 26]. Jurors may assume that the conflicting viewpoints are representative of a general split in the scientific community, even if the vast majority supports one position. They may also choose to ignore evidence because of sympathy for the plaintiff or defendant.

Juries are, in the end, sometimes asked to resolve questions in the light of scientific uncertainty. The problem for the courts seemingly is to ensure that the uncertainty is real and not manufactured. A second challenge for the court would be to ensure that expert opinion truthfully presents the range of viewpoints on a scientific issue and puts them into a representative context within the relevant scientific community. Simply put, all views may not be equally accepted within the community, and evidence should be permitted to establish how widely different views are held.

Will Silicone Breast Implants Return?

Saline breast implants currently on the market are subject to stringent FDA review of premarket safety testing prior to approval [3, 29]. Nevertheless, about one-third of the recipients experience complications within five years [29]. Scar tissue forms around the implants, as occurs with silicone implants, and the contracture of scar tissue may require removal of the implant. The implants may rupture, causing them to collapse. Women may experience difficulty breast-feeding because of the loss of sensation associated with the implants. Despite improvements, saline breast implants do not mimic breast tissue as effectively as do silicone implants, and some women are dissatisfied with the appearance of their breasts after surgery.

Since 1992, silicone breast implants are available only for reconstructive purposes. In 2003, Inamed Aesthetics (formerly McGhan Corporation), which also manufactures saline implants, petitioned the FDA to approve its silicone implants for cosmetic use as well [30]. It presented evidence that changes in design reduced the rupture rate. Inamed also presented data from its own and other studies further stressing the absence of a link between silicone implants and systemic disease. On October 15, 2003, an FDA advisory review panel voted nine to six to approve the implants, finding that they posed no greater risk than saline ones [31]. In stark contrast to the previous media coverage, journalists were divided on the decision [32, 33, 34].

On January 8, 2004, the FDA announced it would defer a decision on whether to approve Inamed's silicone breast implants and issued new guidelines for information required by manufacturers who wished to market silicone breast implants [35]. Critics suggested that the FDA bowed to political pressure, raising the requirements for scientific data to unprecedented levels [36]. Groups opposed to breast implants praised the decision, arguing that the long-term safety had not been demonstrated. Inamed promised to resubmit its application. In April 2005, a federal advisory panel voted five to four against the approval of Inamed's silicone implants, citing the absence of data concerning the frequency of rupture [37]. The following day, though, the same panel voted to approve implants from a rival company.

Whether silicone breast implants are eventually approved remains to be determined. Implant surgery is still popular, with over three hundred thousand procedures in 2002 [38]. Given the history of litigation regarding silicone breast implants, the FDA's caution is understandable. Yet there are many other products on the market that are demonstrably less safe (see chapter 8), once again suggesting that underlying politics influences regulatory decisions.

Lessons from Breast Implant Litigation

In hindsight, breast implant litigation is a powerful example of how science may be misused in the courts [1, 6]. In the absence of epidemiological evidence, lawyers for plaintiffs presented purely speculative evidence, with no demonstrated causal link [39]. Litigation, however, provided an impetus to conduct epidemiological studies that ultimately demonstrated with high confidence that implants do not cause connective tissue disease [2]. The initial reluctance of juries to accept this evidence suggests that even good science may be ignored when challenged by persuasive attorneys. Ultimately, the flood of litigation slowed to a trickle, as judges rejected evidence linking implants to illness as unsupported by science.

Litigation concerning breast implants also provides a telling example of how some lawyers may capitalize on public fears. The stampede to sue breast implant manufacturers was stimulated by the FDA's decision to ban silicone implants. Kessler admitted that he did not anticipate that the FDA's action would lead to massive tort litigation [1]. Panicked women sought to have their implants removed, in the absence of any apparent problem. Prior to the FDA ban, few women expressed dissatisfaction with their implants; after the ban the percentage increased markedly [7]. Media coverage also increased women's fears, with sensational reports of court cases and exposés on television [1].

Breast implant litigation drove Dow Corning into bankruptcy—a casualty of mass tort litigation. Silicone breast implants represented only 1 percent of its sales, yet the burden of thousands of lawsuits was unmanageable [40]. Did Dow Corning behave unethically in its failure to conduct safety testing prior to marketing its breast implants? In the 1960s, manufacturers were not required to conduct extensive safety testing on new medical devices. Dow Corning had built an extensive literature about the safety of silicone products that dated back to the 1940s. The company's internal documents suggest that there was disagreement about whether leakage of silicone presented a hazard, but no compelling data indicated that the silicone was dangerous [1, 6, 40]. As the risk of rupture became clear, Dow Corning issued new warning directives to physicians, but these largely were ignored [1]. The vindication of Dow Corning's claims that silicone breast implants did not cause connective tissue disease came too late for the company.

Whether Dow Corning should be condemned for failing to conduct direct safety studies of silicone breast implants depends on one's viewpoint. Nevertheless, its bankruptcy has implications beyond breast implants. Dow Corning was a major manufacturer of silicone used in many other medical devices. As a result of litigation and increased FDA testing requirements, Dow Corning stopped supplying silicone materials to manufacturers in March 1993 [4]. The loss of a major manufacturer may impact production of other silicone-containing medical devices, such as shunts, artificial lenses, and heart valves [1, 6, 41]. Indeed, unsuccessful lawsuits claiming systemic illness have been filed on behalf of clients with penile implants and other devices [1, 42]. The possibility of litigation may also make other manufacturers reluctant to produce medical devices. Since these devices contain silicone, manufacturers remain at risk for lawsuits [16]. One other possibility is that liability provides a strong disincentive to large firms to manufacture such devices, inviting smaller, less wealthy, higher risk-taking companies into the market. If these smaller companies produce defective devices, consumers may have little or no legal recourse if injuries occur.

The strongest criticism is directed at a system of tort law that enables mass claims and enormous damage awards. Some critics accuse tort lawyers of unmitigated greed and unethical behavior [1, 6, 15, 26, 40]. They point out that the lawyers for plaintiffs often pocket as much as half of a damage award and in other cases the plaintiffs receive nothing after paying the lawyers' fees. Some tort lawyers advertise for clients; a quick perusal of the Web reveals dozens of sites still seeking clients for breast implant and other injury litigation. There have been repeated calls for "tort reform," to place limits on punitive awards and prevent lawyers from trolling for clients. Among the suggestions are calls to eliminate the contingency-fee system for lawyer compensation—a system unique to the United States [43]. Nevertheless, tort law is needed to hold corporations accountable for what they produce, and protect individuals who are genuinely injured by defective and hazardous products. Large multinational corporations that market extremely high-volume products can do untold harm to consumers and others. As such, finding mechanisms to accomplish the goals of compensating injured parties, promoting product safety, and ensuring fairness for manufacturers and others in the supply chain is a real public policy challenge.

References

1. Angell, M. *Science on trial: The clash of medical evidence and the law in the breast implant case*. New York: W. W. Norton and Company, 1996.

2. Hersch, J. Breast implants: Regulation, litigation, and science. In *Regulation through litigation*, ed. W. K. Viscusi, 142–182. Washington, DC: AEI–Brookings Joint Center for Regulatory Studies, 2002.

3. U.S. Food and Drug Administration. Breast implants: Chronology of FDA breast implant activities. *FDA*. August 29, 2003. Available at <http://www.fda.gov/cdrh/breastimplants/bichron.htm> (accessed February 16, 2004).

4. Munsey, R. R. Trends and events in FDA regulation of medical devices over the last fifty years. *Food and Drug Law Journal 50* (1995): 163–177.

5. Kessler, D. The basis for the FDA's decision on breast implants. *New England Journal of Medicine 326* (1992): 1713–1715.

6. Fumento, M. *Silicone breast implants: Why has science been ignored?* Washington, DC: American Council on Science and Health, 1996.

7. Bondurant, S., V. Ernster, and R. Herdman, eds. *Safety of silicone breast implants*. Washington, DC: National Academy Press, 1999.

8. Gabriel, S. E., J. E. Woods, M. O'Fallon, C. M. Beard, L. T. Kurland, and J. M. Melton III. Complications leading to surgery after breast implantation. *New England Journal of Medicine 336*, no. 10 (March 1997): 677–682.

9. Gabriel, S. E., M. O'Fallon, L. T. Kurland, C. M. Beard, J. E. Woods, and L. J. Melton 3rd. Risk of connective-tissue diseases and other disorders after breast implantation. *New England Journal of Medicine 330*, no. 24 (June 16, 1994): 1697–1702.

10. Sanchez-Guerrero, J., G. A. Colditz, E. W. Karlson, D. J. Hunter, F. E. Speizer, and M. H. Liang. Silicone breast implants, and the risk of connective tissue diseases and symptoms. *New England Journal of Medicine 332*, no. 25 (June 22, 1995): 1666–1670.

11. Janowsky, E. C., L. L. Kupper, and B. Hulka. Meta-analyses of the relation between silicone breast implants and the risk of connective tissue diseases. *New England Journal of Medicine 342*, no. 11 (March 16, 2000): 781–790.

12. *Black's Law Dictionary*. 5th ed. Saint Paul, MN: West Publishing Co., 1979.

13. Page Keeton, S., D. Dobbs, R. Keeton, D. Owen, and W. Prosser. *Prosser and Keeton's hornbook on torts*. 5th ed. Eagan, MN: West Publishing Co., 1984.

14. American Law Institute. *Restatement of torts third: Products liability.* Saint Paul, MN: American Law Institute Publishers, 1998.

15. Huber, P. W. *Galileo's revenge: Junk science in the courtroom.* New York: Basic Books, 1991.

16. Orr, D. A. Saline breast implants: A tricky solution. *Rutgers Law Record* 25 (March 1, 2001): 1.

17. *Mariann Hopkins v. Dow Corning Corporation*, 33 F. 1116, (U.S. App. 9th Cir. 1994).

18. *In re Silicone Gel Breast Implant Products Liability Litigation*, MDL No. 926 (N. D. Ala. 1994).

19. MDL 926 Breast implant litigation. *Federal Judicial Center.* September 25, 2003. Available at <http://www.fjc.gov/BREIMLIT/md926.htm> (accessed February 16, 2004).

20. Sanders, J., and D. H. Kaye. Expert advice on silicone implants: Hall v. Baxter Healthcare Corp. *Jurimetrics Journal* 37 (1997): 113–128.

21. *Hall v. Baxter Healthcare Corp*, 947 F. Supp. 1387 (D. Or. 1996).

22. National Science Panel. *Silicone breast implants in relation to connective tissue diseases and immunologic dysfunction.* Washington, DC: Federal Judicial Center, 1998.

23. *Brenda G. Meister v. Medical Engineering Corp. and Bristol-Myers Squibb Company*, No. 00–7241 (U.S. App. DC 2001).

24. Phillips, C. V., and K. J. Goodman. The missed lessons of Sir Austin Bradford Hill. *Epidemiologic Perspectives and Innovations* 1, no. 1 (October 4, 2004): 3–7 (published online October 4, 2004).

25. Rothman, K. J., and S. Greenland. Causation and causal inference in epidemiology. *American Journal of Public Health* 95, no. S1 (July 2005): S144–S150.

26. Faigman, D. L. *Legal alchemy: The use and misuse of science in the law.* New York: W. H. Freeman and Co., 1999.

27. Jasanoff, S. *Science at the bar: Law, science, and technology in America.* Cambridge, MA: Harvard University Press, 1995.

28. Science, Technology, and Law Panel. National Research Council. *The age of expert testimony: Science in the courtroom; report of a workshop.* Washington, DC: National Academy Press, 2002.

29. U.S. Food and Drug Administration, Center for Devices and Radiological Health. Breast implant risks. *FDA.* October 23, 2003. Available at <http://www.fda.gov/cdrh/breastimplants/breast_implant_risks_brochure.html> (accessed March 2, 2004).

30. Kolata, G. Company making case to allow breast implants. *New York Times,* October 11, 2003, A10.

31. Dembner, A. Panel would lift ban on silicone breast implants: 9–6 vote urges changes to safety restrictions put into effect in 1992. *Boston Globe,* October 16, 2003, A3.

32. Kolata, G. How do I look? A sexual subtext to the debate over breast implants. *New York Times,* October 19, 2003, sec. 4, 4.

33. Beam, A. Scientifically, implant ban was a bust. *Boston Globe,* October 28, 2003, E1.

34. Jacobson, N. No clearer than it was before. *Washington Post,* October 26, 2003, B01.

35. U.S. Food and Drug Administration. FDA news: FDA provides pathway for sponsors seeking approval of breast implants. *FDA.* January 8, 2004. Available at <http://www.fda/gov/bbs/topics/NEWS/2004/NEW01003.html> (accessed March 2, 2004).

36. Kolata, G. F.D.A. defers final decision about implants. *New York Times,* January 9, 2004, A1.

37. Harris, G. Citing safety concerns, panel rejects bid to sell silicone breast implants more widely. *New York Times,* April 13, 2005, A16.

38. Dembner, A. US keeps limits on implants. *Boston Globe,* January 9, 2004, A1.

39. Epstein, R. A. Implications for legal reform. In *Regulation through litigation,* ed. W. K. Viscusi, 310–335. Washington, DC: AEI–Brookings Joint Center for Regulatory Studies, 2002.

40. Driscoll, D. M. The Dow Corning case: First, kill all the lawyers. *Business and Society Review* 101, no. 101 (1998): 57–63.

41. Renwick, S. B. Silicone breast implants: implications for society and surgeons. *Medical Journal of Australia* 165 (1996): 338–341.

42. Highlights of recent federal court rulings. *Dow Corning Corporation.* Available at <http://www.implantclaims.com/rulings.html> (accessed February 18, 2004).

43. Brent, R. L. What the legal system can do. *Scientist* 9, no. 24 (December 11, 1995): 11.

Further Reading

Angell, M. Science on trial: The clash of medical evidence and the law in the breast implant case. New York: W. W. Norton and Co., 1996.

• A detailed study of the breast implant case by an insider, and an indictment of the misuse of science in the courts

Huber, P. W. Galileo's revenge: Junk science in the courtroom. New York: Basic Books, 1991.

• A readable overview of the history of abuse of science in the courts

Jasanoff, S. Science at the bar: Law, science, and technology in America. Cambridge, MA: Harvard University Press, 1995.

• An interesting view of the intersection between science and the law

Setting Limits on Expert Testimony: Bendectin and Birth Defects

A Supreme Court decision, *Daubert v. Merrell Dow Pharmaceuticals* [1], changed the way that the courts handle expert testimony. The case was one of many filed against the manufacturer of an anti–morning sickness drug, Bendectin. Merrell-National Laboratories began the manufacture of Bendectin in 1956; the company was sold to Dow Chemical Company in 1981, when it became Merrell Dow Pharmaceuticals.

Bendectin, a combination drug used to relieve the nausea associated with severe morning sickness in pregnant women, went on the market in 1957 [2]. Over the next twenty years, it became the leading prescribed drug for morning sickness; in some foreign countries it was available over the counter [3]. Beginning in 1969, the FDA received anecdotal reports that Bendectin might be associated with birth defects in children of women who took it while pregnant. Yet the number of reports was low compared to the millions of prescriptions filled over the years, and thus the FDA did not initiate any action against the drug. The FDA asked Merrell to conduct additional studies, and the company did so.

In 1975, David Mekdeci was born with serious birth defects—a shortened right arm, missing muscles, and a malformed hand. His mother had taken Bendectin, along with other medications, during her pregnancy. In 1977, the family filed suit against Merrell, claiming that Bendectin caused David's birth defects. After a long and expensive trial in Florida, a divided jury awarded $20,000 to David's parents, but nothing to David, the injured party. The judge voided the decision on the basis of inconsistency [2, 3]. On retrial, the jury found for the defendant, Merrell; the evidence linking Bendectin to birth defects presented by the plaintiffs was unconvincing. This decision was upheld on appeal (*Mekdeci v. Merrell-National Laboratories*, 711 F 2nd 1510 ([11th Cir. 1983]).

The *Mekdeci* case received extensive media coverage and generated additional lawsuits against Merrell. While the case was in litigation, the FDA held a hearing to review the safety data on Bendectin; while it did not rule out the possibility of a link with birth defects, the FDA kept the drug on the market. The FDA also recommended that additional studies be conducted. By 1984, most scientists agreed that there was no animal or epidemiological evidence that suggested a link between Bendectin and birth defects [2, 3]. The number of new lawsuits continued to grow, however, with plaintiffs' experts continuing to insist that Bendectin might cause birth defects.

In the midst of litigation, Merrell announced it would cease to produce and market Bendectin in 1983. Companies may voluntarily withdraw FDA-approved drugs for a variety of reasons; in this case, reduced sales as a result of bad publicity and rising costs was the major reason [2, 3]. Critics viewed Merrell's action as an indirect admission that the drug was unsafe, even though the drug was still FDA approved. Ironically, some physicians suggested that the withdrawal of Bendectin might *increase* the number of birth defects, since severe morning sickness poses a risk for the fetus [2, 3].

In 1981, Merrell asked that forty-eight pending lawsuits be consolidated into a class action suit (MDL–486) [4]. After an attempt to negotiate a settlement failed, the case went to trial in 1985 to determine the central question of whether Bendectin was likely to have caused birth defects. The jury took only 4.5 hours to render a verdict in favor of Merrell [2]. The failure to win the class action suit, that by then involved over eight hundred plaintiffs, did not deter lawyers, who continued to file individual lawsuits. By the late 1980s, the number of new cases dwindled to a trickle. No plaintiff ever received damages after a trial. The consistent failure of juries to believe the plaintiff's expert witnesses undoubtedly discouraged further litigation.

Daubert v. Merrell Dow Pharmaceuticals and Rules of Evidence

One Bendectin case, appealed to the U.S. Supreme Court, led to a landmark decision regarding the admissibility of expert scientific testimony in the courts. Prior to the *Daubert* decision, most federal and state courts followed the "general acceptance test" established in *Frye v. United States* (54 App. D.C. 47, 293 F. 1013, 1014 [1923]). In *Frye*, the court ruled that expert testimony involving the use of a crude lie detector device was inadmissible because it failed to demonstrate that the conclusions were based on reliable methods that have gained general acceptance in the scientific field [1]. In *Daubert*, the judge summarily found for the defendant, after excluding the plaintiffs' expert witnesses using the general acceptance standard. The plaintiffs appealed the ruling. The Court of Appeals affirmed the lower court decision, citing the *Frye* rule. The case was appealed to the Supreme Court, which issued a ruling in June 1993.

The Supreme Court reversed the lower courts on the grounds that the *Frye* rule was superseded by the revised Federal Rules of Evidence (FRE; Pub. L. 93–595, Jan. 2, 1975, 88 Stat. 1926) enacted in 1975 [1, 5]. Rule 702 of the FRE states that any scientific or technical knowledge that might assist the court in trying the case is admissible, and that experts who are qualified by their knowledge, experience, or education may testify. The Supreme Court concluded that a "generally accepted standard" is not part of the FRE, and that the FRE allows considerable flexibility in determining the admissibility of evidence. The court, however, did not entirely liberalize the standard for scientific evidence. The opinion emphasized that testimony must stem from "scientific knowledge," and to qualify the evidence must be derived by the scientific method and be supported by "appropriate validation" [1]. The following criteria for determining the admissibility of expert testimony were outlined: whether theories or techniques on which the testimony relies are based on testable hypotheses; whether they have been subject to peer review; whether the method has a known or predicted error rate (to allow for statistical analysis); and whether the method is generally accepted in the relevant scientific community [1, 6]. The *Daubert* decision stresses methodology over conclusions; the judge should review expert testimony based on its methods or techniques, not on the conclusions drawn from the data [5]. Finally, the court indicated that the "helpfulness" of scientific testimony requires that it has a valid connection to the case—that is, it must be relevant [1]. The Supreme Court gave the trial judge considerable responsibility as a gatekeeper, who must determine whether an expert's testimony is both reliable and relevant, and also adheres to the FRE standards as defined by the *Daubert* decision.

The *Daubert* case itself was remanded to the Court of Appeals, where a three-judge panel undertook to review the plaintiff's evidence using criteria established in the Supreme Court decision. The presiding judge, Alex Kosinski, dismissed the case against Merrell Dow, arguing that the plaintiffs' expert testimony was not admissible under the guidelines spelled out in the Supreme Court decision [7].

The *Daubert* decision triggered lively discussion in both legal and scientific circles. Some expressed concern about the ability of individual judges to adequately assess the validity of the scientific methods employed by experts [8, 9, 10]. Others worried that the Supreme Court's focus on methods, not conclusions, was not necessarily followed, even by the appeals court judges who ruled on the remanded *Daubert* case [11].

The Trilogy: Supreme Court Rulings and Expert Testimony

Daubert was the first of three opinions by the U.S. Supreme Court in the 1990s that focused on expert testimony. The trilogy established rules for expert testimony in both criminal and civil federal courts, and expanded the guidelines to include all expert testimony, not just scientific material [12].

The second case, *General Electric v. Joiner* (522 U.S. 136 [1997]), involved a longtime smoker who claimed that work-related exposure to polychlorinated biphenyls (PCBs) caused his lung cancer. The Supreme Court upheld the trial judge's decision to exclude evidence from animal studies in which a different kind of cancer was linked to PCBs, and affirmed that the judge had gatekeeper responsibility. The third case, *Kumho Tire v. Carmichael* (526 U.S. 137 [1999]), involved expert testimony as to whether a tire defect led to a blowout and a fatal accident. The trial judge excluded evidence from a "tire-failure" expert, who concluded based on a visual inspection that the tire had not been misused, and dismissed the case. The appellate court reversed the decision, claiming the judge's application of *Daubert* was inappropriate since *Daubert* applied only to scientific evidence. The Supreme Court ruled that the rules applied to all expert testimony, whether scientific or not. The court also suggested that the *Daubert* rules might be applied in a flexible manner, in which some criteria might not be met, as long as a degree of "intellectual rigor" is maintained [12].

Science in the Courts: Post-*Daubert*

In the federal courts, the *Daubert* decision changed the way expert testimony is used. The FRE were rewritten in 2000 to reflect new guidelines [5]. Judges commonly hold pretrial *Daubert* hearings to assess the validity of proposed testimony [13]. Judges also may create panels of neutral experts, called masters, to advise them on whether certain testimony should be admissible. Professional organizations are developing lists of neutral experts on whom judges may call for advice [14].

A superficial examination might suggest that junk science has been eliminated from federal courtrooms. By its nature, though, scientific evidence is always incomplete, and any absolute standard of "reliability" and "relevance" is unattainable [6, 15]. If "appropriate" data are unavailable, how should the courts proceed? Critics suggest that an overly rigid application of the *Daubert* rules deprives the jury of evidence that may help determine the outcome of a trial [9, 16, 17]. Critics also claim that defendants may manipulate the review process by challenging a plaintiff's evidence as incomplete and requesting a *Daubert* review with the goal of rendering the evidence inadmissible [18].

Physicians are particularly concerned about inconsistencies in the handling of clinical evidence by the courts [19]. Judges may exclude cause-and-effect clinical evidence if epidemiological evidence (the gold standard) is not available. Such a standard exceeds that normally used in good clinical practice. For example, in *Moore v. Ashland Chemical Company*, the plaintiff, a smoker with a history of asthma, developed a respiratory disorder after he was ordered to clean up solvent containing toluene that had leaked from drums in his truck. Two pulmonologists (lung specialists) conducted a battery of lung function tests after reviewing the plaintiff's medical history and carrying out a physical examination. They concluded that solvent exposure likely precipitated Moore's illness, given the short time between the exposure and the illness. Toluene also is known to irritate the lungs. But the judge ruled that the evidence was inadmissible because the pulmonologists' conclusion was not based on an "objectively validated method" [19]. Moore lost his case.

Conclusions based on "differential diagnosis" are especially problematic [20]. Such diagnoses are based on causality, the elimination of other possible causes, and information from animal studies and other data. For the courts, the ability to "rule in" the alleged cause of injury is important too. The quality of a differential diagnosis may be highly variable, as in the examples presented here. Nevertheless, differential diagnosis is commonly used in medicine. A blanket rejection may unjustly deny plaintiffs restitution in the courts.

Daubert in State Courts

The impact of the *Daubert* decision is uneven since it applies to federal cases only [21]. The states may or may not choose to follow federal rules. Thirty-eight states accept the Uniform Rules of Evidence, which are similar to the FRE [22]. By implication, these states accept the *Daubert* guidelines since they are written into the FRE. The remaining states operate under the older *Frye* standards for evidence or have their own tests [23]. Many state judges remain suspicious of *Daubert,* feel that court-appointed experts are unnecessary, and evince trust in the ability of juries to be fact finders [24]. The state judges believe that scientific evidence is not a major problem in the state courts, and see little reason to change their practices.

Conclusion

The application of *Daubert* remains challenging for the courts. The judge's role is a difficult one, and inconsistencies in practice are inevitable [13]. The conflict between the "cultures" of science and the law remain; no single court ruling will change this. Still, the *Daubert* decision has reduced the flood of junk science in tort cases. Whether ultimately the system of tort law in the United States is further changed awaits future developments.

References

1. *Daubert v. Merrell Dow Pharmaceuticals Inc.,* 509 U.S. 579, 1993.

2. Green, M. D. *Bendectin and birth defects: The challenges of mass toxic substances litigation.* Philadelphia: University of Pennsylvania Press, 1996.

3. Sanders, J. *Bendectin on trial: A study of mass tort litigation.* Ann Arbor: University of Michigan Press, 1998.

4. *In re Richardson-Merrell, Inc.* "Bendectin" Prods. Liability Litig., 624 F. Supp. 1212 (S.D. Ohio 1985), aff'd, 857 F. 2d 290 (6th Cir. 1988), cert. denied sub nom. *Hoffman v. Merrell Dow Pharmaceuticals,* Inc., 488 U.S. 1006 (1989).

5. Samelman, T. R. Junk science in federal courts: Judicial understanding of scientific principles. *University of Florida Journal of Law and Public Policy* 6 (2001): 263–271.

6. Science, Technology, and Law Panel. National Research Council. *The age of expert testimony: Science in the courtroom; report of a workshop.* Washington, DC: National Academy Press, 2002.

7. *Daubert v. Merrell Dow Pharmaceuticals, Inc.* (on remand), 43 F. 3d. 1311 (9th Cir. 1995).

8. Brent, R. L. What the legal system can do. *Scientist* 9 (December 11, 1995): 24.

9. Cecil, J. S. Ten years of judicial gatekeeping under *Daubert. American Journal of Public Health* 95, no. S1 (July 2005): S74–S80.

10. Michaels, D. Scientific evidence and public policy. *American Journal of Public Health* 95, no. S1 (July 2005): S5–7.

11. Chesbro, K. J. Taking Daubert's "focus" seriously: The methodology/conclusion distinction. *Cardozo Law Review* 15, nos. 6–7 (April 1994): 1745–1753.

12. Berger, M. A. Expert testimony: The Supreme Court's rules. *Issues in Science and Technology* 16, no. 4 (Summer 2000): 57–63.

13. Krafka, C., M. A. Dunn, M. Treadway Johnson, J. A. Cecil, and D. Miletich. Judge and attorney experiences, practices, and concerns regarding expert testimony in federal civil trials. *Psychology, Public Policy, and Law* 8, no. 3 (2002): 309–332.

14. American Association for the Advancement of Science. New AAAS project links judges to experts in science and engineering. *AAAS.* February 14, 2001. Available at <http://www.aaas.org/news/releases/2001/case.html> (accessed February 18, 2004).

15. Breyer, S. Science in the courtroom. *Issues in Science and Technology* 16, no. 4 (Summer 2000): 52–56.

16. Brown, E. D., B. Snider, and N. V. Svilik. Ruling on reliability and relevance in a *Daubert* hearing: The methodology-conclusion debate and other issues. *The Judicial Gatekeeping Project.* April 25, 2003. Available at <http://cyber.law.harvard.edu/daubert/ch4.htm> (accessed February 23, 2004).

17. Vidmar, N. Expert evidence, the adversary system, and the jury. *American Journal of Public Health* 95, S1 (July 2005): S137–S143.

18. Melnick R. L. A *Daubert* motion: A legal strategy to exclude scientific evidence in toxic tort litigation. *American Journal of Public Health* 95, S1 (July 2005): S30–S34.

19. Kassirer, J. P., and J. S. Cecil. Inconsistency in evidentiary standards for medical testimony: Disorder in the courts. *JAMA.* 288, no. 11 (September 18, 2002): 1382–1387.

20. Akers, N., and N. Scott. Differential diagnosis under *Daubert. The Judicial Gatekeeping Project.* April 25, 2003. Available at <http://cyber.law.harvard.edu/daubert/ch7.htm> (accessed February 23, 2004).

21. Walsh, J. T. Keeping the gate: The evolving role of the judiciary in admitting scientific evidence. *Judicature* 83, no. 3 (November–December 1999): 140–143.

22. Uniform rules of evidence locator. *Legal Information Institute.* March 5, 2003. Available at <http://www.law.cornell.edu/uniform/evidence.html> (accessed March 2, 2004).

23. Conley, J. M., and S. W. Gaylord. Scientific evidence in the state courts: Daubert and the problem of outcomes. *American Bar Association Judges Journal* 44, no. 5 (Fall 2005): 6–15.

24. Forum for State Court Judges. *Scientific evidence in the courts: Concepts and controversies.* Washington, DC: Roscoe Pound Foundation, 1998.

Further Reading

Green, M. D. Bendectin and birth defects: The challenges of mass toxic substances litigation. Philadelphia: University of Pennsylvania Press, 1996.

• A detailed analysis of Bendectin cases, focusing on the legal process

Sanders, J. Bendectin on trial: A study of mass tort litigation. Ann Arbor: University of Michigan Press, 1998.

• Another detailed study, with a discussion of the impact of *Daubert*

DNA Forensics in Criminal Trials: How New Science Becomes Admissible

The goal of the prosecution in a criminal case is to develop evidence linking the defendant to the crime, while the defense seeks evidence that demonstrates that the accused is innocent. Physical evidence linking the defendant to the crime is most valuable, while circumstantial evidence makes for a weaker case. The standard of proof is higher: the jury is required to determine whether the prosecution has made its case "beyond a reasonable doubt." This difference was highlighted by the murder trial of O. J. Simpson, in which he was found not guilty. He later was held civilly liable for the death of his ex-wife, Nicole.

It is not surprising that the advent of DNA "fingerprinting" in the 1980s had an enormous impact on the conduct of criminal trials. Comparing DNA found at a crime scene with that of the accused might incontrovertibly link the suspect with the crime (a boon for prosecutors) or demonstrate the opposite (a benefit for the defense and the wrongly convicted). The courts did not immediately accept DNA fingerprinting; exploring the process provides insight into how the law deals with new technology.

A Primer on DNA Forensics

The rationale behind DNA forensics (see also chapter 2) is based on the natural variation (polymorphisms) in DNA sequence between individuals [1]. Stretches of DNA between genes may include long sets of repeats of two nucleotides known as variable nucleotide tandem repeat sequences (VNTRs) or shorter sets of similar repeats called short tandem repeats (STRs). The number of repeats differs markedly in different individuals, producing fragments of DNA that are longer or shorter. These may be identified using gel electrophoresis and probes directed at the repeats. DNA collected at a crime scene, from sources such as blood, semen, or hair, may be compared with a suspect's DNA. If only a small amount of DNA is recovered, it can be amplified using the polymerase chain reaction (PCR) to provide a larger sample for analysis. A number of different probes are used to compare a series of VNTRs or STRs. If the bands on the gel consistently line up, the DNA samples are considered a match. If, on the other hand, there is no consistent alignment, then the suspect may be excluded. When properly done, this kind of analysis is both reproducible and reliable.

Although simple in principle, the analysis of DNA fragments is subject to a number of potential errors [2]. DNA samples may be contaminated during their collection or processing, producing spurious results. Gel electrophoresis requires considerable care in order to obtain reproducible results. Data interpretation is often difficult, because bands on a gel may not line up perfectly; individuals analyzing the gel must use their judgment in determining whether samples match.

The analysis of DNA matches also requires statistical analysis, since similar patterns may appear within subpopulations of individuals. What is the possibility that the DNA pattern obtained might come from another person? Until the mid-1990s, many expressed concern that racial or ethnic groups might share particular variations in DNA sequence; DNA fingerprinting might falsely identify an individual. But as the human genome project provided more sequence information (see chapter 2), these concerns declined. By applying the statistical principles of population genetics, a "likelihood ratio" may be calculated by comparing the frequency of a given pattern to a larger database. This ratio is the probability that the DNA evidence sample and that of the suspect came from the same person compared to the probability of a match if they came from randomly chosen people [1]. Ratios of thousands to millions are commonly calculated.

Likelihood ratios are subject to misinterpretation. The first error, the "prosecutor's fallacy," says that if the likelihood ratio is one thousand, then it is one thousand times as likely that the DNA came from the same person as from different persons. Statistical analysis cannot prove that the DNA sample is from the suspect—that the DNA pattern is unique to a single individual. The "defendant's fallacy," on the other hand, assumes that anyone with the same DNA profile as the suspect is equally likely to have committed the crime; the other 999 people in the ratio might equally be suspects. DNA evidence is rarely the only piece of evidence available, however. Other physical or circumstantial evidence may link the suspect to the crime. Nevertheless, a likely defense strategy is to suggest that the DNA match is not foolproof. This worked in the aforementioned Simpson trial.

DNA in the Courts

The first use of DNA forensics in the United States occurred in 1988, in a rape trial in Florida (*Florida v. Tommy Lee Andrews*, 533 So. 2d 841 [Fla. Dis. Ct. App], aff. 533 So. 2d 851 [1988]). DNA evidence linking the suspect to the crime was offered, and it helped to convict the suspect. In some other early cases, though, the court rejected the use of DNA evidence on the grounds that the technology was not generally accepted [3, 4]. Since these cases preceded the *Daubert* decision, most challenges invoked *Frye* as a standard for whether the expert testimony should be admissible (see previous section). The courts expressed concern that DNA forensics was not subject to general standards for practice, placing its analysis in question [5]. The courts also doubted the significance of a DNA match, questioning the basis of population statistics and therefore the relevance of the evidence.

Leading molecular biologists also expressed concern that without quality controls, DNA forensics might be subject to misuse [6]. Some wondered whether the loci used to uniquely identify individuals might not be selective enough. The NRC was asked to study the issue of DNA forensics, and formed a panel in late 1989. Its report, issued in 1992, affirmed the use of DNA analysis for forensic purposes and made a series of recommendations to establish quality-control guidelines [7]. The panel also recommended further studies to elucidate the population genetics of DNA testing; in the interim, it proposed a set of statistical assumptions to be used until further data were obtained. These assumptions were criticized as arbitrary. The NRC convened a second panel to further study the issue. The second report, issued in 1996, again affirmed the value of DNA forensics and resolved many of the statistical issues raised in the first report [1]. In response to increasing concerns about quality controls in forensic laboratories, the NRC report presented a new series of stricter guidelines.

By the mid-1990s, most courts ruled that DNA testing was admissible, under the standards established by either *Frye* or *Daubert*. By 1999, twelve states had enacted statutes regarding the admissibility of DNA evidence [8]. Pretrial DNA testing is used to eliminate suspects prior to trial, and DNA evidence is viewed as the gold standard for physical evidence in criminal trials. At trial, however, defense attorneys often challenge the inclusion of DNA test results on the grounds of questionable methodology and poor quality controls. A series of scandals involving crime labs and misconduct by investigators provided ample ammunition for such claims [9, 10]. A study by the RAND Science and Technology Policy Institute reported that most police departments lack both the staff and the equipment to conduct high-quality DNA analysis, and forensic labs are plagued with huge backlogs [11].

Postconviction Exoneration

Without doubt, the most dramatic application of DNA testing has been the reversal of wrongful convictions. A study for the U.S. Department of Justice in 1996 reported on twenty-eight cases in which postconviction DNA testing revealed that the person convicted of the crime, frequently murder or rape, was innocent [12]. By 1999, the Innocence Project, founded by a group of lawyers to help wrongly convicted prisoners gain new hearings, described sixty-two cases where DNA evidence exonerated the prisoners [13]. A third of the cases involved "tainted or fraudulent science" [14].

The exoneration of individuals using DNA forensics calls into question many previously accepted types of scientific evidence, including hair comparison, bite-mark analysis, and fingerprint examination [14]. The findings also reinforce the unreliability of eyewitness identifications [12, 13]. Defense attorneys are beginning to challenge the scientific validity of such evidence, with some success, using *Daubert* as a justification. Yet the U.S. Supreme Court has not applied *Daubert* in its rulings on cases relating to expert testimony in criminal trials [14].

Conclusion

DNA analysis has been hailed as "the most important tool for human identification since Francis Galton developed the use of fingerprints for that purpose" [1]. Its power to clear suspects and exonerate those already convicted is unmistakable. DNA forensics provides a powerful tool to link suspects to crime scenes and more reliable evidence than that from most eyewitnesses. The reliability of DNA testing is dependent on the quality of the analysis, however, and ample evidence suggests that many forensic labs are understaffed, overburdened, and prone to error. Until quality control is assured, its application will continue to be questioned by the courts.

References

1. Committee on DNA Forensic Science. National Research Council. An Update. *The evaluation of forensic DNA evidence*. Washington, DC: National Academy Press, 1996.

2. U.S. Congress Office of Technology Assesment. *Genetic witness: Forensic uses of DNA tests*. OTA–BA–38. Washington, DC: U.S. Government Printing Office, July 1990.

3. King, J. Disorder in the court when science takes the witness stand. *Scientist* 4, no. 21 (October 29, 1990): 15.

4. Sylvester, J. T., and J. H. Stafford. Judicial acceptance of DNA profiling. *Federal Bureau of Investigation.* Available at <http://www.textfiles.com/law/judna.txt> (accessed March 8, 2004).

5. Kaye, D. H. The admissibility of DNA testing. *Cardozo Law Review* 13 (November 1991): 353–360.

6. Lander, E. S. DNA fingerprinting on trial. *Nature.* 339 (1989): 501–505.

7. National Research Council. *DNA technology in forensic science.* Washington, DC: National Academy Press, 1992.

8. National Conference of State Legislatures. State statutes regarding admissibility of DNA evidence. *NCSL.* Available at <http://www.ncsl.org/programs/health/genetics/DNAadmiss.htm> (accessed March 8, 2004).

9. McDonald, R. Juries and crime labs: Correcting the weak links in the DNA chain. *American Journal of Law and Medicine* 24 (1998): 345.

10. Giannelli, P. C. Crime labs need improvement. *Issues in Science and Technology* 20, no. 1 (Fall 2003): 55–58.

11. Schwabe, W., L. M. David, and B. A. Jackson. *Challenges and choices for crime-fighting technology.* MR 1349–OSTP/NIJ. Arlington, VA: RAND, 2001.

12. Connors, E., T. Lundegran, N. Miller, and T. McEwen. *Convicted by juries, exonerated by science: Case studies in the use of DNA evidence to establish innocence after trial.* NCJ 161258. Washington, DC: U.S. Department of Justice, June 1996.

13. Scheck, B., P. Neufeld, and J. Dwyer. *Actual innocence: Five days to execution and other dispatches from the wrongly convicted.* New York: Doubleday, 2000.

14. Giannelli, P. C. Expert admissibility symposium: Reliability standards—too high, too low, or just right? *Seton Hall Law Review* 33 (2003): 1071.

Further Reading

Committee on DNA Forensic Science. National Research Council. An Update. *The evaluation of forensic DNA evidence.* Washington, DC: National Academy Press, 1996.

• Contains an excellent summary of the science behind DNA forensics and its applications

Scheck, B., P. Neufeld, and J. Dwyer. *Actual innocence: Five days to execution and other dispatches from the wrongly convicted.* New York: Doubleday, 2000.

• A description of convictions reversed by DNA evidence with interesting insights into the misuse of other evidence

7

Selling Science: New Cancer Treatments and the Media

Public opinion has a strong impact on policymaking. In order to form an opinion, individuals first must become aware of an issue and gain enough information to develop a viewpoint. Citizens may call on their legislators for action on a concern—and few elected officials can afford to ignore their constituents. Most Americans have limited training, if any, in science, and the vast majority of legislators at both the federal and state level also lack training in science. Where does the public get its information? Most citizens rely on the media (newspapers and newsmagazines, television and radio news programs, and Web sites) to learn about science. Different news sources may vary in the accuracy of the information they present, leaving individuals vulnerable to having their opinions swayed by distorted or biased reports. Understanding who reports on advances in science and their influences on public opinion is important for policymaking. This chapter explores the relationship between scientists, journalists, and the public by focusing on cancer treatments. Few diseases have more impact on the public; cancer is the second-leading cause of death for Americans after heart disease, and is perhaps the most feared. Thousands of reports about cancer are released every year, and many announce the discovery of new "miracle cures." What is the responsibility of journalists to assure that the information they present is accurate?

Media Coverage of Cancer Cures

In May 1998, noted science writer Gina Kolata wrote a story that appeared on page one of the Sunday *New York Times* [1]. She reported that studies by M. Judah Folkman, a researcher at Harvard Medical School, demonstrated dramatic cancer cures in mice, using a new class of experimental drugs called antiangiogenic agents. These drugs prevented the growth of new blood vessels into tumors without the side effects associated with chemotherapy; the tumors withered and disappeared after weeks of treatment. Because the target cells were normal, rather than cancer cells, no resistance to the drugs developed. Nobel laureate James Watson was quoted in the article as promising "Judah will cure cancer in two years." Kolata's article made it clear that no tests had been done on humans, and treatments used on mice might not work with humans. Yet the overall tone was enthusiastic. An editorial accompanied the article [2], somewhat more cautious but still laudatory. Other media outlets released similarly enthusiastic reports [3, 4, 5, 6].

The public responded predictably. Harvard, other research centers, and individual physicians were flooded with calls from desperate patients asking for treatment [7]. EntreMed, a small biotech company that held the license on the two drugs, angiostatin and endostatin, saw its stock rise from $12 to over $80 a share in one day [7, 8, 9]. The demand for treatment, even though not a single trial had been conducted on humans, continued for months.

Both the scientific community and journalists criticized Kolata for hyping the story [10, 11, 12]. They pointed out that Folkman's research had been reported in the media earlier; the *New York Times* published a report in November 1997, following the publication of promising studies on mice [13]. An unintended outcome of Kolata's article was the response to later reports that indicated other labs were having difficulty replicating Folkman's results [14]. Some news coverage hinted at fraud by Folkman [7]. Further work revealed that both angiostatin and endostatin were sensitive to changes in pH, and had been damaged during shipment. Nevertheless, the inflated promises in Kolata's article contributed to strong criticism of Folkman and his colleagues when additional tests ran into problems.

Since 1998, research continues to assess the potential value of angiogenesis drugs as cancer treatments. Angiostatin and endostatin began clinical trials in 2000, with modest success in reducing tumor growth. They clearly are not the magic bullets touted in 1998, though. An alternative approach involves using antibodies directed against vascular endothelial growth factor (VEGF), which stimulates the growth of new blood vessels, rather than the natural antiangiogenic agents themselves. In clinical trials, the antibody block of VEGF prolongs the lives of late-stage colon cancer patients when combined with chemotherapy [15]. The FDA approved Avastin, the trade name for the antibody, in February 2004 for colon cancer treatment. The full value of antiangiogenesis drugs for the treatment of cancer, of which some sixty are currently under study, will not be known for several more years [16, 17].

Cancer Biology

All cancers are characterized by a loss of cellular growth control and genetic instability [18]. The proliferation of cells, their division to produce more cells, is normally tightly controlled. Different cell types divide at different rates, and some, such as neurons, do not divide at all after reaching maturity. The central regulatory network for cell division is the cell cycle, a series of phases that allows cells to accurately replicate (copy) their DNA and assemble the machinery to divide. This cycle is precisely regulated in healthy cells; cascades of interacting proteins determine whether a given cell will proceed to the next step of the cycle. For example, normal cells will not begin to replicate their DNA unless they have grown large enough to generate viable offspring. Signaling proteins called cyclins must reach a critical level before the cell proceeds to the DNA synthesis phase. After DNA is replicated, the process pauses to determine whether the DNA is an accurate copy. If it is not, either the cell repairs the errors or it is directed to die, being too defective to be allowed to divide.

How does a cell "know" when to divide? Cells may receive signals from the outside that trigger the intracellular changes ultimately leading to cell division. The pathways responding to external signals are termed signal transduction pathways. These pathways are extraordinarily complex; cell biologists continue to discover new components of growth control. Signals finally reach the cell nucleus and cause changes in gene expression that lead to the accumulation of the cyclin proteins that in turn trigger cell division.

Promoting Cancer: Oncogenes

In 1912, Peyton Rous reported that a virus could cause cancer in chickens; Rous sarcoma virus generated a particular kind of cancer, sarcoma, in infected animals [19]. A number of other cancer-causing (oncogenic) viruses were identified in other animals. How can a virus cause cancer? Researchers reasoned that the virus contained a gene that when expressed, caused the cells it infected to "transform" or become cancerous. The transforming gene present in the Rous sarcoma virus was named v-src (viral sarcoma-causing gene). Scientists identified a number of other transforming genes in other viruses in the 1970s and early 1980s, but found no common characteristic shared by all. The putative genes were dubbed oncogenes (cancer-causing genes). Some of these genes had homologues (similar genes) in normal cells, but their functions were unknown.

In 1984, Julian Downward and his colleagues reported a link between oncogenes and normal cell signal transduction pathways [20]. An oncogene called v-erb-B was shown to be similar to a normal cellular protein, the cell surface receptor for a growth factor called the epidermal growth factor. The two proteins differed in their structure in the cell cytoplasm. The protein produced by v-erb-B seemed able to activate signal transduction pathways even when the epidermal growth factor was absent.

Researchers then determined in the 1980s and 1990s that all oncogenes were related to normal cellular proteins associated with signal transduction pathways and cell cycle control. A new view of cancer emerged: cancer results from the mutation of these normal cellular proteins (or proto-oncogenes) [18, 19]. A small number of cancers might be linked to cancer-causing viruses, but the vast majority are not. The rest come from changes in the cells themselves as a result of the random mutation of the genes controlling growth. Since signal transduction pathways are complex, the loss of growth control can result from changes in a large number of genes. In addition, most cancers have mutations in several different genes, suggesting that changes may accumulate over time. Some cancers also result from large errors in reproduced DNA, such as translocations of major chromosomal segments (as in chronic myelogenous leukemia).

The Flip Side: Tumor Suppressors

With the characterization of oncogenes, researchers thought that they had solved the puzzle of cancer. A problem persisted, however; in many cases, even when cells exhibited mutations in proto-oncogenes, the cells did not become cancerous. Another factor appeared to be necessary.

In the 1980s, a different class of genes was discovered. These genes appeared normally to suppress cell division, rather than promote it; they were named tumor suppressors. Among them was a gene dubbed p53, based on the size of the protein it encoded [21]. In 1990, Stephen Friend and his colleagues reported patients with Li-Fraumeni syndrome, a familial disorder involving a high frequency of occurrence of many different cancers at a young age, consistently had a mutation in p53 [22]. The cancers also contained mutations in proto-oncogenes, but not always the same ones. Based on additional studies of colorectal cancer (cancer of the colon, a portion of the large intestine), Bert Vogelstein and his colleagues suggested that cancerous transformation is a process of accumulation of mutations in proto-oncogenes and then finally p53. When p53 is mutated, all growth control is lost. More and more errors in DNA accumulate as the cancer cells proliferate [23].

What does p53 do? Cells must not only respond to growth-promoting signals. They must also be able to stop dividing if the conditions are not right. Cells contain tumor suppressors that block passage through the cell cycle. The p53 protein serves as a master gatekeeper in the cell cycle: it blocks movement through the cell cycle if DNA is damaged and also directs irreparably damaged cells into a pathway leading to their death. Other tumor suppressors also contribute to controlling the cell cycle, but p53 appears to be the most critical component. When p53 is inactivated, it is unable to stop damaged cells from dividing. Cells accumulate more and more errors with each division.

Mismatch Repair Genes

A third category of genes has been found to play a role in cancer—those involved in the repair of DNA [19]. The proteins encoded by repair genes correct any damage to DNA after it is replicated during the cell cycle. When repair genes do not function, cells may accumulate more and more errors in DNA. In 1993, two laboratories reported that one such gene, MSH2, is mutated in patients with early onset colon cancer, a syndrome called hereditary nonpolyposis colorectal cancer. As many as 10 to 15 percent of colon cancers are associated with a defect in the gene.

What Causes Mutations?

Most cancers occur in older people, with the exception of cancers of rapidly dividing cells, such as blood cells, that may affect children. Scientists postulate that errors in DNA accumulate over one's life span as a result of exposure to chemicals, ultraviolet radiation from the sun, environmental exposures to ionizing radiation, other lifestyle and environmental factors, and simple bad luck. A small number of mutations that cause or make individuals susceptible to cancer may be inherited. Many carcinogenic (cancer-causing) agents have a strong association with particular kinds of cancer. For example, most skin cancers are caused by DNA damage from exposure to the sun. These cancers are among the most preventable; if skin is protected from ultraviolet radiation, DNA damage does not occur. The majority of lung cancers are associated with tobacco smoking (see chapter 8). Many mutations are benign, but if they affect proto-oncogenes and tumor suppressors, the loss of growth control may occur. A small number of human cancers are associated with viruses;

for instance, cervical cancer is associated with infection by a particular strain of human papilloma virus, which causes genital warts [18].

Fortunately, most transformed cells are recognized by the immune system as "foreign" and are killed (see chapter 12 for more details). The immune system function tends to decline with age, contributing to the development of cancer in older people.

Why Does Cancer Kill?

Cancer is not a single disease; it can affect nearly all cells of the body. Some cancers directly cause the function of a critical organ to fail. For example, pancreatic cancer disrupts the crucial functions of the pancreas, and patients die quickly. In many cases, however, the cancerous cells that arise at a single site exert their damaging effects by moving to other parts of the body—by metastasizing. Breast cancer does not kill by growing in the breast alone. It metastasizes to other organs such as the liver, lungs, bones, and brain, and disrupts critical organ function.

Cancer disrupts organ function in several ways. The growing tumors may place physical pressure on an organ and mechanically disrupt its function. The tumors also secrete factors that stimulate blood vessels to grow into the tumors (angiogenesis), further depriving the organ of oxygen and nourishment [7, 18]. Cancer cells, because they divide more frequently than normal cells, can effectively starve the rest of the body of nourishment.

Treating Cancer

General approaches to cancer treatment involve surgery to remove tumors, and radiation or chemical treatment (chemotherapy) to kill the remaining cells. The strategy involves using radiation or toxic chemicals to kill rapidly dividing cancer cells. Unfortunately, radiation and chemotherapy also kill normal dividing cells. This is why cancer patients lose their hair, suffer gastrointestinal difficulties, and become so sick when undergoing treatment. Unfortunately, cancer cells, as they mutate further, may become resistant to chemotherapy drugs. Pharmaceutical companies and the National Cancer Institute (NCI) continue to search for more effective chemotherapy agents.

Because chemotherapy is a sledgehammer approach, researchers continue to search for more targeted ways to selectively kill cancer cells. These include using antibodies directed against proteins expressed on the surface of cancerous cells and agents to block blood vessel growth into tumors (antiangiogenesis agents) [7]. Efforts to develop new treatments for cancer are discussed below.

Is There a Magic Bullet for Cancer?

Because cancer can arise from the mutation of a large number of genes in different cells, it is not surprising that finding a cure for cancer is maddeningly elusive. Cancer cells may exhibit different features depending on their parental cell type; an effective treatment for one type of cancer may have no effect on another. Because cancer cells are genetically unstable, a treatment may be effective at one stage but not at another, and a treatment that is effective in one patient will be less so in another patient. As the complexities of cancer are recognized, scientists are opting for a variety of approaches for treatment, to be used in combination, in hopes of making cancer survivable, if not curable. This has met with substantial success in the last thirty years, with survival rates for many cancers increasing markedly.

A History of Media Hype in Cancer Treatment

One impetus for the FDA's formation was the blatant hucksterism of supposed miracle cures in the early twentieth century (see chapter 5). Newspapers published advertisements touting cures and announced upcoming visits by experts peddling their latest miracle drug.

In 1971, President Nixon announced a "war on cancer." Congress passed the National Cancer Act, doubling the budget of the NCI and giving it autonomy from the other institutes of the NIH [24]. The stated goal was to cure cancer by 1976, the bicentennial year [25]. Although critics asserted it was not possible to cure cancer by a mandate, the NCI focused its efforts on developing effective treatments for cancer. Although new chemotherapeutic agents were developed, most were too toxic for general use. Others proved to be transiently effective; cancer cells became resistant with continued treatment. Effective treatments for a small number of cancers were developed; the death rates from testicular cancer and Hodgkin's disease (a cancer of white blood cells) dropped dramatically.

Interferon, a chemical-signaling molecule (chemokine), is released from cells infected with viruses and helps other cells to resist viral infection. Since many scientists in the 1970s believed that viruses might cause human cancers, interferon became an attractive target [26]. Scattered treatment of a small number of cancer patients reportedly led to miracle cures. But interferon is produced in such small amounts by cells that controlled testing of its action in cancer proved impractical. The drug's rarity added to its mystique. Articles touting the promise of the drug appeared widely in the early 1980s [27, 28]. The gene encoding interferon was cloned in the early 1980s, allowing the production of large amounts of the chemokine for the first time. Newspaper articles even promised that enough interferon would be produced to cure the common cold [25]. Almost immediately, however, interferon's limited efficacy against cancer was reported; some patients died [29]. Ultimately, the FDA approved interferon for use against a single rare cancer, hairy cell leukemia [30]. Interferon proved to be useful for treating a number of viral diseases, such as hepatitis C and some AIDS-related diseases, but its promise as a cure for cancer came up empty.

The second "blockbuster" drug of the 1980s was interleukin-2, a member of a class of hormones called cytokines that serve to activate immune responses. Again, interleukin was promoted as a miracle drug that caused tumors to melt away [25]. In the mid-1980s, the initial reports of success treating cancers using interleukin-2 in experiments were tempered by the drug's extreme toxicity [31, 32, 33, 34]. An editorial in *JAMA* lambasted the media for overselling the rather modest success of the trials [35]. Continued research demonstrated that interleukin-2 benefited a small number of cancer patients and had serious side effects. The promised cure failed to materialize.

The most recent approach to show promise (and generate hype) is molecular targeting. Drugs are developed that target the specific signaling pathway altered in a certain kind of cancer. Therapies such as Herceptin, which blocks a receptor on breast cancer cells, and Gleevec, which prevents the action of an enzyme in leukemia and other cancers, are touted as the holy grail [36, 37]. Gleevec, in particular, dramatically reduced tumors in clinical trials, and was approved for use by the FDA in only three months after the release of the results of the trials. Unfortunately, these targeted therapies appear to lose effectiveness after eighteen to twenty-four months [38].

Most scientists argue that a cure for cancer will not be found, given the complexity of the disease. Cancer is not the result of a single agent, as is the case for polio or smallpox [25]. Promising the public that cancer will be cured adds to both the confusion about the disease and the intense frustration that many feel about the failure to overcome it. Current researchers prefer to suggest that treatments may make cancer a chronic, but survivable disease. Research must continue to develop therapies to achieve this goal.

Who Reports on Science?

Media coverage of science has increased markedly over the last thirty years [26, 39]. Yet there are at most a few hundred professional science journalists in the world. The quality of science coverage depends partially on what sector of the media is examined. Major newspapers such as the *New York Times* and the *Washington Post*, leading newsmagazines, and network news organizations have full-time science reporters on staff [40]. Many of these writers have science backgrounds or gained expertise over years of experience. They may specialize in covering particular science disciplines. Smaller media outlets, such as regional and local newspapers, seldom have dedicated science writers available. Reporters cover whatever topics are current. These outlets may rely on wire services or pick up material from major newspapers. After rewriting, distortions of original articles may occur, akin to the old children's game "telephone," in which a whispered message becomes progressively more confused with each communication.

The behavior of science journalists, of course, is driven by factors in addition to the simple presentation of scientific information. The science must be newsworthy—it must be new and likely to interest general readers by having some impact on their lives or imagination [40, 41]. In the interests of space, time, and readability, journalists must simplify complex science to make it accessible to the public. Finally, journalists need to present material that will help the bottom line of the news organization.

The source of information also affects how a story is presented [42]. If a journal article is the source, the reporter is at least assured that other scientists reviewed it prior to publication. On the other hand, reports at scientific meetings may be preliminary. Press releases from universities or companies may present an overly optimistic interpretation of new results. Journalists may also receive information by mail or e-mail purporting to represent new scientific findings. Journalists need to be careful of their sources; misrepresentation is not solely the fault of the media [19, 26, 42]. A continuing challenge is the pressure of publication deadlines—reporters may not be able to contact reliable sources to confirm information prior to its release.

Science journals themselves can be blamed for some of the hype. Leading journals such as *Science*, *Nature*, *JAMA*, and the *New England Journal of Medicine* have public relations offices that promote forthcoming articles, ensuring the widespread public dissemination of new findings. Science writers are given advanced notice of articles; they will typically speak to the investigators and other researchers working in the field; and they will always look for a public-interest aspect, normally a personal story, to highlight the human side of the disease or treatment. Given that the journals are competing for readership and prestige both in the scientific community as well as the public eye, there is a palpable pressure for the journals, as well as the authors, to make the science appear important and on the cutting edge.

The advent of the Internet adds another factor: anyone can post information on a Web site. Internet sites contain claims about science that reflect the views of the individuals creating them. It is common to find sites revealing government conspiracies and cover-ups for (or against) certain scientific claims, promising new cures for illnesses, and dismissing scientific reports as biased or immoral. Since no control is exercised over the content of Web sites, the amount of misinformation about science that is available to the public seems to have grown enormously.

What Is the Effect of Media Hype?

Although an analysis of statistics from the NCI suggests that significant improvement in the cancer survival rate has occurred, cancer is still very much with us [43]. Some critics claim that the overall death rates from cancer have remained relatively constant over the past twenty years [12, 19], and interpret the statistics as an indication of a complete failure of the war on cancer. Some suggest that there has been no success in treating or curing cancer, despite notable successes in treating many forms of leukemia as well as testicular and cervical cancer. The cancer research community is subject to blistering criticism from many quarters [19]. The NCI is accused of pandering to pharmaceutical companies that seek large profits, instead of focusing on cancer prevention. The public wonders if the billions of dollars directed at the war on cancer have been wasted. Government regulatory agencies are accused of ignoring environmental and industrial hazards that might contribute to cancer because of their close ties to industry.

As pharmaceutical companies become more involved in cancer research, press releases aimed at bolstering the bottom line contribute to overly optimistic predictions for new therapies. Each failed promise adds to the frustration of the public, and its increasing suspicion of the motives of scientists. Because the public often fails to recognize the complexity of cancer, it may focus more on hype than on the quiet success of cancer treatments.

Web sites now urge cancer sufferers not to believe what their doctors tell them, and instead to opt for diets and "natural treatments" (available for purchase on the Web sites) that do not "poison the body." Alternative approaches for treating cancer (and many other diseases) have been offered for decades, in some ways mimicking the snake oil sales of yesteryear. Among the most notorious of these was laetrile, a B vitamin derived from apricot pits, touted in the 1970s as an inexpensive cure for cancer that was "covered up" by the government [19]. Purveyors of such "treatments" often set up clinics or go offshore to avoid FDA regulations. The availability of ineffective treatments has two effects: it causes cancer patients to waste money on pointless treatments, and it may deter patients from making use of chemotherapy and other treatments that may prolong life. Cancer treatments may not cure cancer, but for many they allow for the extension of a good quality of life.

The Scientist's Role in Improving the Public Understanding of Science

The relationship between scientists and the media is an uneasy one. In the past, many scientists felt that it was not their responsibility to communicate the results of their work to the public; since the scientists knew it was valuable, it was expected that the public would take it on faith that science was important [26, 40, 41]. This attitude has changed slowly, as the public has become more suspicious of science and scientists. On the one hand, scientists rely on the media to present their work to the public, in order to maintain public support for science and the funding of future research. On the other hand, scientists complain that the media misrepresent and oversimplify their work, misleading the public in order to tell a good story [26, 44, 45]. Preliminary work is represented as fact, omitting or "burying" the caveats associated with the findings. Modest but potentially important new observations are hyped as extraordinary new discoveries.

Many researchers continue to maintain that any interactions with the media are pandering to the public and detract from the scientific enterprise [7]. Some scientists view even highly successful communicators of science to the public such as the late astronomer Carl Sagan and evolutionary biologist Stephen J. Gould as less-than-serious scientists. Yet both made significant contributions to their scientific disciplines.

While it is easy to blame the media for the public's confusion about cancer and cancer treatment, responsibility also lies with the scientific community. Scientists need to become more engaged in explaining the complexities of biology to the public. The scientific community needs to recognize its critical role in educating the public, rather than viewing such activity as demeaning or unimportant. If researchers engage more fully with the public, some of the mystery, and hence suspicion, of science may be removed. This might allow greater tolerance of the limits of scientific knowledge and the recognition of the need for further study.

References

1. Kolata, G. Hope in the lab: A special report; a cautious awe greets drugs that eradicate tumors in mice. *New York Times*, May 3, 1998, sec. 1, 1.

2. The new cancer hope. *New York Times*, May 3, 1998, A30.

3. Radford, T. Cancer drug discovery hailed by scientists. *Guardian* (London), May 5, 1998, 3.

4. SoRelle, R. Cancer weapon has wide application; protein pair may fight heart-related problems. *Houston Chronicle*, May 5, 1998, A1.

5. Sternberg, S., and A. Manning. Cancer war may have a new weapon. *USA Today*, May 4, 1998, 1A.

6. van Rijn, N. Cancer fight's new magic bullet drug treatment stems cancer in mice. *Toronto Star*, May 4, 1998, 1.

7. Cooke, R. *Dr. Folkman's war: Angiogenesis and the struggle to defeat cancer.* New York: Random House, 2001.

8. Investors get excited about report of promising cancer treatment. *San Francisco Chronicle*, May 5, 1998, B1.

9. Gillis, J. Of mice and med; for little EntreMed, the problem will be living up to the sudden promise of its cancer research. *Washington Post*, May 11, 1998, F12.

10. Markman, M. The cruel "cancer cure" hype. *Cleveland Clinic Journal of Medicine* (July–August 1998): 386.

11. Shapiro, M. Pushing the "cure": Where a big cancer story went wrong. *Columbia Journalism Review* 37, no. 2 (July/August 1998): 15.

12. Groopman, J. Too-high hopes. *New York Times*, May 6, 1998, A23.

13. Boehm, T., J. Folkman, T. Browder, and M. S. O'Reilly. Antiangiogenic therapy of experimental cancer does not induce acquired drug resistance. *Nature* 390 (1997): 404–407.

14. Grady, D. Optimistic results in battling tumors face fresh doubts. *New York Times*, November 13, 1998, A1.

15. McCarthy, M. Antiangiogenesis drug promising for metastatic colorectal cancer. *Lancet* 361 (June 7, 2003): 1959.

16. Marx, J. A boost for tumor starvation. *Science* 301 (July 25, 2003): 452–454.

17. Ferrara, N., and R. Kerbel. Angiogenesis as a therapeutic target. *Nature* 438 (December 15, 2005): 967–974.

18. Varmus, H., and R. A. Weinberg. *Genes and the biology of cancer.* Vol. 42. New York: Scientific American Library, 1993.

19. Proctor, R. N. *Cancer wars: How politics shapes what we know and don't know about cancer.* New York: Basic Books, 1995.

20. Downward, J., Y. Yarden, E. Mayes, et al. Close similarity of epidermal growth factor receptor and v-erb-B oncogene protein sequences. *Nature* 307 (1984): 521–527.

21. Finlay, C. A., P. W. Hinds, and A. J. Levine. The p53 proto-oncogene can act as a suppressor of transformation. *Cell* 57 (1989): 1083–1093.

22. Malkin, D., F. P. Li, L. C. Strong, et al. Germ line p53 mutations in a familial syndrome of breast cancer, sarcomas, and other neoplasms. *Science* 250, no. 4985 (1990): 1233–1238.

23. Hollstein, M., D. Sidransky, B. Vogelstein, and C. C. Harris. P53 mutations in human cancers. *Science* 253 (1991): 49–53.

24. The National Cancer Act of 1971. 92nd Cong., 1971.

25. Groopman, J. The thirty years' war. *New Yorker* (June 4, 2001): 32.

26. Nelkin, D. *Selling science: How the press covers science and technology.* Rev. ed. New York: W. H. Freeman and Company, 1995.

27. Edelhart, M. Putting interferon to the test. *New York Times*, April 26, 1981, sec. 6, 33.

28. Fenyvesi, C. Beyond interferon. *Washington Post Magazine*, June 14, 1981, 28.

29. Reuters. France halts interferon study. *New York Times*, November 6, 1982, sec. 1, 14.

30. Associated Press. F.D.A. to approve interferon for use against rare cancer. *New York Times*, June 5, 1986, A20.

31. Associated Press. Successes seen in experimental cancer therapy. *New York Times*, December 5, 1985, A19.

32. Kolata, G. Study finds hope in immune therapy for cancer. *New York Times*, March 23, 1984, A15.

33. Schmeck, H. M., Jr. Cautious optimism is voiced about test cancer therapy. *New York Times*, December 5, 1985, A1.

34. Sullivan, W. Warrior: Research rooted in surgery and immunology. *New York Times*, December 6, 1985, A23.

35. Moertel, C. G. On lymphokines, cytokines, and breakthroughs. *JAMA* 256, no. 22 (December 12, 1986): 3141.

36. Wade, N. Powerful anti-cancer drug emerges from basic biology. *New York Times*, May 8, 2001, F1.

37. Mishra, R. Turning the cancer switch off: Unlike today's often brutal cancer treatment, a new generation of drugs aim to shut the disease down at its source. *Boston Globe*, June 19, 2001, C4.

38. Foreman, J. A cycle of hope, heartbreak; new cancer therapies do extend lives, but the drugs' effectiveness may only be temporary. *Los Angeles Times*, August 18, 2003, sec. 6, 9.

39. Pellechia, M. G. Trends in science coverage: a content analysis of three US newspapers. *Public Understanding of Science* 6 (1997): 49–68.

40. Friedman, S. M., S. Dunwoody, and C. L. Rogers, eds. *Scientists and journalists: Reporting science as news*. New York: Free Press, 1986.

41. Friedman, S. M., S. Dunwoody, and C. L. Rogers, eds. *Communicating uncertainty: Media coverage of new and controversial science*. Mahwah, NJ: Lawrence Erlbaum Associates, 1999.

42. Allan, S. *Media, risk, and science*. Buckingham, UK: Open University Press, 2002.

43. Ries, L., M. Eisner, C. Kosary et al. *SEER cancer statistics review, 1975–2001*. Available at <http://seer.cancer.gov/csr/1975_2001/> (accessed March 30, 2005).

44. Murray, D., J. Schwartz, and S. R. Lichter. *It ain't necessarily so; how media make and unmake the scientific picture of reality*. Lanham, MD: Rowan and Littlefield, 2001.

45. Moynihan, R., L. Bero, D. Ross-Degnan et al. Coverage by the news media of the benefits and risks of medications. *New England Journal of Medicine* 342, no. 22 (June 1, 2000): 1645–1650.

Further Reading

Cooke, R. *Dr. Folkman's war: Angiogenesis and the struggle to defeat cancer*. New York: Random House, 2001.

• A profile of Judah Folkman and angiogenesis research; the book also provides an interesting view into the culture of scientific research

Nelkin, D. *Selling science: How the press covers science and technology*. New York: W. H. Freeman and Co., 1995.

• An overview of the relationship between science and the media

Proctor, R. N. *Cancer wars: How politics shapes what we know and don't know about cancer*. New York: Basic Books, 1995.

• A highly critical analysis of the war on cancer, with excellent explanations of basic science

Responsible Journalism: Are There Health Risks from Electromagnetic Fields?

We live in a world where electricity is ubiquitous, and ignore it except when we lose access in blackouts or other power failures. Could our dependence on electricity place us at health risk? Most Americans are aware of the hazards from electric shock, yet are there other dangers? Scientists began studying the biological effects of electric fields in the 1950s. In 1979, research raised the possibility that the electromagnetic fields (EMFs) produced by a sixty hertz alternating current (common household electricity) might be associated with an increased risk of childhood cancer [1]. Both the scientific community and the public largely ignored this work until it received major media coverage in 1989.

For some, the benign view of electricity was changed by the work of a single journalist, Paul Brodeur, a staff reporter for the *New Yorker* magazine. In 1989, the *New Yorker* published a three-part article by Brodeur that became a book titled *Currents of Death* [2]. In his book, Brodeur presented a chilling story of dedicated researchers who discovered a link between EMFs and childhood leukemia and brain cancer. Children who lived in neighborhoods near electric power distribution lines were more likely to develop cancer than those who did not. But a wide conspiracy involving power companies, government agencies, other scientists, professional scientific societies, and the media suppressed the findings. Further studies showed that adults who worked closely with EMFs (electric power line workers, for example) as well as individuals who spent many hours in front of computer video display terminals or used electric blankets also had higher rates of cancer or reproductive problems. Brodeur suggested that manufacturers actively covered up this information also. It was only through the continued efforts of a few scientists, local activists, concerned citizens, and brave reporters that the information came to the public's eye.

The culprits, according to Brodeur, were the magnetic fields generated by the flow of electric current, not the electricity itself. Brodeur presented information from studies demonstrating that EMFs caused a variety of changes in living cells, including alterations of the movements of ions across cell membranes and changes in hormone production. He presented a number of epidemiological studies showing a link between EMFs and cancer. Brodeur suggested that studies finding no evidence of a link between EMFs and cancer had methodological flaws or were biased because of funding sources.

Unfortunately, the story presented by Brodeur is incomplete [3]. There is no cover-up; scientists and policymakers simply question the validity and reproducibility of the studies. A large number of studies find no link between EMFs and cancer, or with the use of video display terminals. While biological responses to EMFs are observed, no clear mechanism of how the observed cellular changes might lead to cancer is apparent. Finally, even the evidence linking EMFs to cancer suggests that the risk, if any, is low—much less than the risks posed by smoking, air pollution, and other identified carcinogens [4]. Scientists and other journalists criticized Brodeur for his biased, alarmist approach.

The media and public response to Brodeur's articles and book was intense, and continued for several years [5, 6, 7, 8, 9, 10, 11]. Local governments sought to block new power lines, and parents demanded that power lines be moved away from schoolyards [12, 13, 14]. Some states set limits on exposure to EMFs [15]. Power companies found themselves subject to lawsuits claiming that their power lines caused disease [16].

How should policymakers respond? The potential impact on health was enormous, with the risk of increasing cancer across the large population of the United States. On the other hand, the possible costs of protecting the public, by rerouting power lines or burying them, rewiring homes, and requiring shielding for all electric appliances to reduce EMFs was in the billions of dollars. Should taxpayer dollars be spent to protect against a risk whose magnitude was undetermined? The government sought to find a balance: to take time to assess the evidence that EMFs posed a health hazard, while reassuring the public that the exposure risks were small. Some European countries tried to follow the "precautionary principle," which argues that in the face of incomplete evidence, policy should err on the side of caution [17]. Some countries set mandatory exposure limits, and others recommended the "prudent avoidance" of unnecessary exposure, where practical [18].

The U.S. government ordered a number of studies to assess the degree of risk from EMFs. An Office of Technology Assessment (OTA) report in 1989 suggested that the evidence that EMFs posed a serious hazard was weak [18], but recommended more research. The OTA proposed the prudent avoidance of EMFs, where practical. For example, the OTA report suggested that electric blanket users might use the blankets to warm the empty bed and then unplug them when entering the bed. Children might play in a schoolyard away from power lines, rather than in one near power lines. A 1990 EPA report also found that evidence supporting a link between EMFs and cancer was uncertain, and recommended that additional research should be conducted [19].

In 1996, the NRC released a report that concluded there was no compelling evidence that EMFs are associated with any human health risks, based on the available scientific data [20]. A central issue was the use of "proxies" such as wire ratings (as an indirect way to measure how much current is carried by a power line) to assess the strength of magnetic fields. While the NRC found a slight association of cancer with wire ratings, no relationship was discovered when actual magnetic fields were measured. Such findings called into question the validity of the original epidemiological studies [21].

Cellular Phones and Cancer

The NRC study did not assuage the fears of those who worry about EMFs. While discussion continued, a new component was added. In January 1993, David Reynard alleged on CNN that his wife's brain tumor had been caused by her heavy cellular phone use. Although he presented no medical evidence to support his claim, the media covered the story heavily [22, 23, 24, 25, 26, 27]. Reynard repeated his allegations on other television programs and other "victims" came forward. The stock value of cellular phone companies dropped dramatically, despite industry denials of any risk. Wisely, the industry requested that the federal government fund additional research on cell phone safety. In the meantime, the federal government and industry recommended that cellular phone users limit their use [28].

Cellular phones operate on a different range of the electromagnetic spectrum from household currents. Cell phones use radio frequencies in the microwave range. At a high current, microwaves can heat tissue—the principle behind microwave ovens. The amount of energy emitted by the antennas of cellular phones, however, is small. Numerous scientific studies failed to find a link between cellular phones and brain cancer [29, 30, 31].

The absence of a scientific link between cellular phones and cancer did not deter individuals from filing lawsuits against cell phone manufacturers. In one highly publicized case, Christopher Newman sued Motorola in 2000, alleging that his use of a cellular phone from 1992 until his diagnosis with brain cancer in 1998 caused his cancer. His case was dismissed in October 2002 for lack of scientific evidence [32]. Newman appealed in January 2003; his appeal was denied in fall 2003 [33]. The courts rejected other class action suits alleging cover-ups by the cellular phone manufacturers as well [34].

Does Electromagnetic Radiation Pose a Hazard?

Most scientists agree that there is no compelling evidence that EMFs pose a health hazard. Although some studies suggest a link between EMFs exposure and cancer, the relationship breaks down with further analysis. In 2001, the International Agency for Research on Cancer placed EMFs in the "possible carcinogen" range, akin to coffee and tea [17]. If there is a risk from exposure to EMFs, it is extremely small. For cellular phones, no evidence links usage to brain cancer.

The lack of evidence for links between EMFs and cancer has not quieted the fears of many. Groups continue to oppose new power lines and to file lawsuits against power companies. The conspiracy theory first voiced by Brodeur remains very much alive on the Web.

Responsible Journalism?

There is no doubt that Brodeur's writings increased awareness of the EMFs controversy. Prior to his work, the public was largely unaware of even the potential hazards of EMFs. In addition, only limited scientific study of the biological effects of EMFs was carried out. Brodeur's efforts led to a burst of research on EMFs; unfortunately, the results are inconsistent. It remains unresolved whether EMFs pose a health hazard or not. This lack of consistency makes policymaking difficult, for the reasons spelled out above.

Was Brodeur's approach overly alarmist? Brodeur's articles succeeded in unnecessarily frightening many citizens. Newspaper articles describe families moving out of their homes for fear of exposing their children to EMFs and frantic parents demanding that schools be moved away from power lines. Litigation costs against both power and cellular phone companies run into the millions of dollars. Concerns about EMFs may also distract the public from other clearly defined health hazards.

If we want the media to accurately inform the public of emerging health hazards, then Brodeur (and the *New Yorker*) failed in his role as a responsible journalist. Given the challenges of explaining science to the public in a balanced manner, journalists who seriously overstate the evidence do the profession a disservice.

References

1. Wertheimer, N., and E. Leeper. Electrical wiring configurations and childhood cancer. *American Journal of Epidemiology* 113 (1979): 273–284.

2. Brodeur, P. *Currents of death*. New York: Simon and Schuster, 1989.

3. Morgan, M. G. Exposé treatment confounds understanding of a serious public health issue. *Scientific American* 262, no. 4 (April 1990): 188–123.

4. Moulder, J. E., and K. R. Foster. Biological effects of power-frequency fields as they relate to carcinogenesis. *Proceedings of the Society for Experimental Biology and Medicine* 209 (1995): 309–324.

5. Kong, D. Scientists warn against panic over electromagnetic fields. *Boston Globe*, November 13, 1992, 11.

6. Specter, M. Electromagnetism: Can human disorders flow from waves of household current? *Washington Post*, May 7, 1990, A3.

7. Stevens, W. K. Scientists debate health hazards of electromagnetic fields. *New York Times*, July 11, 1989, C1.

8. Suplee, C. Electromagnetic fields: A highly charged issue; draft report on hazards generates controversy. *Washington Post*, December 20, 1990, A21.

9. Hilts, P. J. Study says electric fields could be linked to cancer. *New York Times*, December 15, 1990, sec. 1, 12.

10. Broad, W. J. In short: Science books. *New York Times*, April 8, 1990, sec. 7, 21.

11. Chandler, D. L. Research effort intensifies on power line link to cancer. *Boston Globe*, December 31, 1990, 33.

12. Carlton, S. Power line opponents rally: Hearing on Hillsborough regulations drawing statewide attention. *St. Petersburg Times* (Florida), January 18, 1989, 1.

13. Rogers, D. K. Electric lines spur a debate on safety. *St. Petersburg Times* (Florida), September 15, 1989, 1A.

14. Farber, M. A. Town astir as critics link cancer to electricity. *New York Times*, August 20, 1990, B1.

15. Harvard School of Public Health. Electric and magnetic fields. *Rapid Science Publishers*. Available at <http://www.hsph.harvard.edu/cancer/publications/reports/ vol1_full_text/ vol1_emfields_html> (accessed January 1, 2004).

16. Hoversten, P., and B. Ross. Lawsuit calls power lines cancer threat. *USA Today*, April 6, 1993, 1A.

17. Foster, K. R. The precautionary principle: Common sense or environmental extremism. *IEEE Technology and Society Magazine* (Winter 2002–2003): 9–13.

18. Nair, I., M. G. Morgan, and H. K. Florig. *Biological effects of power frequency electric and magnetic fields*. OTA–BP–E–53. Washington, DC: U.S. Office of Technology Assessment, 1989.

19. U.S. Environmental Protection Agency. *Evaluation of the potential carcinogenicity of electromagnetic fields: Review draft*. EPA/600/6–90/005B. Washington, DC: Environmental Protection Agency, 1990.

20. National Research Council. *Possible health effects of exposure to residential electric and magnetic fields*. Washington, DC: National Academy Press, 1997.

21. Murray, D., J. Schwartz, and S. R. Lichter. *It ain't necessarily so: How media make and unmake the scientific picture of reality*. Lanham, MD: Rowan and Littlefield, 2001.

22. Burgess, J. Cancer scare leaves cellular phone owners, firms with frayed nerves; stock prices fall than rally as companies deny health risk. *Washington Post*, January 26, 1993, C1.

23. Burgess, J. Cellular phone industry fights cancer allegation; new study planned; investors unload shares. *Washington Post*, January 30, 1993, C1.

24. Angier, N. Cellular phone scare discounted. *New York Times*, February 2, 1993, C1.

25. Ramirez, A. Health claims cause turmoil in cellular-phone market. *New York Times*, January 30, 1993, sec. 1, 1.

26. Skrzycki, C. Scientists urge more cellular phone studies; no proof of cancer link, Hill panel told. *Washington Post*, February 3, 1993, A1.

27. Skrzycki, C. Cram course in crisis management; the cellular industry struggles to contain a health scare. *Washington Post*, February 9, 1993, D1.

28. Nordenberg, T. Cell phones and cancer: No clear connection. *FDA Consumer Magazine* 34, no. 6 (November/December 2000). Available at <www.fda.gov/fdac/fectures/2000/ 600_phone.html> (accessed July 6, 2006).

29. Moulder, J. E., L. S. Erdreich, R. S. Malyapa, J. Merritt, W. F. Pickard, and Vijaylaxmi. Cell phones and cancer: What is the evidence for a connection? *Radiation Research* 151 (1999): 513–531.

30. Muscat, J. E., M. G. Malkin, S. Thompson, et al. Handheld cellular phone use and risk of brain cancer. *JAMA* 284, no. 23 (December 20, 2000): 3001–3007.

31. Inskip, P. D., R. E. Tarone, E. E. Hatch, et al. Cellular-telephone use and brain tumors. *New England Journal of Medicine* 344, no. 2 (January 11, 2001): 79–86.

32. Silva, J. Newman's lawyers to file appeal on cancer suit. *RFSafe*. January 20, 2003. Available at <http://www.rfsafe.com/articles/1_20.htm> (accessed December 18, 2003).

33. US court throws out cellphone cancer case. *Cellular-News.* October 24, 2003. Available at <http://www.howandforums.com/archive/topic/226191-1.html> (accessed June 14, 2006).

34. Cell phone radiation suit dismissed. *Consumer Affairs.Com, Inc.* March 8, 2003. Available at <http://consumeraffairs.com/news03/cell_suit.html> (accessed December 18, 2003).

Further Reading

Brodeur, P. *Currents of death.* New York: Simon and Schuster, 1989.

• The book that suggests a link between EMFs and cancer

Concealing Evidence: Science, Big Business, and the Tobacco Industry

Science is a central player in business, where it may come into conflict with corporate goals. If revealing scientific information will compromise the earning potential of a business, what happens? This part will explore the conflict between science and business by examining the tobacco industry along with efforts to regulate it. Humans began to cultivate tobacco (*Nicotiana sp.*) in the Western hemisphere over two thousand years ago; early Mayan art from AD 600 to AD 1000 pictures tobacco smoking [1, 2]. Native North and South Americans used tobacco for its psychoactive properties in rituals, and smoked for pleasure. Native Americans introduced smoking to Europeans beginning in the late sixteenth and early seventeenth centuries. Tobacco rapidly became a popular product and valued commodity for international trade. It was touted as a panacea for a variety of illnesses and for generally boosting health, although social disapproval was voiced broadly as early as the seventeenth century [2]. The view of tobacco as a healthful substance began to change in the nineteenth century, when the first reports of toxic effects appeared.

The Economics of Tobacco in the United States

Tobacco was a major commercial crop in the American colonies even before they secured independence from England. The wealth generated by the export of tobacco to Europe provided one means to finance the Revolutionary War [1]. Production boomed particularly in the southern colonies because of the highly favorable climate for cultivation; in New England, efforts were made to ban both the cultivation and consumption of tobacco on moral grounds [3]. Until the late nineteenth century, most tobacco was consumed via hand-rolled cigarettes, pipes, or in smokeless form. The development of the first cigarette machines in 1880 changed patterns of consumption beginning in the early twentieth century, and commercial cigarettes became cheap enough for even the poor.

The tobacco industry today is a worldwide enterprise. The United States is the second-largest producer (after China) [4, 5]. Eighty percent of tobacco is now consumed as cigarettes. The total annual revenue by U.S. tobacco companies is around $150 billion [6]. Although U.S. consumption has declined since the mid-1980s, overseas sales allowed tobacco companies to remain financially healthy; the United States is the world's leading exporter of cigarettes [4, 5]. U.S. consumers spent an estimated $86.7 billion on tobacco products in 2003 [7]. Tobacco companies sold nearly four hundred billion cigarettes in that same year [7]. In response to the declining numbers of smokers in the United States, the tobacco industry increased its advertising and promotion expenditures, spending $12.5 billion in the United States in 2002 [8].

The tobacco industry and its ancillary marketing operations is a major employer in the United States. There are over sixty-five thousand farms in the United States, mostly in southern states, whose primary product is tobacco, and an estimated one to three million people are employed in jobs related to tobacco [9].

Taxation

The federal government proposed a tax on tobacco in 1794 as a means to generate revenue, but the first excise tax was not enacted until the Civil War in 1862 [3]. Taxes on tobacco remain a major source of federal and state revenue; an estimated $13.5 billion in tax revenue, or twenty-six cents for each dollar spent on tobacco products, was generated in 1997 [5, 9]. In recent years, most states increased taxes on tobacco products as part of attempts to discourage teenage smoking; the move also enhances states' revenue. Nonetheless, according to the World Health Organization, the tobacco tax rate in the United States is among the lowest in the developed world [5, 10].

The Biology of Nicotine and Tobacco

Nicotine is the main active ingredient in tobacco responsible for the pleasurable sensations experienced by smokers and other tobacco users. Nicotine, a natural component of tobacco, mimics the action of acetylcholine (ACh), a chemical used for communication between nerve cells (neurotransmitter) in the brain. The high levels of nicotine present in tobacco act as a poison protecting the plant against insect attack. ACh is used broadly across the nervous system of humans and other mammals, and is the primary neurotransmitter for controlling the activation of the so-called pleasure centers of the brain, the mesolimbic reward system. The mesolimbic system itself uses another neurotransmitter, dopamine, for communication. ACh stimulates the release of dopamine, which in turn activates brain cells that produce sensations of pleasure.

Nicotine binds to the same cell surface receptor as ACh, the nicotinic cholinergic receptor. The binding of nicotine to this receptor stimulates the release of dopamine in the limbic system. The dopamine activates nerve cells producing feelings of calm, alertness, reduced stress, and reduced appetite [11].

Is Nicotine Addictive?

Much of the controversy about the tobacco industry centers on whether or not nicotine is an addictive drug. Addiction, or substance dependence, is recognized as a compulsion to continue to use the drug and difficulty in controlling its use, and is a major social problem. The American Psychiatric Association developed criteria to define addictive substances: substance dependence is associated with tolerance, a need to increase the amount of the drug in order to have the same effect; characteristic symptoms of withdrawal when the drug is not taken; continued use in larger amounts or over longer periods than intended; unsuccessful attempts to cut down the usage; considerable time spent in obtaining or consuming the agent; abandonment of other activities in order to continue using the substance; and continued use despite knowledge that it contributes to ongoing medical problems [12]. Any three of these characteristics is considered diagnostic for addiction.

Nicotine fulfills most criteria for an addictive substance, although many require additional explanation. For example, tolerance is demonstrated by enhanced physiological effects of the first cigarette of the day (including dizziness and nausea), with a less profound response from cigarettes smoked later in the day [12]. This differs from other addictive substances, such as heroin, where users must consume larger amounts for any effect at all. Withdrawal symptoms include depression, irritability, anxiety, difficulty concentrating, a decreased heart rate, and weight gain. Smokers continue to use tobacco despite knowledge that it is harmful (see below). It is difficult to quit smoking: about 35 percent of smokers attempt to quit each year, and fewer than 7 percent are successful without intervention [12, 13]. Most return to smoking after a few days. Even if the pharmacological symptoms of withdrawal are managed, most smokers find that there are strong psychological components to addiction—the smoker finds the rituals of lighting up in particular situations as well as the feel and sight of cigarettes pleasurable. This psychological addiction may lead to persistent cravings for tobacco long after the physiological dependence has eased.

The Cigarette as a Nicotine Delivery System

The typical cigarette contains from five to ten milligrams of nicotine [11, 13]. When tobacco is burned, the nicotine becomes gaseous and is inhaled with smoke. About one to two milligrams of nicotine are taken into the body with each cigarette. The nicotine is absorbed through the linings of the lungs, and reaches the brain within seconds through the bloodstream. Nicotine may also be absorbed through the skin and the linings of the mouth and throat; this is the route taken by nicotine from pipe and cigar smoking (in which smoke is usually not inhaled), and from smokeless tobacco such as chewing tobacco and snuff. The effects of nicotine are delayed using these methods.

The effects of nicotine are short-lived, leading the smoker to consume additional cigarettes to continue dosing with the drug. Nicotine is rapidly broken down in the bloodstream, limiting its effect on the brain. A typical heavy user may consume one to two packs of cigarettes (twenty to forty cigarettes) a day to maintain pleasurable sensations. This use can disrupt normal daily activities, particularly in the many environments where smoking is now restricted.

The Office of the Surgeon General of the United States, the chief medical officer of the country, formally concluded that nicotine was addictive in a 1979 report and issued a second, stronger report in 1988 [14]. The delayed recognition of the addictive nature of nicotine provided important ammunition for those opposed to tobacco regulation, as described below.

The Dark Side of Tobacco Use

Were nicotine the only active component of tobacco, the pattern of consumption might not generate such concern and controversy. But nicotine is only one of approximately four thousand chemicals naturally present in tobacco that affect the body. Many of the chemicals are toxic or carcinogenic. Throughout the 1950s and 1960s, accumulating scientific evidence made the link clear between smoking and disease. The link between cigarette smoking and illness was recognized formally in 1964 when the Office of the Surgeon General of the United States issued a report linking cigarette smoking to lung cancer, heart disease, and chronic bronchitis and emphysema [15]. This report, however, suggested that nicotine was "habituating," but not necessarily addictive [15].

About 90 percent of lung cancers are now linked to cigarette smoking [13]. Smoking is also associated with cancers of the mouth, throat, pancreas, kidney, cervix, and other organs. The most recent report from the surgeon general's office linked cigarette smoking to a broad range of cancers, cardiovascular diseases, respiratory disorders, reproductive problems including low birth-weight babies, and additional health effects [16]. Smoking is linked to approximately 438,000 premature deaths in the United States each year; the economic costs from the health effects and loss of productivity attributable to smoking were over $167 billion annually between 1997 and 2001 [16].

The Path to Addiction

Over 80 percent of the current adult smokers began smoking regularly in their teens, with over 30 percent beginning before age sixteen [17]. By age twenty, the majority of these smokers will continue to smoke into adulthood, finding it difficult to quit. In 2004, over 28 percent of high school students and almost 12 percent of middle school children used tobacco [18]. These statistics suggest that tobacco use will continue to be a major health problem in the future.

Although the number of adult smokers in the United States declined over the past forty years, about 21 percent of adults were cigarette smokers in 2004 [19]. While the number of male smokers declined, the numbers of female smokers increased; women now make up close to 50 percent of the consumers.

Is There a Safe Cigarette?

With increasing recognition that smoking is hazardous, the tobacco industry has focused efforts to create a "safe" cigarette—one that delivers the pleasurable effects of nicotine without the toxins and carcinogens present in smoke. Early efforts included the addition of filters to cigarettes to absorb particulate matters ("tars"). Cigarettes formulated to be "low-tar/nicotine" have no apparent health benefits; smokers simply consume more cigarettes or inhale more deeply to maintain their nicotine levels [20]. The smokeless cigarette, a cigarette-shaped nicotine delivery system with almost no tobacco, was introduced in 1987 as a supposedly safe alternative [2, 21]. It proved a commercial failure, though, lacking the taste and sensations of smoke so desired by smokers.

Secondhand Smoke and the Risk to Nonsmokers

Smokers are not the only ones affected by cigarette smoke. Environmental tobacco smoke (ETS, or secondhand smoke) consists of smoke exhaled by smokers and the smoke that is released by burning tobacco when the smoker is not inhaling (side-stream smoke). Both sources of smoke contain similar levels of toxins and carcinogens as does inhaled smoke; some argue that side-stream smoke may contain higher levels of chemicals because of the lower temperature at which the tobacco is burning [22]. A 1986 surgeon general's report concluded that exposure to cigarette smoke by nonsmokers contributes to lung cancer and other disease in healthy adults [22]. The children of smokers have an increased frequency of respiratory infections and retarded lung development. The report also argued that simple separation of smokers and nonsmokers does not eliminate the exposure of nonsmokers to ETS. A second report, released by the EPA in 1992, echoed these findings and labeled ETS as a class A carcinogen, indicating that there was sufficient evidence to conclude that it caused cancer in humans [23].

Early reports of the effects of ETS were treated with skepticism by critics. Yet controlled studies measuring the levels of nicotine and conitine (a breakdown product of nicotine) in the blood of study participants revealed widespread exposure to ETS in the U.S. population between 1988 and 1991 [24]. The American Heart Association also links ETS to an increased risk for cardiovascular disease, citing studies that suggest that nonsmoking spouses of smokers have a 25 percent increased risk of heart disease [25].

The surgeon general's 1986 report and subsequent studies led to stiffer regulations intended to limit nonsmokers' exposure to ETS in the workplace, classroom, and some public areas. Of course, limiting exposure in private homes is beyond the direct reach of regulatory agencies (see below), so exposure continues for the family members of smokers.

A History of Tobacco Regulation

Given that smoking is a recognized health hazard, why is the sale of tobacco permitted? The history of tobacco regulation is a long and convoluted one, dating back to the seventeenth century in Europe [1]. Efforts to control tobacco usage in Europe varied from country to country, but in general were not successful.

In the late nineteenth century, an antitobacco movement developed parallel to the antialcohol temperance movement in the United States. Some states barred smoking in public; others banned the sale of cigarettes [3]. Much of the rhetoric was aimed at preventing young people from smoking, as is the case today. The failure of Prohibition, the constitutional amendment banning alcohol sales, led to the repeal of most state statutes banning tobacco sales by the late 1920s. The Pure Food and Drug Act of 1906 also omitted tobacco from the products subject to regulation (see chapter 5). Instead, states began taxing tobacco products and continued to limit sales of tobacco to minors.

Tobacco consumption increased dramatically in the twentieth century, particularly during World War II, and then peaked in the 1960s. Cigarettes were viewed as indispensable by the military; tobacco farmers received deferments in order to aid the war effort [21]. Consumption was boosted by the high visibility of smoking by actors, athletes, and other public figures. Smoking was portrayed as healthy, stylish, and sophisticated.

Despite this, many people referred to cigarettes as "coffin nails" and "cancer sticks" in the early and mid-twentieth century, even though manufacturers did not acknowledge the health risks [26]. Cigarette manufacturers began reformulating cigarettes in the 1950s in response to highly publicized reports linking smoking to cancer. Filter cigarettes were advertised as "smoother" and "milder" [21]. As discussed below, tobacco companies were aware of the increasing evidence linking smoking to disease, but publicly continued to deny that smoking posed a health risk [21, 27].

Regulation after the Surgeon General's Report of 1964

In 1962, Luther Terry, the surgeon general, appointed an expert committee to examine the accumulating data on the health effects of smoking. The report, released in January 1964, unequivocally linked smoking to lung cancer and other ailments (see above). The report's release led to the first federal regulation of the tobacco industry.

In June 1964, the Federal Trade Commission (FTC) ruled that cigarette advertising was deceptive, and ordered that all packaging and advertising carry a statement linking cigarette smoking to death from cancer and other diseases. A strong lobbying effort by tobacco companies, however, persuaded Congress to delay the implementation of any warning and also to weaken the statement to read: "Cigarette Smoking May Be Hazardous to Your Health" [3]. Tobacco companies presented their own medical experts disagreeing with the conclusions of the surgeon general's commission. Congress passed the Cigarette Labeling and Advertising Act of 1965

(P.L. 89–92, 79 Stat. 282), which required the warning label on packages, but in response to lobbying by the tobacco industry, limited the FTC or other federal or state agencies from requiring other labeling in advertising until after 1969 [21]. Antismoking advocates criticized the federal law as toothless and a capitulation to the tobacco industry. The addition of a warning label had little effect on cigarette sales, which continued to increase.

In 1967, the Federal Communications Commission (FCC), which has oversight of broadcast media, ruled that television and radio stations had to provide free advertising to groups with antismoking messages to counteract the widespread cigarette commercials on air. Its rationale grew from the "Fairness Doctrine," under which the public has a right to hear differing viewpoints on issues [3]. The U.S. Court of Appeals upheld the FCC decision, and the Supreme Court declined to hear the appeal filed by cigarette manufacturers.

After the expiration of the Cigarette Labeling Act in 1969, Congress enacted stronger legislation, the Public Health Cigarette Smoking Act of 1969 (P.L. 91–222, 84 Stat. 87). It required a more prominent health warning—"Warning: the Surgeon General has determined that cigarette smoking is dangerous to your health"—on packages and advertising, and banned cigarette advertising on radio and television beginning in 1971. Yet the bill also limited the power of the FTC to regulate other advertising [3, 21].

Congress enacted the Comprehensive Smoking Act (P.L. 98–474, 98 Stat. 2200) in 1984, aimed at increasing the public's awareness of the hazards of smoking. It mandated a federally sponsored program of research, education, and information; required more explicit warning labels on cigarette packs and in advertising; and required that manufacturers provide the government with a list of ingredients added to tobacco in the manufacture of cigarettes [21].

Regulation was not limited to advertising. Berkeley, California, became the first municipality to restrict smoking in public indoor places in 1977 [21]. Through the 1980s, municipalities and states passed legislation limiting smoking in indoor public spaces. By 2003, all states had regulations in place limiting smoking in indoor public spaces. In 1987, as part of a transportation appropriations bill, Congress banned smoking on domestic commercial flights of two hours or less, to begin in 1988; this was expanded in 1990 to a total smoking ban on almost all domestic flights.

The federal government also demanded greater oversight of youth smoking. In 1992, the Synar Amendment (a provision of P.L. 102–321, 106 Stat. 394) required that state governments must have a law prohibiting the sale of tobacco products to persons under eighteen, in order to receive federal grants for the control of alcohol or drug abuse. States also must establish a system of yearly inspections and report annually to the Department of Health and Human Services (DHHS).

The FDA Steps In

In 1995, David Kessler, head of the FDA, announced that the FDA would begin to regulate nicotine as a drug, and also cigarettes and other tobacco products as "drug delivery systems." The FDA issued broad proposed regulations to discourage youth consumption of tobacco, requiring photo ID verification of a purchaser's age (over eighteen), prohibiting cigarette vending machines except in adults-only areas, limiting advertising and promotional items such as T-shirts and hats, and barring tobacco company sponsorship of sporting and musical events [21]. The move was unprecedented. The FDA had declined previously to claim jurisdiction over tobacco products—a decision dating back to 1906.

The FDA does not have broad powers to protect public health; it may regulate only certain classes of products dictated by its enabling legislation (see chapter 5) [28]. Two events changed the FDA's approach. The surgeon general's report in 1988 concluding that nicotine was an addictive substance provided a central justification for FDA action [29]. Kessler also argued that tobacco manufacturers had concealed their knowledge of the addictive nature of nicotine and deliberately designed cigarettes to deliver nicotine efficiently in order to promote addiction in consumers [30].

The tobacco manufacturers filed suit to block the FDA regulations, arguing that the FDA did not have the jurisdiction to regulate tobacco products and that the broad restrictions on advertising violated the companies' First Amendment free speech rights. They also contended that the FDA would have the authority, and perhaps the obligation, to ban cigarettes as an unsafe drug. Since the FDA was not proposing to ban tobacco products, its application of law was inconsistent with its own policies [28]. An initial ruling in favor of the FDA was reversed on appeal. In 2000, the U.S. Supreme Court ruled that previous congressional actions precluded the FDA from exercising jurisdiction over tobacco products [31]. The court majority concluded that the FDA's stance on regulation, controlling but not banning the product, was inconsistent with FDA policy and hence lay outside its jurisdiction. An FDA ban on tobacco products also would contradict Congress, whose own legislation was more measured.

The court left open further congressional action to give the FDA the authority to regulate tobacco. In the meantime, many states passed their own laws limiting advertising and youth access to tobacco products. In July 2004, the Senate voted to allow the FDA to regulate tobacco in exchange for an industry-sponsored buy out of tobacco farmers, who were suffering from a decline in tobacco demand, but the bill failed to reach a vote in the House [32].

Litigation and the Concealing of Evidence

Individuals began filing lawsuits against tobacco companies in the 1950s. The plaintiffs claimed that smoking caused their illnesses and that cigarette companies were negligent in not warning them of the risks. The tobacco industry strongly contested over eight hundred suits; until the mid-1990s, all but two were decided in the industry's favor. The remaining cases resulted in no damages [21]. The industry's initial defense was that it could not be held liable for dangers that were not known. It also argued that the plaintiffs had chosen to smoke, and were not forced to do so. In the alternative, the industry also asserted that the risks were broadly known, and that smokers accepted the risks by beginning and continuing to smoke. In addition, the industry employed numerous legal delaying tactics; many plaintiffs ran out of money before the cases ever came to trial [21].

Behind the scenes, however, the tobacco industry was well aware of the dangers of its own products, based on its own in-house research, by the late 1950s [21]. Publicly, the industry continued to challenge the increasing evidence of the link between smoking and disease, and most strongly contested the claim that smoking was addictive. The front collapsed in 1994, when thousands of internal industry documents were leaked to the public. An analysis of the documents revealed that industry scientists concluded that smoking caused cancer and that nicotine was addictive long before the surgeon general drew similar conclusions. Tobacco companies suppressed their own data on the risks of smoking. The industry also explored the effects of manipulating nicotine content on the behavior of smokers, concluding that nicotine was the primary reason why people kept smoking. Industry spokespersons issued public statements, placed advertisements, and gave testimony to Congress in direct contradiction to this evidence—they lied [27]. People died.

Following these revelations, the tide began to turn toward the plaintiffs in tort lawsuits against the tobacco industry. In 1994, the first class action suit was filed against the tobacco industry, charging that the tobacco companies had knowingly marketed an addictive product [21]. The suit, while unsuccessful, provided an impetus for a series of lawsuits filed by the states' attorneys general seeking the recovery of Medicaid expenses for smoking-related illnesses. The first of these, *Moore v. American Tobacco Co. et al*, was filed in Mississippi in 1994; the state asked for $1 billion in compensation [21]. The tobacco companies made attempts to deflect the cases by legal and political manipulations, but were not successful.

In August 1996, a Florida court awarded $750,000 in damages to a sick smoker. The Liggett Corporation, the smallest of the major tobacco companies, offered to settle with the plaintiffs in the class action suits in order to avoid insolvency [21]. This action marked the first capitulation on the part of the tobacco industry. Four states reached separate settlements with the tobacco companies [21].

By 1997, thirty-one other states had filed suit against the tobacco industry. Rather than face bankruptcy, the four largest tobacco companies—Philip Morris, R. J. Reynolds, Brown and Williamson, and Lorillard—offered to settle with the states' attorneys general. Additional states also joined the settlement, for a total of forty-six states and territories. Industry payments of $368.5 billion (including $60 billion in punitive damages) over twenty-five years were proposed. In return, the industry would be protected from future class action lawsuits and settlements of individual suits would be capped at no more than $5 billion per year. The agreement also included the FDA's proposed restrictions on advertising that targeted children, banned outdoor advertising, and called for explicit warning labels on packages. The industry would bear the costs of enforcing these regulations. The FDA would have the responsibility for regulating the components of cigarette smoke. The agreement also called for a reorganization of the tobacco industry itself in order to assure the information would not be concealed in the future [21, 33].

The agreement required that Congress enact the appropriate legislation, because it included federal provisions. Congress failed to do so, facing criticism from both the Left and the Right. Tobacco farmers argued that their interests had not been considered; many representatives from tobacco-growing states objected to the proposed settlement on the grounds that it damaged their states' economies. Lawyers objected because the agreement placed limits on future lawsuits and protected the industry from further liability. Antismoking advocates argued that the agreement was too kind to the industry [21].

A bill was introduced in 1998 that called for industry payments of $516 billion over twenty-five years and removed many of the industry protections. The industry counterattacked with an aggressive ad and lobbying campaign accusing the federal government of greed [33]. The bill died in the Senate in June 1998, on procedural votes [21].

The Master Tobacco Settlement Agreement of 1998

In November 1998, the attorneys general of forty-six states reached an agreement, titled the Master Tobacco Settlement Agreement, with the four largest tobacco companies ostensibly for reimbursement of Medicaid expenses and punitive damages. The remaining states negotiated individual settlements with the tobacco manufacturers. The tobacco companies would pay $206 billion over twenty-five years; the states would use the funds at their own discretion. Tobacco companies were prevented from targeting advertising to youth by banning the use of cartoon characters in advertising, the sponsorship of events with significant youth audiences, outdoor advertising, or promotional materials. The companies were required to create a national foundation and public education fund aimed at discouraging youth smoking. The agreement also called for restricting industry lobbying, disbanding trade associations, and opening industry records and research to the public as means to change corporate culture [34].

Missing from the settlement was the federal regulation of tobacco and other federal programs, thus dropping the requirement for congressional action [35]. The agreement dropped the requirement for FDA regulation of tobacco (a proposal abrogated by the U.S. Supreme Court in 2000), and also weakened provisions for stronger warnings on packages, narrowed limitations on advertising and promotional activities, weakened enforcement requirements for rules regarding the sale of tobacco to minors, and sharply cut back on clean air requirements [35]. The tobacco industry also lost provisions contained in the 1997 proposed settlement. The new agreement afforded no protection from or limits on private individual or class action suits. The agreement did not preclude the federal government from filing suit against the tobacco companies as well; the federal government filed a civil suit in 1999 against the tobacco companies using the provisions of an antiracketeering law [21]. The case went to trial in September 2004. In February 2005, a federal appeals court ruled that the government may not seek $280 billion in past profits as part of its case, thereby limiting the potential size of any payments by the industry if the suit is successful [36]. In October 2005, the Supreme Court declined to hear the government's appeal [37].

Many criticized the Master Settlement Agreement. Public health advocates argued that the settlement did not go far enough. Others charged that the states were merely grabbing for money from a vulnerable industry and that the payments represented an additional tax on smokers [38]. Critics also argued that the settlement violated antitrust laws by allowing the major tobacco companies to shield themselves from further state lawsuits, leaving smaller manufacturers still vulnerable [21].

Has the Master Settlement Agreement Helped Reduce Smoking?

The Master Settlement Agreement went into effect in November 1999 with up-front payments of $2.4 billion to the states and industry-sponsored antismoking campaigns directed at youth [21]. What has been the effect of these programs?

Many states failed to apply industry payments directly to covering medical costs or reducing smoking [39]. Instead, the payments were directed to general funds, used to reduce budget shortfalls or to secure bond issues for other projects. A General Accounting Office study in 2003 reported that a declining amount was directed toward youth smoking cessation programs or other control measures, and that states increasingly borrowed on future payments to balance their budgets [40].

The Master Settlement Agreement also failed to limit advertising targeted at teenagers. An analysis of advertising in "youth-oriented" magazines in 2001 revealed that cigarette manufacturers continued to spend well over $200 million per year [41]. More than 80 percent of children age twelve to seventeen were reached by these advertisements, especially those that marketed "youth" brands of cigarettes, which are smoked by more than 5 percent of smokers in grades eight to twelve. Since the Master Settlement Agreement barred tobacco companies from marketing in some venues, they simply shifted to other areas: youth magazines, free gifts, and discounts on brands popular with children [42]. In 2004, over 34 percent of middle school students reported seeing advertisements for tobacco products on the Internet [18]. Smoking also continues to be depicted in movies and on television programs; almost 78 percent of middle school students reported seeing actors smoking [18].

A small number of states developed aggressive programs to limit and discourage smoking, and evidence suggests that these state approaches are proving somewhat successful [42, 43]. State efforts to limit advertising, however, were also met with lawsuits filed by tobacco manufacturers. In 1999, Massachusetts issued regulations that broadened bans on advertising not included in the Master Settlement Agreement. These included bans on advertising on billboards and other open-air advertising that might be seen by children [21, 42]. The tobacco companies and distributors filed suit, claiming that the new regulations infringed on their First

Amendment rights and were inconsistent with federal laws regulating tobacco advertising. In 2001, the U.S. Supreme Court ruled that portions of the Massachusetts regulations violated First Amendment rights and were superseded by federal law, including the regulations passed in 1969 (*Lorrilard v. Reilly, Attorney General of Massachusetts*) [44]. In December 2005, the Illinois Supreme Court, in a split vote, reversed a $10.1 billion lower court ruling (*Price v. Philip Morris, Inc.*) against Philip Morris USA for misleading advertisements about its "light cigarettes" because such ads are permitted by the FTC [45]. These rulings demonstrate the continuing conflict between federal and state regulations.

Lessons from Tobacco

The challenges of regulating a legal product that carries with it significant health problems is exemplified by tobacco. The federal government has a responsibility to take action in the name of public welfare. Despite unquestionable evidence that nicotine is addictive and smoking is linked to over four hundred thousand deaths each year, the federal government failed to step in to regulate tobacco products, beyond limiting some advertising and barring access by children. States' efforts to tighten regulations have met with mixed success. The tobacco industry is a formidable lobby that exerts a strong influence on legislators. The impact of the industry on the country's economy—and critically that of several southern states—also makes it difficult for legislators to move to eliminate smoking. Recent litigation and legislation suggests that things may change in the future.

The deeper story lies in the conflict between free enterprise, profits, and scientific knowledge. The federal government, in general, has tended to limit oversight of corporate activities, except when the abuses are so egregious that public outcry forces government action. Documents from the tobacco industry make it clear that the scientists employed by tobacco companies knew that cigarettes were dangerous and reported their data to executives. Industry executives decided to preserve markets and concealed damaging evidence. Why did the scientists fail to independently report their findings? Employees are bound by corporate rules of disclosure, which bar independent publication without approval. Going public would mean a scientist loses his or her job; the scientist might also be liable for contract violation, and could be criminally charged under state and federal trade secrecy laws.

Scientists did speak out; whistle-blowers provided invaluable information leading up to the Master Settlement Agreement. In 1994, Jeffrey Wigand, a biochemist who was the head of research and development at Brown and Williamson from 1989 to 1993, told the FDA of the industry's attempts to boost the effectiveness of nicotine using additives [21]. Other former research employees also provided information to the FDA. Whistle-blowing, or an express duty to act to prevent risks to the public even when corporate secrets might be jeopardized, is not an ethical requirement in the sciences. Because whistle-blowing involves an express breach of loyalty, it raises suspicions about a whistle-blower's motives [46].

Is the story of the tobacco industry an isolated case, or is conflict of interest common? The evidence suggests that conflicts between scientific goals and corporate interests are not rare. When data indicate that a promising new drug is associated with side effects or is not as effective as hoped, one instinct is to conceal the evidence in order to protect the company. In June 2004, New York State filed suit against GlaxoSmithKline, the manufacturer of the popular antidepressant drug, Paxil. The suit charged that the company had committed fraud by knowingly concealing clinical evidence that the drug was ineffective in teenagers [47, 48]. The publicity led to renewed calls to establish a database of all drug trials in order to block attempts to bury negative data [49, 50]. In August 2004, GlaxoSmithKline settled with the state and agreed to post the results of all drug trials online [51]. More recent cases involving a class of popular painkillers, the Cox-2 inhibitors, also support the contention that concealment of evidence is common (see chapter 5). Efforts to find a balance between the two competing interests are ongoing, yet the central challenges are structurally ingrained in science performed for commercial interests. Requiring full disclosure by corporations is an important step, but enforcing the rule is another matter.

References

1. Gately, I. *Tobacco: A cultural history of how an exotic plant seduced civilization*. New York: Grove Press, 2001.

2. Hughes, J. *Learning to smoke: Tobacco use in the West*. Chicago: University of Chicago Press, 2003.

3. McGrew, J. L. History of tobacco regulation. In *National commission on marihuana and drug abuse*. 1972. Available at <http://www.druglibrary.org/schaffer/LIBRARY/studies/nc/nc2b.htm> (accessed September 18, 2004).

4. van Liemt, G. *The world tobacco industry: Trends and prospects*. Sector WP 179–2002. Geneva: International Labour Office, 2002.

5. World Health Organization. Tobacco atlas. *WHO*. Available at <http://www.who.int/tobacco/statistics/tobacco_atlas/en/> (accessed December 24, 2005).

6. U.S. Department of the Treasury. *Testimony by assistant secretary for financial markets Gary Gensler before the Senate Democratic Task Force on Tobacco.* RR–2439. Washington, DC: U.S. Department of the Treasury, 1998.

7. Capehart, T. *The changing tobacco user's dollar.* TBS257–01. Washington, DC: U.S. Department of Agriculture, October 2004.

8. Frieden, T. R., and D. E. Blakeman. The dirty dozen: 12 myths that undermine tobacco control. *American Journal of Public Health* 95, no. 9 (September 2005: 1500–1505).

9. Gale, J., H. Frederick, L. Foreman, and T. Capehart. *Tobacco and the economy: Farms, jobs, and communities.* AER–789. Washington, DC: Economic Research Service, U.S. Department of Agriculture, September 2000.

10. World Health Organization. Price policy. *WHO.* Available at <http://www.who.int/tobacco/en/atlas35.pdf> (accessed March 31, 2005).

11. Ember, L. R. The nicotine connection. *Chemistry and Engineering News* 72, no. 48 (November 28, 1994): 8–18.

12. American Psychiatric Association. *Diagnostic and statistical manual of mental disorders.* DSM IV–TR ed. Washington, DC: American Psychiatric Association, 2000.

13. National Institute on Drug Abuse. *Nicotine addiction.* NIH 01–4342. Bethesda, MD: National Institute on Drug Abuse, August 2001.

14. Advisory Committee to the Surgeon General. *The health consequences of smoking: Nicotine addiction.* Washington, DC: U.S. Department of Health and Human Services, May 3, 1988.

15. Advisory Committee to the Surgeon General. *Smoking and health.* PHS–1103. Washington, DC: U.S. Public Health Service 1964.

16. Centers for Disease Control. Annual smoking-attributable mortality, years of potential life lost, and productivity losses—United States, 1997–2001. *Morbidity and Mortality Weekly Report* 54, no. 25 (July 1, 2005): 625–628.

17. Schoenborn, C., P. Adams, P. Barnes, J. Vickerie, and J. Schiller. Health behaviors of adults: United States, 1999–2001. *Vital Health Statistics* 10, no. 219 (2004): 33.

18. Centers for Disease Control. Tobacco use, access, and exposure to tobacco in media among middle and high school students—United States, 2004. *Morbidity and Mortality Weekly Report* 54, no. 12 (April 1, 2005): 297–301.

19. Centers for Disease Control. Cigarette smoking among adults—United States, 2004. *Morbidity and Mortality Weekly Report.* 11 November 2005;54(44): 1121–1124.

20. Advisory Committee to the Surgeon General. The health consequences of smoking: A report of the Surgeon General. *DHHS.* 2004. Available at <http://www.cdc.gov/tobacco/sgr/sgr_2004/index.htm> (accessed August 31, 2004).

21. Derthick, M. A. *Up in smoke: From legislation to litigation in tobacco politics.* 2nd ed. Washington, DC: CQ Press, 2004.

22. U.S. Department of Health and Human Services. *The health consequences of involuntary smoking: A report of the surgeon general.* Rockville, MD: Department of Health and Human Services, December 15, 1986.

23. U.S. Environmental Protection Agency. Setting the record straight: Secondhand smoke is a preventable health risk. *EPA.* June 1994. Available at <http://www.epa.gov/smokefree/pubs/strsfs.html> (accessed September 18, 2004).

24. Pirkle, J., K. Flegal, J. Bernert, D. Brody, R. Etzel, and K. Maurer. Exposure of the US population to environmental tobacco smoke: The third national health and nutrition examination survey, 1988–91. *JAMA* 275, no. 16 (April 24, 1996): 1233–1240.

25. Brook, R., B. Franklin, W. Cascio et al. Air pollution and cardiovascular disease: A statement for healthcare professionals from the expert panel on population and prevention of the American Heart Association. *Circulation* 109 (June 1, 2004): 2655–2671.

26. R. J. Reynolds Tobacco Company. Tobacco risk awareness timeline. *Brown and Williamson.* 2005. Available at <http://www.brownandwilliamson.com/smoking/awareTimeline.aspx> (accessed April 3, 2005).

27. Glantz, S. A., J. Slade, L. A. Bero, P. Hanauer, and D. E. Barnes. *The cigarette papers.* Berkeley: University of California Press, 1996.

28. Glantz, L. H., and G. J. Annas. Tobacco, the Food and Drug Administration, and Congress. *New England Journal of Medicine* 343 (December 14, 200): 1802–1806.

29. Goitein, L., G. S. Chernack, G. Liu, and M. T. Davis. Developments in policy: The FDA's tobacco regulation. *Yale Law and Policy Review* 15 (1996): 399–446.

30. Kessler, D. The Food and Drug Administration's regulation of tobacco products. *New England Journal of Medicine* 335, no. 13 (September 26, 1996): 988–994.

31. *Food and Drug Administration, et al. v. Brown and Williamson Tobacco Company, et al.,* 98–1152. U.S. Supreme Court, 2000.

32. Dewar, H. Senate backs compromise on tobacco; regulation of industry paired with buyout. *Washington Post,* July 16, 2004, A1.

33. Kagan, R. A., and W. P. Nelson. The politics of tobacco regulation in the United States. In *Regulating tobacco,* ed. R. L. Rabin and S. D. Sugarman, 11–38. New York: Oxford University Press, 2001.

34. Wilson, J. J. Summary of the attorneys general master tobacco settlement agreement. *National Conference of State Legislatures.* March 1999. Available at <http://academic.udayton.edu/health/syllabi/tobacco/summary.htm> (accessed August 5, 2004).

35. Schroeder, S. A. Tobacco control in the wake of the 1998 Master Settlement Agreement. *New England Journal of Medicine* 350, no. 3 (January 15, 2004): 293–301.

36. Levin, M. U.S. can't go after tobacco's past profits: An appellate court rules that the racketeering statute does not allow the government to claim $280 billion it alleges was earned illegally. *Los Angeles Times,* February 5, 2005, A1.

37. Greenhouse, L. Justices reject appeal in tobacco case. *New York Times,* October 18, 2005, A18.

38. Wagner, R. E. Understanding the tobacco settlement: The state as partisan plaintiff. *Regulation* 22, no. 4 (Winter 1999): 38–41.

39. Gross, C. P., B. Soffer, P. B. Bach, R. Rajkumar, and H. P. Forman. State expenditures for tobacco-control programs and the tobacco settlement. *New England Journal of Medicine* 347, no. 14 (October 3, 2002): 1080–1086.

40. U.S. General Accounting Office. *Tobacco settlement: States' allocations of fiscal years 2002 and 2003 Master Settlement Agreement payments.* GAO–03–407. Washington, DC: General Accounting office, February 2003.

41. King, C., III, and M. Siegel. The Master Settlement Agreement with the tobacco industry and cigarette advertising in magazines. *New England Journal of Medicine* 345, no. 7 (August 16, 2001): 504–511.

42. Kessler, D. A., and M. L. Myers. Beyond the tobacco settlement. *New England Journal of Medicine* 345, no. 7 (August 16, 2001): 535–537.

43. Tauras, J. A., F. J. Chaloupka, M. C. Farrelly et al. State tobacco control spending and youth smoking. *American Journal of Public Health.* 95, no. 2 (February 2005): 338–344.

44. *Lorrilard Tobacco Company, et al. v. Thomas F. Reilly, Attorney General of Massachusetts, et al.,* 00–596. U.S. Supreme Court, 2001.

45. Vise, D. A. Court overturns $10 billion verdict against Philip Morris. *Washington Post,* December 16, 2005, D2.

46. Bok, S. Whistleblowing and professional responsibilities. In *Ethics teaching in higher education,* ed. D. Callahan and S. Bok, 277–295. New York: Plenum Press, 1980.

47. Harris, G. Spitzer sues a drug maker, saying it hid negative data. *New York Times,* June 3, 2004, A1.

48. Marshall, E. Buried data can be hazardous to a company's health. *Science* 304 (June 11, 2004): 1576–1577.

49. Meier, B. Group weighs plan for full drug-trial disclosure. *New York Times,* June 15, 2004, A1.

50. Dickersin, K., and D. Rennie. Registering clinical trials. *JAMA* 290, no. 4 (July 23–30, 2003): 516–523.

51. Harris, G. Glaxo agrees to post results of drug trials on Web site. *New York Times,* August 27, 2004. C4.

Further Reading

Derthick, M. *Up in smoke: From legislation to litigation in tobacco politics.* 2nd ed. Washington, DC: CQ Press, 2004.

• A detailed report on tobacco regulation

Gately, I. *Tobacco: A cultural history of how an exotic plant seduced civilization.* New York: Grove Press, 2001.

• A lively account of the history of tobacco

The Academic-Industrial Complex and Conflict of Interest

Far from being ivory towers, academic institutions have a long history of working relationships with industry. Scientific research in academic institutions supplies the raw material for many industrial applications (technology transfer); industry provides support to scientists in academia to encourage research in areas of interest. In the life sciences, the close ties between pharmaceutical companies, researchers, and clinicians offer both benefits and risks. Support for biomedical research that is then translated into clinical uses boosts the nation's health and economy [1]. These financial arrangements may lead to bias in clinical studies, however, risking the health of human subjects and distorting the results of experiments. Concerns over potential conflicts of interest have grown in recent years, as more and more researchers develop financial relationships with industry.

Academic-industry ties in the life sciences developed in the early twentieth century, but increased dramatically after 1980, as biotechnology began to flourish. The Bayh-Dole Act of 1980 (P.L. 96–517, the Patent and Trademark Law Amendments Act) allowed federal grant recipients to retain patents ownership; this opened the door for increased academic-industry cooperation [2, 3]. One estimate places industry's share of biomedical research support in 2000 at 62 percent (see the following section) [4]. By the mid-1990s, over 90 percent of companies with life science activities had a relationship with at least one university, and 25 percent of life scientists at major research institutions received some research support from industry. Some faculty members held equity in biotechnology companies (7 percent), and over 50 percent reported consulting for industry [1]. Roughly two-thirds of academic institutions hold equity in companies that sponsor research at the same institution [4].

What Is Conflict of Interest?

A conflict of interest is the situation in which a person serves two or more masters. A conflict exists when a scientist's financial relationship with a company risks compromising the researcher's professional judgment and independence in the design, conduct, or publication of research [5, 6, 7]. The conflict is a source of bias that may be intentional or unintentional, and the company may contribute to the problem by further manipulating results or establishing criteria for selecting subjects. Among the manipulations are the selective use of data that provide positive results ("data dredging"), building bias into a study by preselecting the population examined, or shortening a trial when the long-term results appear less promising [8, 9]. Conflict of interest is different from scientific misconduct, when an investigator knowingly and deliberately falsifies or fabricates data or plagiarizes (copies the work of another without attribution) (see the following section).

Numerous studies demonstrate that conflict of interest exerts effects on research outcomes. Studies paid for by industry, or conducted by researchers with financial ties to a company, are more likely to report positive outcomes in clinical trials [4, 10, 11]. Scientists whose data do not support the company may be barred from publishing their negative data [1, 10]. For example, in 1987, Betty Dong, a University of California at San Francisco researcher, was denied permission by the sponsoring company to publish results indicating that the company's commercially leading synthetic thyroid hormone was no more effective than competing preparations [10]. Other cases involved a researcher at the University of Toronto [12] and another at Brown University [13], both of whom were blocked from publishing their findings about the health risks to patients or employees, respectively, that were inimical to the financial interests of the companies that, respectively, sponsored the research or permitted the research to be done at their facility. Notably, in both of these latter cases, the company executives had strong ties to the universities (Toronto was courting a multimillion dollar gift for building construction from the company executives, and in the last case, the company CEO sat on Brown's board). For different reasons, the administrations in all three cases failed to support the academic freedom of their faculty to publish.

The potential risk to human subjects was dramatized in 1999, when Jesse Gelsinger, a teenager who volunteered for a gene therapy clinical trial at the University of Pennsylvania, died. A review determined that the principal investigator had a strong financial stake in the outcome of the trial and had skirted requirements for trial approval [9]. The Gelsinger case and others led to calls by officials at the FDA and the NIH for better regulation of clinical trials [14].

The close relationship between academics and industry may have other negative effects. Agreements between companies and researchers often include requirements that information be kept secret in order to allow companies to file for patents [4, 15]. Such agreements may produce as much as a six-month additional lag in publication [16]. Agreements may also limit the ability of researchers to collaborate with others by sharing reagents and data [1, 17]. Directions of research may be selected for perceived financial gain, leaving other critical areas underserved. Finally, publicized accounts of bias serve to erode the public's confidence in science [10].

Managing Conflict of Interest

Preventing all conflicts of interest is impossible, given the intertwined relationship between academics and industry. Ensuring that the relationships are made public and transparent, and providing clear guidelines for the ethical conduct of research, are more attainable goals.

The U.S. government instituted a series of requirements for researchers receiving federal support. The NIH requires that researchers reveal all conflicts of interest to their home institutions, and that institutions have conflict of interest policies in place [1]. In 1995, the NIH and NSF instituted new guidelines that gave the primary responsibility for managing conflicts to research institutions [18]. A report in 2001 called for increased oversight by the DHHS to assure that institutions have effective conflict of interest policies [5]. In response, in 2003, the NIH issued a guidance document intended to protect human subjects in cases of financial conflict of interest; the document called for greater oversight by Institutional Review Boards (see chapter 5) [19]. The FDA requires researchers to disclose any financial arrangements when research data are submitted for review [1, 19]. In February 2005, Elias Zerhouni, the NIH director, announced a new policy that barred all researchers at the NIH from any consulting or other activities that might constitute a conflict of interest [20].

Professional associations and academic institutions develop guidelines for the proper conduct of industry-supported research. For example, the American Association of Medical Colleges, the umbrella organization for medical schools, issued new guidelines in 2001, placing the safety of human subjects as paramount and spelling out circumstances under which a researcher with a conflict of interest may participate in human research [2]. Yet, policies developed by institutions suffer from inconsistencies in what constitutes inappropriate activity, and from a lack of oversight and enforcement [1, 4, 5, 18, 21, 22].

Editors of biomedical journals were among the first to establish disclosure policies for authors. Nevertheless, a study in 2001 reported that only 16 percent of over thirteen hundred highly ranked journals had conflict of interest policies in 1997 [23]. Such findings stimulated journal editors to develop conflict policies as part of their "instructions to authors." In general, authors must indicate their ties to commercial enterprises as a condition of manuscript acceptance. Still, critics have repeatedly reported that the disclosures are inadequately policed. The Integrity in Science project, run by the Center for Science in the Public Interest, advocates for transparency in scientific research, publications, and presentations, and the protection of human subjects [24]. In July 2004, the center released a report that criticized several leading journals for failing to enforce their own policies regarding financial conflicts of interest. The center cited several examples of authors who failed to disclose financial ties to companies [25]. It may be obvious that disclosure may not be enough to avoid the risks of bias caused by conflicts of interest, because reviewers and readers cannot reliably assess whether and to what extent the results are in fact influenced (if at all); disclosed conflicts thus result in the disclosure being ignored or the results being totally discounted as hopelessly flawed.

Why Conflict of Interest Matters

Scientists like to believe that research is conducted to further knowledge and benefit the public, and ought to be independent of the more base goals of profit or fame. Yet scientists are prey to ego and greed, as are all humans. The development of close ties between academe and industry adds to the challenges to conduct research dispassionately. Unlike interests that are unavoidably intrinsic to academic science—such as the need to publish, to build an international reputation, and even to cure disease—financial ties may be viewed as avoidable and impure. Because the effects are difficult to measure, financial conflicts can result in effectively unanswerable criticisms that one's work cannot be believed because of the conflict. When financial ties spur an individual to distort data or to allow the suppression of negative results, the damage may be extensive—to health and the enterprise itself. The public's trust in science is fragile. Scientists have both professional and ethical responsibilities to see that the conduct of science is as free of bias as possible.

References

1. Blumenthal, D. Academic-industrial relationships in the life sciences. *New England Journal of Medicine* 349, no. 25 (December 18, 2003): 2452–2459.

2. Kelch, R. P. Maintaining the public trust in clinical research. *New England Journal of Medicine* 346, no. 4 (January 24, 2002): 285–287.

3. Good, M. L. Increased commercialization of the academy following the Bayh-Dole Act of 1980. In *Buying in or selling out? The commercialization of the American research university*, ed. D. G. Stein, 48–55. New Brunswick, NJ: Rutgers University Press, 2004.

4. Bekelman, J. E., Y. Li, and C. P. Gross. Scope and impact of financial conflicts of interest in biomedical research. *JAMA* 289, no. 4 (January 22–29, 2003): 454–465.

5. U.S. General Accounting Office. *HHS direction needed to address financial conflicts of interest*. Washington, DC: General Accounting Office, November 2001.

6. Kassirer, J., and M. Angell. Financial conflicts of interest in biomedical research. *New England Journal of Medicine* 329 (1993): 570–571.

7. Porter, R. Conflicts of interest in research: The fundamentals. In *Biomedical research: Collaboration and conflict of interest*, ed. R. Porter and T. Malone, 121–134. Baltimore, MD: Johns Hopkins University Press, 1992.

8. Angell, M. Time for a drug test registry. *Washington Post*, August 13, 2004, A25.

9. Angell, M. The clinical trials business: Who gains? In *Buying in or selling out? The commercialization of the American research university*, ed. D. G. Stein, 127–132. New Brunswick, NJ: Rutgers University Press, 2004.

10. Bodenheimer, T. Uneasy alliance: Clinical investigators and the pharmaceutical industry. *New England Journal of Medicine* 342, no. 20 (May 18, 2000): 1539–1544.

11. Rennie, D. Pharmacoeconomic analyses: Making them transparent, making them credible. *JAMA* 283, no. 16 (April 26, 2000): 2158–2160.

12. Nathan, D. G., and D. J. Wentherall. Academic freedom in clinical research. *New England Journal of Medicine* 347 (2002): 1368–1371.

13. Shuchman, M. Secrecy in science: The flock worker's lung investigation. *Annals of Internal Medicine* 129 (1999): 341–344.

14. Stolberg, S. G. F.D.A. officials fault Penn study in gene therapy death. *New York Times*, December 9, 1999, A22.

15. Blumenthal, D., N. Causino, E. Campbell, and L. Seashore. Relationships between academic institutions and industry in the life science: An industry survey. *New England Journal of Medicine* 334 (1996): 368–373.

16. Blumenthal, D., E. Campbell, N. Anderson, N. Causino, and L. Seashore. Withholding research results in academic life science: Evidence from a national survey of faculty. *JAMA* 277 (1997): 1224–1228.

17. Rosenberg, S. A. Secrecy in medical research. *New England Journal of Medicine* 334, no. 6 (February 8, 1996): 392–394.

18. McCrary, S. V., C. B. Anderson, J. Jakovlevic et al. A national survey of policies on disclosure of conflicts of interest in biomedical research. *New England Journal of Medicine* 343, no. 22 (November 30, 2000): 1621–1626.

19. Krimsky, S. Reforming research ethics in an age of multivested science. In *Buying in or selling out? The commercialization of the American research university*, ed. D. G. Stein, 133–152. New Brunswick, NJ: Rutgers University Press, 2004.

20. National Institutes of Health. NIH announces sweeping ethics reform. *NIH*. February 1, 2005. Available at <http://www.nih.gov/news> (accessed March 7, 2005).

21. Ehringhaus, S., and D. Korn. Conflict of interest in human subjects research. *Issues in Science and Technology* 19, no. 2 (Winter 2002): 75–81.

22. Cho, M., R. Shohara, A. Schissel, and D. Rennie. Policies on faculty conflicts of interest at U.S. universities. *JAMA* 284 (2000): 2203–2208.

23. Krimsky, S., and L. S. Rothenberg. Conflict of interest policies in science and medical journals: Editorial practices and author disclosures. *Science and Engineering Ethics* 7, no. 2 (2001): 205–218.

24. Sharpe, V. A. Oversight, disclosure, and integrity in science. In *AAAS science and technology policy yearbook 2003*, ed. A. H. Teich, S. D. Nelson, S. J. Lita, and A. E. Hunt, 115–123. Washington, DC: American Association for the Advancement of Science, 2003.

25. Goozner, M. *Unrevealed: Non-disclosure of conflicts of interest in four leading medical and scientific journals*. Washington, DC: Center for Science in the Public Interest, July 12, 2004.

Further Reading

Angell, M. The truth about the drug companies: How they deceive us and what to do about it. New York: Random House, 2004.

• A recent criticism of the pharmaceutical industry

Krimsky, S. Science in the private interest: Has the lure of profits corrupted biomedical research? New York: Rowman and Littlefield, 2003.

• A second viewpoint on the conflict between industry and academe

Stein, D. G., ed. Buying in or selling out? The commercialization of the American research university. New Brunswick, NJ: Rutgers University Press, 2004.

• A collection of short essays on issues of academic-industrial relations

The Darker Side of Science: Scientific Misconduct

When scientists read a new scientific paper, they assume that the methods and data presented are genuine. Science is based on trust [1]. The ethics of science—which include honesty, the obligation to publish, teach, and mentor, and the duty to defend scientific and intellectual freedoms—promote trustworthiness in the scientific enterprise [2]. Nonetheless, it is well established that sometimes, scientists do not act by these lofty ideals [3]. Scientific misconduct refers to those behaviors that undermine the veracity of the scientific record [4]. Misconduct is commonly defined as falsification, fabrication, or plagiarism. Both the scientific community and the public view such misconduct with horror since it runs counter to all the principles on which scientific research is based.

What Is Scientific Misconduct?

Scientific misconduct is not new. A highly publicized case occurred in the early twentieth century, when the Piltdown man, fossilized remains that were purportedly the missing link between apes and modern humans, was determined to be a deliberate hoax by persons unknown. Science misconduct in more modern times gained exposure when psychologists determined after his death in 1971 that Cyril Burt, a celebrated British researcher of intelligence in twins, had fabricated most of his data [5].

A series of cases in the 1980s changed the public's perception of scientific misconduct, leading to conclusions that misconduct was widespread [6, 7]. One notable case involved David Baltimore, who, at the age of thirty-eight, won the 1975 Nobel laureate in physiology and medicine. In 1986, Margot O'Toole, a postdoctoral fellow in the laboratory of Thereza Imanishi-Kari (of MIT and then Tufts University), reported to university authorities that the central conclusions of a paper that

Imanishi-Kari published with Baltimore and other coauthors were not supported by the raw data [6]. The initial investigation found no serious errors and concluded that O'Toole's concerns stemmed from a scientific dispute. O'Toole persisted in her accusations and drew the attention of two self-appointed misconduct watchdogs at the NIH, Walter Stewart and Ned Feder. Stewart and Feder, in turn, reported the case to Congressperson John Dingell (D-MI), chair of the House subcommittee that provided oversight of the NIH. Dingell believed that misconduct was widespread and was an outspoken critic of the scientific community's ability to handle such cases [8].

In 1985, long before the so-called Baltimore case came to light, at Dingell's urging, Congress enacted legislation requiring the NIH to create an office to investigate cases of alleged misconduct; the Office of Scientific Integrity was opened in 1989 [9]. In the Baltimore case, Dingell argued that the NIH's own review processes for investigating misconduct were inadequate, and demanded a major investigation involving the Secret Service and other law enforcement agencies. After a lengthy investigation, the Office of Scientific Integrity found Imanishi-Kari guilty of falsifying data. The government, however, declined to file criminal misconduct charges against Imanishi-Kari, citing inadequate evidence [10]. In the meantime, Baltimore, who defended his colleague throughout the extended investigation, resigned the presidency of Rockefeller University in 1991, as a result of the uproar and a vote of no confidence by the faculty. The media coverage was heavy, and Baltimore was accused of heavy-handed and misguided support for the wrong side [6]. Nevertheless, Imanishi-Kari's fabrication ruling was overturned in 1996 by a review panel at the NIH and she was fully exonerated (guilty of no more than sloppy laboratory notebook practices). Baltimore became the president of the California Institute of Technology in 1997; he stepped down and returned to research in 2006.

Is misconduct widespread? The public perception of misconduct run rampant was encouraged by the large amount of media coverage in the 1980s. The suggestion that scientists might be subject to the same frailties as other humans was viewed with dismay. Most scientists believe that misconduct is relatively rare because of the self-correcting nature of science [11, 12]. They think that fraudulent research is uncovered fairly quickly because the confirmation of results, either directly or through further study in the area, is a central part of the scientific enterprise. Some cases come to light when coworkers call attention to irregularities. But other scientists suggest that more widespread misconduct may occur because most scientists are simply not alert to it and are reluctant to believe misconduct when it happens;

it may take years to detect fraudulent activity [13]. The assertion that science is self-correcting cannot go unchallenged, as scientists rarely duplicate prior studies, but rather build on previous results. Only if studies fail will scientists go back and attempt to reproduce the background research on which their own (failed) research depends. Cases of data manipulation or the doctoring of figures are more difficult to detect, especially now that images are digitized and easily altered. Estimates of the frequency of misconduct vary widely, depending on whether perceptions or actual cases are measured, but it is generally acknowledged that no one knows the number of occurrences [14].

What motivates a scientist to fake or steal data? Career pressure is often cited as a reason why scientists commit misconduct [11, 15]. Most scientists under similar pressure maintain their integrity, though, so simple pressure is insufficient to explain such behavior. Another reason offered is that scientists may believe so strongly in their hypothesis that they feel compelled to alter data that do not support it [11]. Even Gregor Mendel purportedly fit his data to his expectations [3]. Finally, scientists who commit misconduct believe that they will not get caught. Many suggest that the highly competitive nature of science, the lack of training in responsible conduct, and the limited (or wrongheaded) mentoring of junior researchers by senior faculty contribute to an environment that might encourage misconduct [13, 15].

Response to Misconduct

How has the scientific community responded to misconduct? The Office of the Assistant Secretary of DHHS maintains an office, now known as the Office of Research Integrity (ORI), that provides support for research into misconduct, produces educational materials, and investigates reports of misconduct (a similar office exists at the NSF) [16]. The ORI publishes notices about scientists found to have committed misconduct; guilty individuals are generally barred from receiving federal funds or serving on federal panels for a period of several years. The ORI received 987 allegations of misconduct between 1993 and 1997; most were found to be false, and only 19 percent went on to the inquiry stage [15]. Because of its legacy of perhaps overzealous investigations in the past, institutions and scientists often view ORI activities with suspicion [15]. Scientists are accused of blocking attempts to thoroughly investigate the range of fraudulent activities [17].

Gatekeeping responsibility is also assigned to journal editors, who are expected to pursue possible misconduct in submitted or published manuscripts [18]. The

International Committee of Medical Journal Editors developed uniform requirements for manuscripts that cover a broad range of editorial issues, including misconduct. Editors who receive allegations of potential misconduct from reviewers are expected by the ORI to respond promptly and begin an investigation [19]. Should evidence arise that suggests a significant problem, the host institution, which has the primary responsibility for conducting an investigation, and the ORI should be contacted. Editorial responsibility includes the timely retraction of articles whose contents are found to be fraudulent. The delayed retraction of flawed research is a common source of criticism [13]. In addition, researchers may not take note of retraction notices and flawed research may continue to inform additional studies.

A notable change in attitude since 1990 has been the development of programs to educate young scientists about responsible conduct; institutions that receive federal funds are required to have such programs in place [16]. The NRC issued several reports, the most recent in 2002, providing suggestions for programs to educate scientists about the responsible conduct of research and steps that might create an environment that promotes it. The council described integrity as "a commitment to intellectual honesty and personal responsibility for one's actions, and to a range of practices that characterize responsible conduct of research" [20]. Institutions must provide leadership and educational opportunities, thoroughly investigate allegations of misconduct, and regularly assess the institutional environment that supports research activities in order to improve.

The recent NRC report also presented a set of sobering conclusions: that there are no established measures to assess integrity; having policies and procedures will not assure responsible conduct; and unless education is conducted in creative ways, it is unlikely to be effective [20].

Scientific misconduct remains a serious problem for the scientific community; the recent fabricated report of human stem cells from cloned embryos is an example (see chapter 4). Whether perpetrated by a single individual or indicative of a wider problem, allegations of misconduct serve to weaken the public's trust in science, and lead to increased calls for the closer management and oversight of the scientific enterprise. Misconduct represents a challenge to the "free enterprise" of research, and requires a thoughtful response by the scientific community. The public, legislators, and the media, however, must recognize that every scientific disagreement is not misconduct; science thrives on competing hypotheses.

References

1. Barber, B. Trust in science. *Minerva* 25 (Summer–Spring 1987): 123–134.

2. Glass, B. The ethical basis of science. *Science* 150 (1965): 1254–1261.

3. Brush, S. Should the history of science be rated X? The way scientists behave (according to historians) might not be a good model for students. *Science* 183 (1974): 1164–1172.

4. Commission on Research Integrity. *Integrity and misconduct in research*. Washington, DC: U.S. Department of Health and Human Services, 1995.

5. Broad, W. J. Frauds from 1960 to the present: Bad apples or bad barrel? In *The dark side of science*, ed. B. K. Kilbourne and M. T. Kilbourne, 26–33. San Francisco: American Academy for the Advancement of Science, 1983.

6. Kevles, D. J. *The Baltimore case: A trial of politics, science, and character*. New York: W. W. Norton, 1998.

7. Crewdson, J. *Science fictions: A scientific mystery, a massive cover-up, and the dark legacy of Robert Gallo*. Boston: Little, Brown and Company, 2002.

8. Dingell, J. D. Misconduct in medical research. *New England Journal of Medicine* 328, no. 22 (June 3, 1993): 1610–1615.

9. Office of Research Integrity. HHS fact sheet: Promoting integrity in research. *DHHS*. October 22, 1999. Available at <http://ori.dhhs.gov/html/about/HHS.asp> (accessed October 14, 2004).

10. Kassirer, J. P. The frustrations of scientific misconduct. *New England Journal of Medicine* 328, no. 22 (June 3, 1993): 1634–1636.

11. Goodstein, D. Scientific misconduct. *Academe* 88, no. 1 (January–February 2002): 28–31.

12. Shafir, S., and D. Kennedy. Research misconduct: Media exaggerate results of a survey. *Scientist* 12, no. 13 (June 22, 1998): 9.

13. Gunsalus, C. Rethinking unscientific attitudes about scientific misconduct. *Chronicle of Higher Education*, March 28, 1997, B4.

14. Panel on the Responsible Conduct of Science, National Academy of Sciences. Misconduct in science: Incidence and significance. In vol. 1, *Responsible science: Ensuring the integrity of the research process*, 80–97. Washington, DC: National Academy Press, 1992.

15. Shamoo, A. E., and D. B. Resnik. *Responsible conduct of research*. New York: Oxford University Press, 2003.

16. Office of Research Integrity. About ORI. *DHHS*. March 17, 2004. Available at <http://ori.dhhs.gov/html/about/function.asp> (accessed October 14, 2004).

17. Soft responses to misconduct. *Nature* 420, no. 6913 (November 21, 2002): 253.

18. Uniform requirements for manuscripts submitted to biomedical journals. *Annals of Internal Medicine* 126, no. 1 (January 1, 1997): 36–47.

19. Office of Research Integrity. *Managing allegations of scientific misconduct: A guidance document for editors*. Bethesda, MD: Department of Health and Human Services, January 2000.

20. Institute of Medicine, National Research Council. *Integrity in scientific research: Creating an environment that promotes responsible conduct.* Washington, DC: National Academy Press, 2002.

Further Reading

Crewdson, J. Science fictions: A scientific mystery, a massive cover-up, and the dark legacy of Robert Gallo. Boston: Little, Brown and Company, 2002.
• A thorough account of research leading to the discovery of HIV and the behavior of scientists competing for high-stakes discovery

Judson, H. F. *The great betrayal: Fraud in science.* Orlando, FL: Harcourt, Inc., 2004.
• A scathing condemnation of the scientific community, using the Baltimore case as an example of the lack of commitment to prevent fraud

Kevles, D. *The Baltimore case: A trial of politics, science, and character.* New York: W. W. Norton, 1998.
• An account of a celebrated case of alleged misconduct, with a thoughtful analysis of the culture that stimulated the investigation

9

Science in the National Interest: Bioterrorism and Civil Liberties

Bioterrorism is the threat or act of an intentional release of viruses, bacteria, or their toxins with the aim of influencing government conduct and terrorizing a civilian population [1]. Since the terrorist attacks of September 11, 2001, and the subsequent mailings of anthrax-laced letters to public figures and other citizens in the United States, developing policy responses to potential attacks using biological agents such as toxins or pathogens has taken on a new urgency. How might the general public be protected from an attack using biological agents? If protection by means of quarantine, forced vaccination, or other measures results in the loss of some civil liberties, what trade-offs are acceptable? This part will explore the government response to bioterrorism by examining the anthrax letters as well as earlier attempts at bioterrorism.

A Short History of Biological Weapons Programs

The use of biological weapons is not new. Beginning in ancient times, armies deliberately contaminated the water supplies of enemy cities with animal and human carcasses. British troops gave smallpox-contaminated blankets to Native Americans during the French and Indian Wars from 1754 to 1767 to eliminate hostile populations [2]. By the early twentieth century, both the use of chemical agents such as mustard gas and efforts to "weaponize" pathogens were widespread. Both Germany and the Allied forces had programs to infect horses and cattle with anthrax using contaminated feed beginning in World War I [2].

After the horrors of chemical warfare in World War I, the international community negotiated the Geneva Protocol for the Prohibition of the Use in War of Asphyxiating, Poisonous, or Other Gases, and of Bacteriological Methods of Warfare in 1925. The treaty did not prevent the research, production, or stockpiling of

biological weapons, only their use in war [2]. The loopholes allowed many countries to continue to develop and in some cases use biological weapons. The United
States, Japan, and the former Soviet Union had active programs to develop biological weapons by midcentury. Japan allegedly conducted field trials of biological
agents in China during World War II, including anthrax, plague (using infected
fleas), cholera, and other agents.

The United States maintained an offensive biological weapons development
program centered at Fort Detrick, Maryland, from 1942 to 1970, when President
Nixon terminated the program by executive order. Weaponized forms of anthrax
and botulinum toxin were generated, and tests were conducted on a number of other
pathogens, including tularemia, other toxins, and plant pathogens (to damage
enemy crops) [2, 3]. The United States signed the Convention on the Prohibition,
Production, and Stockpiling of Bacteriological (Biological) and Toxin Weapons and
on their Destruction (Biological Weapons Convention, 26 U.S.T. 583) [2, 4]. The
Biological Weapons Convention did not preclude maintaining stocks for "prophylactic, protective, or other peaceful purposes," and the verification mechanisms were
limited. An outbreak of anthrax in the Soviet Union in 1979 near a biological
weapons facility in Sverdlovsk made it clear that efforts to weaponize anthrax were
continuing [2, 4, 5, 6]. The unintentional release sickened as many as 250 people,
of whom 100 died [5].

After the Gulf War of 1991, Iraq, a signatory to the Biological Weapons Convention, admitted to having an offensive biological weapons program involving
anthrax, botulinum toxin, and other agents. The United Nations Special Commission on Iraq oversaw the destruction of the facilities and stockpiles in 1996 [2].
Other nations—including Libya, Iran, and North Korea—are suspected of having
offensive biological weapons [7]. Putting these weapons in context, these countries
have all pursued nuclear weapons as well. Iraq's program was disabled by an Israeli
missile attack on the Osirak reactor site in 1981 and the first Gulf War; Libya abandoned its program in 2004; Iran asserts its nuclear programs (involving the construction of Russian reactors) are for power production only; and North Korea is
suspected of possessing several bombs.

Biological Warfare: Candidate Pathogens

What agents are candidates for biological warfare? Biological agents may be used
in two general ways: first, as infectious agents to cause disease that might disable

or kill enemies; and second, as biologically derived toxins that might directly disable or kill target populations. In addition to their potential to kill large numbers of people, biological agents are also effective in causing panic in unaffected populations because of their potential for attack without detection.

The ideal pathogen will be relatively stable (able to survive under adverse natural conditions), infectious (able to infect individuals), contagious (able to spread from person to person), and able to be dispersed as an aerosol for maximal transmission [6]. The target population should be vulnerable to the agent; the agent is not one for which a vaccine is generally given. The initial symptoms should mimic those of other diseases, confounding early diagnosis. Finally, the resulting disease should be difficult or impossible to treat. For the bioterrorist, it is desirable that the agent be easy to produce and disseminate, and relatively safe to handle.

Among the diseases most frequently cited as likely targets for bioterrorism are smallpox, anthrax, plague, tularemia, and a number of hemorrhagic diseases including Marburg and Ebola viruses [6]. The initial symptoms of all these diseases resemble a number of other flu-like conditions, making early detection difficult. Treatment of these diseases is limited; only anthrax responds to antibiotics in the early stages of the disease. The mortality rate for these diseases is high. Smallpox and anthrax are considered the most attractive target diseases for bioterrorists. Biologically derived toxins include botulinum toxin (the cause of botulism poisoning). These agents are characterized by an extremely high potency; even a tiny dose can be lethal.

Smallpox is a highly contagious illness characterized by a high fever followed by a blistering skin rash leading to general organ failure from the toxic effects; the mortality rate is as high as 30 percent [8]. Smallpox may be confused with the much less serious illness, chicken pox. Smallpox is the only disease to have been eradicated globally by a concerted effort to vaccinate populations against its virus, *Variola major*. The last case of natural smallpox in the world occurred in 1977; in the United States, population-wide vaccinations against smallpox ceased in 1972 (the last U.S. case occurred in 1949). The World Health Organization recommended that worldwide vaccination cease in 1980. Nevertheless, the United States and the former Soviet Union maintained stocks of the smallpox virus, ostensibly for research purposes. With the collapse of the Soviet Union, concerns increased that lapses in security might allow the Russian stocks to be stolen by terrorist groups [9]. The general population is highly vulnerable to infection by the smallpox virus; even individuals vaccinated prior to 1972 most likely have limited immunity because of the long time span since vaccination. A biological attack using smallpox might spread rapidly through populations via direct contact with infected individuals, who can be highly contagious prior to onset of severe symptoms. The smallpox virus, however, is sensitive to both temperature and humidity, and usually is not effectively transmitted in aerosol form [8]. Yet an outbreak in a hospital in Germany in 1970, in which a single patient infected nineteen other patients on three floors, suggests that the smallpox virus can spread through ventilation systems under the right conditions [8, 9].

Anthrax is an animal disease that especially infects herbivores [5]. Unlike the smallpox virus, the anthrax bacillus, *B. anthracis*, is remarkably stable, forming spores that can survive for decades in the soil. Anthrax bacilli are widespread in soil; animals pick up the disease by ingestion. The vaccination of animals may reduce the incidence of anthrax, but cases still appear worldwide. Humans can become infected through contact with contaminated animals, or by exposure to spores in soil or other material. Only one case of natural anthrax occurred in the United States from 1993 to 2001, although epidemics have occurred in other countries, usually associated with disease outbreaks in animals [10].

Anthrax is not spread from person to person; direct contact with the pathogen is required. Anthrax symptoms differ depending on the route of infection. Cutaneous anthrax occurs when bacteria enter the body through breaks in the skin. Skin lesions develop that can ulcerate and lead to generalized illness as the result of toxins produced by the bacteria. Mortality is low if patients are treated with antibiotics such as ciprofloxacin, doxycycline, or penicillin. If bacteria are inhaled, a more serious form of disease, inhalation anthrax, may develop. Patients initially develop flu-like symptoms followed by difficulty breathing, and usually die within days from internal bleeding and fluid loss if untreated. Aggressive treatment with antibiotics may reduce mortality. Gastrointestinal anthrax may be contracted through the ingestion of contaminated meat. Patients develop gastric and intestinal lesions, and die of sepsis (a generalized systemic infection). A vaccine for humans exists, but it is not generally available in the United States outside the military [10]. Anthrax spores might be prepared as a fine powder to be distributed as an aerosol, enhancing the dissemination of the bacteria. This requires advanced equipment, though, because anthrax spores are comparatively large and heavy.

Botulinum toxin, among the most poisonous substances known, is produced by the bacterium, *Clostridium botulinum* [11]. The bacteria thrive in oxygen-free (anaerobic) conditions in improperly prepared foods. The toxin blocks transmission in the nervous system, leading to paralysis and death. An antitoxin is available to help combat the effects of the toxin, but patients may require artificial ventilation and other support to survive. Ironically, the toxin (as the product Botox) is used cosmetically as injections to paralyze muscles in the face to provide temporary removal of lines and wrinkles. This product is extremely dilute, however, and does not pose a bioterrorist hazard. Botulinum toxin is comparatively easy to produce, as the bacteria are common. A concentrated solution of botulinum toxin might be disseminated as an aerosol or in food. Some suggest that the toxin might be used to contaminate public water supplies. But standard water purification procedures inactivate the toxin, making widespread poisoning unlikely using this route [11, 12].

Making a Biological Weapon

How much expertise is needed to generate a biological weapon is a matter of contention among experts [6]. Some believe that certain agents may be prepared with simple equipment and limited knowledge, while others argue that the development of an effective bioweapon requires considerable expertise and access to expensive equipment.

The first step in developing a biological weapon is obtaining a sample of the agent [6, 9]. The ease with which this can be accomplished varies: anthrax may be obtained from soil samples and plague from sick animals fairly easily, but agents such as smallpox would need to be stolen. Once a sample is obtained, the pathogen must be grown to generate enough material for an attack. Some agents—such as anthrax, smallpox, and *C. botulinum*—are relatively easy to maintain in a cell culture with only limited knowledge about microbiology. As stated above, some agents are difficult to grow and maintain, including plague, tularemia, and the hemorrhagic viruses. The terrorists also need to protect themselves, perhaps by vaccination (for smallpox or anthrax), or by using isolation equipment that may be expensive and difficult to obtain.

Once sufficient quantities of a pathogen are generated, a means to disseminate the agent must be developed. Dispersing the pathogen in aerosol form might infect the largest number of people, provided that the agent is stable in this form. Sunlight, heat, and humidity may kill the pathogen. It may also be difficult to convert the pathogen to an aerosol form that will remain airborne for an extended period; if the powder is too heavy, particles will fall to the ground quickly, reducing transmission. Many experts argue that the effective generation of aerosols requires sophisticated equipment and specialized knowledge not available to the general public [13]. Other means of effective dissemination might be through the contamination of food; for the reasons given above, attacks on public water supplies are less likely to be successful. Using infected individuals as "walking disease carriers" is popular among novelists; however, in most cases, these individuals would be too sick during their contagious phases to be mobile.

Bioterrorism Prior to 2001

Attempts by nongovernment organizations to conduct bioterrorism began before 2001. In 1984, the Rajneesh cult in Antelope, Oregon, sickened over 750 people by spraying salmonella bacteria (which cause gastrointestinal illness) on food in local salad bars in an attempt to influence a local election; no one died, fortunately [6, 13]. During the 1990s, a series of attempts at bioterrorism raised awareness of the apparent growing risk of attacks. In June 1993, the Aum Shinrikyo cult in Japan sprayed a suspension of anthrax bacilli off the roof of their headquarters in Kameido. Although there were complaints of strange odors, no one became ill. The release was the culmination of several years of work on a number of infectious agents, including botulinum toxin and the Ebola virus [7, 14]. In 2001, a molecular analysis of the anthrax strain used by the cult revealed that it was a noninfectious strain commonly employed to immunize animals against the disease [15]. In subsequent years, the cult released sarin gas, a lethal chemical agent, on two occasions—first to target judges involved in a case against the cult in 1994, and second, in the Tokyo subway system in March 1995, ostensibly as part of its goal to rule Japan [14]. Seven people died in the first attack, and around five hundred sought medical attention. In the latter incident, twelve people died, and close to thirty-eight hundred were injured. The relatively poor quality of the manufactured sarin prevented much higher casualities [7].

During the 1990s, an apparently failed bioterrorism attempt and a spate of hoaxes further increased concerns about the potential for attacks. In 1995, Larry Wayne Harris, an Ohio microbiologist and member of a survivalist organization, ordered *Yersinia pestis*, the pathogen that causes plague, from a nonprofit tissue repository, the American Type Culture Collection. Harris falsified his credentials to obtain the culture, claiming to be working in a laboratory and licensed to conduct research on pathogens. Shortly after sending the cultures, the American Type Culture Collection informed the Centers for Disease Control (CDC) of its concerns that the bacteria might not be handled safely. Harris was arrested and convicted of wire fraud, but sentenced to probation. Regulations governing the possession of and research on pathogens were tightened after the Harris case [6, 7].

In the late 1990s, over forty incidents involving letters purported to contain anthrax were reported. The targets included courthouses, Planned Parenthood clinics, churches, schools, and other institutions [13]. While no anthrax was found, over twelve hundred individuals were evacuated or temporarily quarantined. Other hoaxes, perhaps numbering a thousand, involved phone calls but no physical evidence. Some perpetrators were caught, but other cases remained unsolved. Beginning in 1998, in response to concerns about bioterrorism, both federal and state governments developed public health response plans in the event of a bioterrorism attack (see below for details). Hoaxes continued at a steady pace until fall 2001, when genuine bioterrorism using anthrax occurred.

The Anthrax Letters of 2001

On October 5, 2001, less than one month after the September 11 terrorist attacks, Bob Stevens, a photo editor for the tabloid publishing company American Media Inc. (AMI), died in Florida from inhalation anthrax. Although his death was initially thought to be the result of natural exposure to anthrax, Stevens was the first of twenty-two individuals later confirmed to be suffering from anthrax due to bioterrorism [13, 16]. Eleven individuals developed cutaneous anthrax; all survived. Five of the eleven contracting inhalation anthrax died. An estimated thirty-two thousand people began to take antibiotics after possible exposure; of these, over ten thousand were told to complete a full course of antibiotics [16]. The Hart Senate Office Building in Washington, DC, was closed for months and postal facilities were closed for over a year for decontamination.

Investigations determined that letters containing *B. anthracis* were sent from New Jersey, and passed through the Hamilton postal facility in September and October 2001. The targets of the anthrax letters were AMI in Florida, the New York offices of the broadcasting companies NBC, ABC, and CBS along with the *New York Post*, and the offices of the then Senate majority leader, Tom Daschle, and Senator Patrick Leahy, in Washington, DC (the Leahy letter did not reach the Senate Office Building because of misrouting). The victims were largely postal workers and office staff who handled incoming mail. Two other individuals, one in New York and one in Connecticut, apparently contracted inhalation anthrax through cross-contaminated mail and died.

Letters containing anthrax sent to targets in New York were postmarked in New Jersey on September 18; these letters held a fairly crude powder containing anthrax and caused only cutaneous anthrax. The first case developed on September 22, although it was not diagnosed until October 19. It is likely that the letter sent to AMI also was sent in mid-September; two individuals, including Stevens, developed inhalation anthrax. Letters to Senators Daschle and Leahy, postmarked October 9, contained a highly refined aerosolized anthrax powder and caused mostly inhalation anthrax. These later letters were the likely source of cross-contamination that led to the "outlier deaths" in New York and Connecticut; the powder was so fine that it leaked through the pores in the paper envelopes. Given the purity of the anthrax in the Daschle letter, it is fortunate that so few cases occurred. Nevertheless, because of earlier cases, heightened awareness led to the immediate use of antibiotics to prevent infection in exposed individuals [17].

Despite the differences in preparation, all the anthrax letters contained the same strain of *B. anthracis*—a strain called Ames that is used in government research labs [18]. This finding led to speculation that the perpetrator had access to government bioweapons labs [13]. Yet the Ames strain is fairly widely distributed in research laboratories in the United States and abroad.

The identification of the strain of anthrax did not help identify the still-unknown perpetrator(s) of the anthrax mailings. A former bioweapons researcher, Steven Hatfill, was investigated by the FBI, but was never charged. He has maintained his innocence, and yet has been unable to find employment since the investigation [13]. Speculation continues as to whether the perpetrator was a "lone wolf" domestic terrorist, associated with the September 11 terrorists, or associated with a rogue state.

Public Health Response to the Anthrax Letters

How well did the public health community respond to the anthrax attacks? The anthrax letters served as a test of the response network developed by the CDC with federal, state, and local public health officials in 1999, after increasing numbers of anthrax hoaxes. Officials had emergency plans in place, in anticipation of a possible bioterrorist event. These plans were found to be lacking in a GAO study released in 2003 [19].

Because anthrax is rare in the United States, physicians are not likely to suspect the disease when a patient visits. Family physicians initially diagnosed several cases of cutaneous anthrax as spider bites; anthrax was found only after heightened suspicion led to a reevaluation of the cases. Since these patients were given broad-spectrum antibiotics to treat their infections, however, they were given the right treatment at the right time [13]. Diagnosis of cases of inhalation anthrax presented more of a challenge, since the last case in the United States occurred in 1976 [16]. Stevens's case took several days to confirm; he died despite receiving antibiotics. Similar lags occurred in the diagnosis of inhalation anthrax in several postal workers.

More worrisome than diagnosis alone was the lack of coordinated information between the responsible agencies [19]. Although most local and state public health officials planned for coordination, they had not anticipated the challenges of working across many jurisdictions. State and local officials as well as the CDC found it difficult to easily communicate with clinicians. Physicians realized that they needed more information about diagnosing anthrax and had difficulty getting the necessary information; the CDC had not prepared materials in advance. Because of ongoing criminal investigations, public health officials found it difficult to obtain needed information from the FBI about the nature of the anthrax contained in the letters. Public officials also found it hard to communicate with the public because of restrictions on what information might be made available. Incorrect information released by the media led to unnecessary concerns by the public, demands for unneeded treatments, and other actions that compromised the public health effort, such as the hoarding of and unnecessary treatment with the antibiotic ciprofloxacin (Cipro).

The report of anthrax in letters served only to increase the number of hoaxes, flooding the public health system. Laboratories were swamped with samples for testing, coming not only from the genuine letters but also from hoaxes around the country [19]. The laboratory response network tested more than 120,000 samples for the presence of anthrax; other agencies also stepped in to do testing. Many of these samples were from the environment, not from patients, and thus there were no published protocols for testing. The number of laboratory staff was also insufficient to meet the demand for testing.

The CDC acknowledged that it was unprepared for the challenge of coordinating all the federal agencies involved; these included the Department of Defense (DOD), the FDA, the NIH, the EPA, the FBI, and the Federal Emergency Management Authority (FEMA) [19]. The CDC and other federal agencies worked to revamp the emergency response network based on the lessons from 2001.

The U.S. Public Health System

The public health system in the United States is a highly complex network involving local, county, state, and federal health authorities, and nongovernmental organizations such as medical associations, physicians, and researchers [20]. State, county, and municipal health departments have overall responsibility for oversight of public health within their jurisdictions. The federal government monitors a list of fifty-eight "reportable diseases and conditions," including HIV/AIDS, venereal diseases, communicable diseases such as measles and hepatitis, and lead poisoning [21]. The specific reportable diseases vary between states since disease incidence may vary regionally. Health care professionals are required to report cases of the diseases to the CDC, which maintains statistics. The power of the reporting mechanism allows for the detection of unusual clusters or increases in the occurrence of a given disease. The CDC and the DHHS also maintain a list of specific diseases for which isolation (segregating a sick patient in a secure location) or quarantine (restricting the activities of probable exposed individuals who are still healthy) applies. The listed diseases include cholera, diphtheria, plague, tuberculosis, smallpox, yellow fever, and viral hemorrhagic diseases such as Ebola [22]. The federal government typically leaves it to the states to manage quarantine, except for interstate and international border control.

For the most part, health issues are handled at the state and local levels, including natural outbreaks of diseases. Public health responses involve disease surveillance (recognizing the appearance of a disease), epidemiological and laboratory studies to confirm occurrences and possible sources, health care delivery to affected individuals and communities, quarantine management (if needed), disease containment, and casualty handling. In the event of a widespread outbreak, states call on federal support services, including the CDC, FEMA, and other agencies [1]. For example, in summer 1999, the first cases of a mosquito-borne illness, the West Nile virus, appeared in the New York City area [23]. The first indicators of the disease were numerous dead crows and other birds. Sixty-two individuals were diagnosed with encephalitis caused by the virus; seven died. Health departments in New York responded by spraying neighborhoods with pesticides to reduce the mosquito population and advising people to remain indoors during peak hours of mosquito activity [24]. The outbreak was reported to the CDC, as required, and the CDC conducted much of the laboratory testing needed to confirm the virus [1]. In 2000, the disease spread to Connecticut and Massachusetts; by 2003, all states except for

Alaska and Hawaii had reported cases. The CDC issued detailed guidelines to health departments for the surveillance, prevention, and control of the virus and animal carriers [23]. Although the virus is a national problem, state and local health departments make their own decisions regarding controls such as pesticide spraying; some municipalities decide that the health risks of pesticides are high enough to preclude spraying [24].

The emergence of the West Nile virus served as a valuable lesson for the CDC. The volume of lab testing strained the available facilities, even though the initial outbreak was limited to a small area [1].

Preparing for Bioterrorism

Planning for a bioterrorist attack in the United States assumed that the same steps of surveillance and response would be involved, but placed more emphasis on the roles of federal agencies. Congress passed a series of laws to increase the federal role in bioterrorism preparedness [1]. Laws aimed at preventing the acquisition and use of chemical or biological weapons were passed from 1989 to 1996 (sections of U.S. Code Title 18). In 1992, the Stafford Act (P.L. 93–288) established a framework for federal responses to domestic events requiring federal disaster relief. The act was amended in 1999, setting up the framework through which the CDC might manage a bioterrorism event. After the Oklahoma City bombing in 1995, President Clinton issued a directive (presidential decision directive 39) that broadly defined the federal government's role in response to the use of weapons of mass destruction; a second directive in 1998 (directive 62) spelled out the roles that federal agencies would play in the event of an attack with weapons of mass destruction [1].

Preparedness takes several forms. Prior to the anthrax attacks of 2001, the DHHS and the DOD created response teams to provide emergency medical treatment and other assistance to local authorities. The CDC and the DHHS's Office of Emergency Preparedness established stockpiles of pharmaceutical supplies that might be rushed to a site. Bioterrorism exercises were conducted to help train local emergency response teams; these simulations revealed both the strengths and weaknesses of response networks. Hospitals in particular are poorly prepared and equipped to deal with mass causalities [25]. Even a particularly bad flu outbreak in a community may tax a local hospital [1].

The events of September 11 and the subsequent anthrax letters mobilized the federal government to further extend its oversight of public health issues. The USA

Patriot Act (P.L. 107–56), passed in October 2001, broadly expanded the authority of the federal government and law enforcement officials to investigate and prosecute potential terrorists, and expanded prohibitions involving biological agents and their possession. The Public Health Security and Bioterrorism Preparedness and Response Act of 2002 was enacted on June 12 (Bioterrorism Act, P.L. 107–188), after the anthrax letter attacks in fall 2001. The statute mandates the development of more streamlined coordination of responses to bioterrorism by creating the position of an assistant secretary for public health preparedness in DHHS to develop and oversee the National Disaster Medical System; this position was filled by Stewart Simonson in April 2004 [26]. The act requires that states develop their own emergency response plans in collaboration with the new assistant secretary. The act also mandates additional funds and facilities for the CDC, and clarifies its role in overseeing the isolation and quarantine of individuals. Regulations for access to potentially hazardous pathogens and toxins are tightened, along with increased criminal penalties for their illegal possession. New authority for the FDA and other agencies to monitor food, agricultural, and water safety is given.

The Model State Emergency Health Preparedness Act

In order to facilitate the coordinated development of state emergency health laws, the Center for Law and the Public's Health at Georgetown and Johns Hopkins universities prepared the Model State Emergency Health Powers Act (MSEHPA) [27]. The model legislation is intended to serve as a guide for states as they develop their own laws; such drafts are not laws themselves. The proposed law recognized that many existing state emergency health response systems are outdated and outmoded. The first draft of the MSEHPA, released in October 2001 during the anthrax letters attacks, proposed a broad range of powers to be vested in state governors in the event of a health emergency. In the event that a bioterrorism attack or other health emergency is declared by the state's governor, the act "authorizes the collection of data and records, the control of property, the management of persons, and access to communications" [27]. Actions might include the collection and reporting of otherwise-private medical information (including the names of individuals); seizure of property such as hospitals, landfills, or mortuaries to control the spread of pathogens; restriction of commercial enterprise; mandatory medical examinations or vaccinations; mandatory quarantine or isolation of exposed individuals or groups; and criminal penalties for noncooperation by medical professionals or the public [28, 29].

The drafters argued that in an emergency, the loss of certain civil liberties is necessary to protect the public. The philosophical "harm principle" seeks to preserve individual autonomy while recognizing that such autonomy may pose a risk to others. In society, the need to protect the public from harm may appropriately trump an individual's autonomy [30, 31]. The MSEHPA framers based the act on previous legal precedent. In the late nineteenth and early twentieth centuries, states passed laws requiring vaccination against smallpox in an attempt to control the outbreaks that occurred yearly [32]. In 1902, Henning Jacobson, a minister in Cambridge, Massachusetts, refused vaccination against smallpox during an outbreak. He was convicted of having violated the state's compulsory vaccination law and fined. He appealed to the U.S. Supreme Court, which ruled in *Jacobson v. Commonwealth of Massachusetts* (197 U.S. 11 [1905]) that the state had the right to order compulsory vaccination, but stopped short of allowing forcible vaccination [33]. In other words, individuals might be punished for refusing to be vaccinated, but could not be vaccinated against their will. The states did have the power to impose other measures such as quarantine or isolation in the context of protecting public health. The *Jacobson* decision established a framework for the development of the states' public health systems, with their reporting requirements, regulations regarding childhood immunization, and other measures [32]. Nevertheless, the emergence of HIV/AIDS in the 1980s led to challenges of many components of the states' testing and reporting systems; rights to privacy gained greater precedence during this period. The events of 2001 contributed to the erosion of privacy protection from growing governmental intrusiveness and the power of computers to ferret out information.

The MSEHPA triggered a storm of protest over its draconian abrogation of civil liberties and broad powers of the state government to act after declaring a health emergency [32, 34, 35, 36]. Critics on both the Left and the Right argued that the MSEHPA infringed on the rights to privacy and to refuse medical treatment, weakened protections against illegal search and seizure, and criminalized the practice of ordinary civil rights. Quarantine or isolation might be used as a punishment instead of only when medically warranted. Critics also suggested that the definition of a health emergency is too broad; "nonemergency conditions" such as a flu epidemic might also trigger a declaration of emergency [34]. Others claimed that large-scale quarantine is unlikely to be the most effective way to handle disease containment because of logistic problems as well as the unintended negative effects on commerce, public trust, and civil rights [37]. The revised version of the MSEHPA released in

December 2001 responded to some degree to these criticisms by requiring legislative review prior to some actions and the possibility of appeal [38], but critics remained concerned.

Critics contend that the system outlined in the MSEHPA represents an overreaction to a hypothetical problem and assumes that the public will fail to cooperate in an emergency [32, 34]. The MSEHPA's principles are based on outdated attitudes toward public health, and fail to recognize that measures that might preserve individual rights have been demonstrated to be feasible and effective.

States have given mixed responses to the draft legislation. By August 2003, forty-three states had introduced legislation based wholly or partly on the MSEHPA; thirty-two states passed bills that include at least some of its provisions [39]. But many of the most controversial components of the model legislation were rejected [40].

To Vaccinate or Not to Vaccinate: Smallpox and Anthrax

Concerns over the risks of bioterrorism involving smallpox led the federal government to propose vaccinating target populations against the disease. Models suggest that exposure of a vulnerable population to smallpox might have catastrophic outcomes [8, 41]. Studies determined that diluting the current stocks of vaccine still allowed for effective vaccination [42]. Sufficient stocks would be available to vaccinate the U.S. population. In December 2002, President Bush announced a plan for the voluntary vaccination of health care workers and other critical personnel against smallpox [43, 44]. Military personnel, such as soldiers involved in the conflict in Iraq, also were vaccinated; over four hundred thousand were vaccinated by early 2003 [45]. The program is a compromise between no pre-event preparation and the mass voluntary vaccination of the entire U.S. population [46, 47, 48].

Having a pool of vaccinated health care personnel would serve to reduce the spread of smallpox from infected individuals in the event of a bioterrorism attack. The smallpox vaccine, however, carries with it significant known risks of adverse effects, including severe infection with smallpox, encephalitis, and death [44]. Individuals with certain skin disorders such as eczema and those with weakened immune systems such as organ transplant recipients and HIV/AIDs patients are at particular risk. Persons receiving the smallpox vaccine actively shed the virus and may infect those with whom they come in contact, requiring careful management. For example, a vaccinated individual might infect family members [49].

Despite the report that finding volunteers to participate in the vaccine dilution study was easy [42], health care workers have not rushed to get vaccinated. Of the 440,000 targeted, fewer than 40,000 had been vaccinated as of February 2005 [50]. Unanticipated adverse effects, including possible heart problems, in a small number of vaccinated military and health care workers received heavy media coverage. In June 2003, the CDC's Advisory Committee on Immunization Practices recommended against further expansion of the vaccination plan because of the adverse side effects of the vaccine [51].

The U.S. military's policy on the vaccination of troops is controversial. The military carries out mandatory vaccination in order to protect troops against identified hazards. Whether military personnel are fully informed of the risks is unclear, and critics suggest that the practice violates requirements for informed consent. In the 1991 Gulf War, troops were vaccinated with the anthrax vaccine because of fears that Iraq might use biological weapons against them [40, 52]. The anthrax vaccine, although FDA-approved to prevent cutaneous anthrax, was untested for inhalation anthrax. The DOD began vaccinating all military personnel against anthrax in 1998; over one million had received the vaccine by 2002. Some soldiers refused to be vaccinated, though, and were either dismissed from the military or court-martialed for refusing a direct order. Soldiers filed lawsuits opposing the practice just prior to the 2003 Iraq invasion, arguing that the vaccine was an investigational new drug untested for efficacy against inhalation anthrax [53]. Vaccination proceeded until halted by court injunctions in 2003 and 2004.

What Price Security?

The risks of bioterrorism cannot be assessed directly, and thus present a serious challenge for policymakers. If the response infringes heavily on civil liberties in the name of public safety, one might argue that the bioterrorists have achieved their goals of disrupting society. Critics of the MSEHPA suggest that its framers assume that citizens will fail to cooperate in a national emergency, and therefore strong measures are necessary. Heavy-handed responses, such as widespread quarantine, only increase distrust of government and make noncooperation by citizens more likely [34]. The recent response to severe acute respiratory syndrome (SARS; see the following section) suggests that it is more likely that individuals will cooperate in times of crisis [40]. A system of measured responses that might progressively reduce civil liberties in proportion to the severity of the situation is preferable to a one-size-fits-all system.

It is also important to remember that the anthrax attacks of 2001 led to only eleven deaths; about twenty thousand people die from influenza each year. The public needs to be able to see events in context in order to respond appropriately. Widespread panic might have been reduced in 2001 had information been made more readily available to the population. Improved coordination within the public health system is clearly needed, and more direction from the federal government may be useful given the fragmented nature of the current system.

It is impossible to anticipate if and when another bioterrorism event will occur. Increased vigilance to prevent such an occurrence is the best strategy. Tightened controls on access to pathogens are already in place. What is needed is effective intelligence gathering that still protects the rights of individuals.

References

1. U.S. General Accounting Office. *Bioterrorism: Federal research and preparedness activities.* GAO–01–915. Washington, DC: General Accounting Office, September 2001.

2. Christopher, G. W., T. J. Cieslak, J. A. Pavlin, and E. M. Eitzen Jr. Biological warfare: A historical perspective. *JAMA* 278, no. 5 (August 6, 1997): 412–417.

3. Bernstein, B. J. The birth of the U.S. biological-warfare program. *Scientific American* 256 (1987): 116–121.

4. Scharf, M. P. Clear and present danger: Enforcing the international ban on biological and chemical weapons through sanctions, use of force, and criminalization. *Michigan Journal of International Law* 20 (Spring 1999): 477–521.

5. Inglesby, T. V., T. O'Toole, D. A. Henderson et al. Anthrax as a biological weapon, 2002. *JAMA* 287, no. 11 (May 1, 2002): 2236–2252.

6. Kellman, B. Biological terrorism: Legal measures for preventing catastrophe. *Harvard Journal of Law and Public Policy* 24 (Spring 2001): 417–488.

7. Cole, L. The specter of biological weapons. *Scientific American* 275 (December 1996): 60–65.

8. Henderson, D. A., T. V. Inglesby, J. G. Barlett et al. Smallpox as a biological weapon. *JAMA* 281, no. 22 (June 9, 1999): 2127–2137.

9. Henderson, D. A. Bioterrorism as a public health threat. *Emerging Infectious Diseases* 4, no. 3 (July–September 1998): 488–492.

10. Ashford, D. A., and B. Perkins. Use of anthrax vaccine in the United States: Recommendations of the advisory committee on immunization practices. *Morbidity and Mortality Weekly Report* 49, RR–15 (December 15, 2000): 1–20.

11. Arnon, S. S., R. Schechter, T.V. Inglesby et al. Botulinum toxin as a biological weapon. *JAMA* 285, no. 8 (February 28, 2001): 1059–1070.

12. Harvard University. Botulinum toxin as a potential weapon. *University Operations Services.* Available at <http://www.uos.harvard.edu/ehs/fs_botulinum.shtml> (accessed April 8, 2004).

13. Cole, L. *The anthrax letters: A medical detective story.* Washington, DC: Joseph Henry Press, 2003.

14. Olson, K. B. Aum Shinrikyo: Once and future threat? *Emerging Infectious Diseases* 5, no. 4 (July–August 1999): 513–516.

15. Keim, P., K. L. Smith, C. Keys et al. Molecular investigation of the Aum Shinrikyo anthrax release in Kameido, Japan. *Journal of Clinical Microbiology* 39, no. 12 (December 2001): 4566–4567.

16. Jernigan, D. B., P. Raghunathan, B. P. Bell et al. Investigation of bioterrorism-related anthrax, United States 2001: Epidemiological findings. *Emerging Infectious Diseases* 18, no. 10 (October 2002): 1019–1028.

17. Peters, C., R. Spertzel, and W. Patrick III. Aerosol technology and biological weapons. In *Biological threats and terrorism: Assessing the science and response capabilities*, ed. S. L. Knobler, A. A. F. Mahmoud, and L. A. Pray, 66–77. Washington, DC: National Academy Press, 2003.

18. Hoffmaster, A. R., C. C. Fitzgerald E. Ribot, L. W. Mayer, and T. Popovic. Molecular subtyping on Bacillus anthracis and the 2001 bioterrorism-associated anthrax outbreak, United States. *Emerging Infectious Diseases* 8, no. 10 (October 2002): 394.

19. U.S. General Accounting Office. *Bioterrorism: Public health response of anthrax incidents of 2001.* GAO–04–152. Washington, DC: General Accounting Office, October 2003.

20. Centers for Disease Control and Prevention. Achievements in public health, 1900–1999: Changes in the public health system. *JAMA* 283, no. 6 (February 9, 2000): 735–738.

21. Roush, S., G. Birkhead, D. Koo, A. Cobb, and D. Fleming. Mandatory reporting of diseases and conditions by health care professionals and laboratories. *JAMA* 281, no. 2 (July 14, 1999): 164–170.

22. Misrahi, J. J., J. A. Foster, F. E. Shaw, and M. S. Cetron. HHS/CDC legal response to SARS outbreak. *Emerging Infectious Diseases* 10, no. 2 (February 2004): 353–355.

23. Centers for Disease Control and Prevention. *Epidemic/epizootic West Nile virus in the United States: Guidelines for surveillance, prevention, and control.* 3rd ed. Fort Collins, CO: Department of Health and Human Services, 2003.

24. Tickner, J. A. Epidemiology and science: The precautionary principle and public health trade-offs: Case study of West Nile virus. *Annals of the American Academy of Political and Social Science* 584 (November 2002): 69–79.

25. Macintyre, A. C., G. W. Christopher, E. M. Eitzen Jr. et al. Weapons of mass destruction events with contaminated casualties: Effective planning for health care facilities. *JAMA* 283, no. 2 (January 12, 2000): 242–249.

26. U.S. Department of Health and Human Services, Office of Public Health Emergency Preparedness. *DHHS*. April 1, 2005. Available at <http://www.hhs.gov/ophep/index.html> (accessed April 3, 2005).

27. The Center for Law and the Public's Health at Georgetown and Johns Hopkins universities. The Model State Emergency Health Powers Act. October 23, 2001. Available at <http://www.publichealthlaw.net> (accessed March 26, 2004).

28. Gostin, L. O., J. Sapsin, S. P. Teret et al. The Model State Emergency Health Powers Act: Planning for and response to bioterrorism and natually occurring infectious diseases. *JAMA* 288, no. 5 (August 7, 2002): 622–628.

29. Hodge, J., G. James, and L. O. Gostin. Protecting the public's health in an era of bioterrorism: The Model State Emergency Health Powers Act. In *In the wake of terror: Medicine and morality in a time of crisis*, ed. J. D. Moreno, 17–32. Cambridge, MA: MIT Press, 2003.

30. May, T. Harm, public health threats, and the Model State Emergency Health Powers Act: Bio-terror defense and civil liberties. *Joint Services Conference on Professional Ethics*. 2003. Available at <http://www.usafa.af.mil/jscope/JSCOPE03/May03.html> (accessed March 26, 2004).

31. Gostin, L. O. Dunwody distinguished lecture in law: When terrorism threatens health: How far are limitations on personal and economic liberties justified? *Florida Law Review 55* (December 2003): 1105–1170.

32. Bayer, R., and J. Colgrove. Rights and dangers: Bioterrorism and the ideologies of public health. In *In the wake of terror: Medicine and morality in a time of crisis*, ed. J. D. Moreno, 51–74. Cambridge, MA: MIT Press, 2003.

33. Joseph, D. G. Uses of *Jacobson v. Massachusetts* in the age of bioterror. *JAMA* 290, no. 17 (November 5, 2003): 2331.

34. Annas, G. J. Bioterrorism, public health, and civil liberties. *New England Journal of Medicine* 346, no. 17 (April 25, 2002): 1337–1342.

35. Reich, D. S. Modernizing local responses to public health emergencies: Bioterrorism, epidemics, and the Model State Emergency Health Powers Act. *Journal of Contemporary Health Law and Policy* 19 (Spring 2003): 379–414.

36. American Civil Liberties Union. Q&A on the Model State Emergency Health Powers Act. Available at <http://archive.aclu.org/issues/privacy/Model_health_feature.html> (accessed March 26, 2004).

37. Barbera, J., A. C. Macintyre, L. Gostin et al. Large-scale quarantine following biological terrorism in the United States: Scientific examination, logistic and legal limits, and possible consequences. *JAMA* 286, no. 21 (December 5, 2001): 2711–2717.

38. The Center for Law and the Public's Health at Georgetown and Johns Hopkins universities. The Model State Emergency Health Powers Act (revised). December 21, 2001. Available at <http://www.publichealthlaw.net> (accessed March 26, 2004).

39. The Center for Law and the Public's Health at Georgetown and Johns Hopkins universities. Model public health laws. August 11, 2003. Available at <http://www.publichealthlaw.net/Resources/Modellaws.htm> (accessed April 16, 2004).

40. Annas, G. J. Blinded by terrorism: Public health and liberty in the 21st century. *Health Matrix* 13 (Winter 2003): 33–69.

41. Ferguson, N. M., M. J. Keeling, W. J. Edmunds et al. Planning for smallpox outbreaks. *Nature* 425 (2003): 681–685.

42. Frey, S. E., R. B. Couch, C. O. Tacket et al. Clinical responses to undiluted and diluted smallpox vaccine. *New England Journal of Medicine* 346, no. 17 (April 25, 2002): 1265–1274.

43. Centers for Disease Control. Protecting Americans: Smallpox vaccination program. *CDC.* December 13, 2002. Available at <http://www.bt.cdc.gov/agent/smallpox/vaccination/guidelines.asp> (accessed April 16, 2004).

44. Wharton, M., R. A. Strikas, R. Harpaz et al. Recommendations for using smallpox vaccine in a pre-event vaccination program. *Morbidity and Mortality Weekly Report* 52 (February 26, 2003): 1–16.

45. Institute of Medicine. *Review of the Centers for Disease Control and Prevention's Smallpox vaccination program implementation letter report #3.* Washington, DC: National Academy of Sciences, 2003.

46. Bicknell, W. J. The case for voluntary smallpox vaccination. *New England Journal of Medicine* 346, no. 17 (April 25, 2002): 1323–1325.

47. May, T., and R. D. Silverman. Should smallpox vaccine be made available to the general public? *Kennedy Institute of Ethics Journal* 13, no. 2 (2003): 67–82.

48. Fauci, A. S. Smallpox vaccination policy: The need for dialogue. *New England Journal of Medicine* 346, no. 17 (April 25, 2002): 1319–1320.

49. Sankar, P., C. Schairer, and S. Coffin. Public mistrust: The unrecognized risk of the CDC smallpox vaccination program. *American Journal of Bioethics* 3, no. 4 (Fall 2003): W22–W25.

50. Centers for Disease Control. Smallpox vaccination program status by state. *Office of Communication.* January 31, 2005. Available at <http://www.cdc.gov/od/oc/media/spvaccin.htm> (accessed March 8, 2005).

51. Centers for Disease Control. Advisory Committee on Immunization Practices (ACIP) statement on smallpox preparedness and vaccination. *ACIP* June 18, 2003. Available at <http://www.bt.cdc.gov/agent/smallpox/accination/acipjun2003.asp> (accessed June 14, 2006).

52. Katz, R. D. Note: Friendly fire: The mandatory military anthrax vaccination program. *Duke Law Journal* 50 (April 2001): 1835–1865.

53. Michels, J. J., and M. S. Zaid. Anthrax vaccine lawsuit filed by military servicemen against Pentagon. *Alliance for Human Research Protection.* 18 March 2003. Available at <http://www.ahrp.org/infomail/0303/18.php> (accessed June 14, 2006).

Further Reading

Cole, L. *The anthrax letters: A medical detective story.* Washington, DC: Joseph Henry Press, 2003.

• A detailed account of the bioterrorism of 2001, with a thoughtful discussion of its wider implications

McBride, D., ed. *Bioterrorism: The history of a crisis in American society, vol. 1.* New York: Routledge, 2003.

• A collection of articles on history and policy regarding biological weapons and warfare

Moreno, J. D., ed. *In the wake of terror: Medicine and morality in a time of crisis.* Cambridge, MA: MIT Press, 2003.

• A series of essays on ethical issues in bioterrorism

Emerging Diseases: SARS and Government Responses

It is far more likely that public health systems will be challenged by new illnesses than by a bioterrorism attack; new infectious diseases appear yearly, including new variants of influenza. Emergent diseases pose a broad challenge because of their potential for worldwide effects. In a world where air travel can spread a new disease within hours, vigilance is required not only nationally but globally as well. The outbreak of SARS in winter 2002–2003 provides an example of how the international mobilization of public health resources can both succeed and fail.

In late November 2002, an unusual form of pneumonia appeared in Guangdong Province, China. The new disease spread rapidly as infected individuals traveled internationally. In February 2003, the disease spread to Hong Kong, when a Chinese physician fell ill there and infected a dozen people in the hotel where he was staying [1, 2]. These infected individuals then spread the disease to Vietnam, Singapore, Taiwan, and Canada, with a small number of cases in other countries, including the United States. The World Health Organization (WHO) declared a global alert in March 2003, warning of a growing health threat [3]. The epidemic was declared over in July 2003, after a total of over 8,000 cases and 774 deaths worldwide [4]. Despite its short duration, the SARS epidemic had serious effects on the economies of affected countries; WHO's travel advisories led to billions of dollars in lost revenue from declines in commerce and tourism.

The rapid containment of a new epidemic disease might be viewed as a triumph of global cooperation and communication [5, 6]. An international collaboration of researchers identified a novel coronavirus (a family of viruses that includes the common cold virus) responsible for SARS within two weeks of the first public reports of the disease [7, 8]. Screening tests for the presence of the virus followed, enabling public health officials to identify the disease in patients. With the exception of China, governments acted quickly to contain the spread of the SARS virus, instituting isolation and quarantine, tracking contacts of patients, taking precautions to protect health care workers, restricting travel, and screening passengers at airports [9, 10, 11, 12, 13]. After a belated start, Chinese officials instituted strong measures to contain the disease that had spread explosively during the period when China denied there was a problem [10].

The impact of public health actions on citizens varied from country to country. In authoritarian Singapore, the mandatory isolation of infected individuals along with the identification and quarantine of contacts left little room for civil liberties [11]. Home-quarantined contacts were fitted with ankle monitors to assure compliance. In China, government officials refused to acknowledge the existence of an outbreak of SARS in the south until after it had spread to Hong Kong. By then, the number of cases had grown to hundreds. In Beijing, over thirty thousand people were quarantined, patients were isolated in fever hospitals, and public facilities such as schools, theaters, libraries, and sports facilities were closed [10]. Travelers were examined for fevers at train stations, on roads, and at airports. Even in Canada, a country with clearly delineated civil rights, the identities of contacts were not kept confidential. In one case, two thousand students and staff from a single school in Toronto were quarantined after a suspicious illness in one student [14].

In the aftermath of the epidemic, several features suggest that the world was lucky that a new pandemic did not emerge. The pattern of SARS transmission is unusual; a small number of infected individuals are "superspreaders," such as the Chinese physician in Hong Kong, while most victims do not seem to be very contagious [1, 15]. This heterogeneity in response makes modeling a future outbreak difficult. Identifying the animal reservoir of the disease has been difficult too; recent evidence indicates that bats, which are eaten and used in traditional medicine, carry the virus [16]. Despite concerns that SARS might reappear in winter 2003–4, only four cases appeared in Guangdong during that December and January [17]. A second outbreak in China, the result of a laboratory accident, occurred in April 2004. Whether SARS will reappear in the future cannot be predicted.

SARS in the United States

The United States escaped the worst of the epidemic; of nearly fifteen hundred suspicious respiratory illnesses reported to the CDC, seventy-two cases were probable SARS, with eight cases confirmed by laboratory testing to be harboring the SARS virus [18]. No fatalities occurred. The number of cases was low because the CDC instituted surveillance procedures within three days of the WHO declaration of health risk; SARS was added to the list of quarantinable diseases in April 2003 [9]. By setting broad screening criteria, the CDC sought to ensure that no cases would be missed. Individuals with illnesses other than SARS were ruled out by additional screening and testing. Infected individuals were isolated voluntarily in hospitals or at home; only one patient required involuntary isolation [19]. Probable contacts of patients were traced and voluntarily quarantined [20]. Information about SARS was developed and transmitted to both health care workers and the public. The CDC developed response guidelines for future SARS outbreaks, covering surveillance, hospital practices, and containment [21].

Despite the successful containment of SARS in the United States, worrisome issues emerged as the epidemic developed. Reports of discrimination against Asian-appearing individuals appeared; the CDC received phone calls from people fearful of any contact with Asians [22]. Immigration and customs officials detained incoming travelers who appeared to have any respiratory symptoms [9]. The University of California at Berkeley barred all students from China, Singapore, and Taiwan from attending a summer program because of the SARS epidemic [23].

Lessons from SARS

The success of strict isolation and quarantine measures in other countries raises concerns about future intrusions into privacy and other rights [9]. Countries had difficulties finding a balance between public health requirements and individual rights. This suggests that in the event of future terrorist attacks, both overreaction and discrimination are real possibilities.

The failure of China to respond to the initial SARS outbreak allowed the disease to spread to other countries [24]. This failure of disease surveillance exposes the vulnerability of the global community to emerging diseases. The rapid spread of a virus that turned out to be not very infectious is sobering; had SARS been as contagious as influenza, a global pandemic would be unavoidable. A higher degree of preparedness is needed worldwide to prevent disaster the next time around. As concern over SARS began to wane, a new global threat emerged—avian influenza. A particularly virulent strain is killing thousands of domestic fowl in Southeast Asia, and a number of humans around the world have contracted the disease and died.

In January 2005, the WHO issued a report urging governments to prepare for a possible avian flu pandemic [25]. In November 2005, President Bush called for $7.1

billion to be used to prepare for a possible pandemic. The plans include developing vaccines and stockpiling antiflu drugs [26]. The WHO convened a meeting in November 2005 to develop global planning for a possible pandemic. Nations are taking the possibility seriously—but whether preparations will prove adequate cannot be determined in advance.

References

1. Pearson, H., T. Clarke, A. Abbott, J. Knight, and D. Cyranoski. SARS: What have we learned? *Nature* 424 (July 10, 2003): 121–126.

2. Wenzel, R. P., and M. B. Edmond. Managing SARS amidst uncertainty. *New England Journal of Medicine* 384, no. 20 (May 15, 2003): 1947–1948.

3. Altman, L. K., and K. Bradsher. Rare health alert is issued by W.H.O. for mystery illness. *New York Times*, March 16, 2003, sec. 1, 1.

4. Revised U.S. surveillance case definition for severe acute respiratory syndrome (SARS) and update on SARS cases—United States and worldwide, December 2003. *Morbidity and Mortality Weekly Report* 52, no. 49 (December 12, 2003): 1202–1206.

5. Gerberding, J. L. Faster . . . but fast enough? Responding to the epidemic of severe acute respiratory syndrome. *New England Journal of Medicine* 348, no. 20 (May 15, 2003): 2030–2031.

6. Heymann, D. L., and G. Rodier. Global surveillance, national surveillance, and SARS. *Emerging Infectious Diseases* 10, no. 2 (February 2004): 173–175.

7. Grady, D., and L. K. Altman. The SARS epidemic: The researchers; global collaboration bears fruit as SARS gives up some secrets. *New York Times*, May 26, 2003, A1.

8. Rota, P. A., M. S. Oberste, S. S. Monroe et al. Characterization of a novel coronavirus associated with severe acute respiratory syndrome. *Science* 300 (May 30, 2003): 1394–1399.

9. Gostin, L. O., R. Bayer, and A. L. Fairchild. Ethical and legal challenges posed by severe acute respiratory syndrome: Implications for the control of severe infectious disease threats. *JAMA* 290, no. 24 (December 24–31, 2003): 3229–3237.

10. Pang, X., Z. Zhu, F. Xu et al. Evaluation of control measures implemented in the severe acute respiratory syndrome outbreak in Beijing, 2003. *JAMA* 290, no. 24 (December 24–31, 2003): 3215–3221.

11. Arnold, W. In Singapore, 1970's law becomes weapon against SARS. *New York Times*, June 10, 2003, F5.

12. Bradsher, K., and L. K. Altman. Isolation, an old medical tool, has SARS fading. *New York Times*, June 21, 2003, A1.

13. Lau, J. T., H. Tsui, M. Lau, and X. Yang. SARS transmission, risk factors, and prevention in Hong Kong. *Emerging Infectious Diseases* 10, no. 4 (April 2004): 587–592.

14. Krauss, C., and L. K. Altman. Quarantine hits Toronto school in SARS fight. *New York Times*, May 29, 2003, A1.

15. Low, D. E., and A. McGeer. SARS: One year later. *New England Journal of Medicine* 349, no. 25 (December 18, 2003): 2381–2382.

16. Normile, D. Researchers tie deadly SARS virus to bats. *Science* 309 (September 2005): 2154–2155.

17. Associated Press. China reports first SARS death since July. April 23, 2004. Available at <http://www.nytimes.com/aponline/international/AP-China-SARS.html?hp> (accessed April 23, 2004).

18. Schrag, S., J. T. Brooks, and C. Van Beneden et al. SARS surveillance during emergency public health response, United States, March–July 2003. *Emerging Infectious Diseases* 10, no. 2 (February 2004): 185–194.

19. Broder, J. M. The SARS epidemic: The American response; aggressive steps, and luck, help U.S. avoid SARS brunt. *New York Times*, May 5, 2003, A1.

20. Wilgoren, J. From China Sea to U.S. campus, quarantined. *New York Times*, May 12, 2003, A1.

21. Centers for Disease Control. *Severe acute respiratory syndrome: Public health guidance for community-level preparedness and response to severe acute respiratory syndrome (SARS) version 2.* Washington, DC: Department of Health and Human Services, January 8, 2004.

22. Person, B., F. Sy, K. Holton, B. Govert, and A. Liang. Fear and stigma: The epidemic within the SARS outbreak. *Emerging Infectious Diseases* 10, no. 2 (February 2004): 358–363.

23. Murphy, D. E., and K. W. Arenson. The SARS epidemic; precautions; students in SARS countries banned for Berkeley session. *New York Times*, May 6, 2003, A12.

24. We have been warned. *Nature* 424 (July 10, 2003): 113.

25. World Health Organization. *Avian influenza: Assessing the pandemic threat.* WHO/CDS/2005.29. Geneva: World Health Organization, January 2005.

26. Kaiser, J. Pandemic or not, experts welcome Bush plan. *Science* 310 (November 11, 2005): 952–953.

Further Reading

Centers for Disease Control Web site, <http://www.cdc.gov>.

• A source for information about SARS and other infectious diseases

Fidler, D. P. SARS: Governance and the globalization of disease. New York: Palgrave Macmillan, 2004.

• Provides a comprehensive overview of the governmental responses to the SARS outbreak

Limiting Research in an Age of Bioterrorism

Advances in molecular biology and genetics have led to remarkable developments in medicine, agriculture, and industry. Yet as the ability to manipulate the genomes of bacteria and viruses increase, so does the potential for terrorists to create biological weapons using that information. Given this threat, should limits be placed on the conduct of research or its publication? Can knowledge be contained?

Bioterrorism Research Funding

Even before the terrorist attacks of 2001, the federal government began to increase funding for research related to bioterrorism [1]. The 2001 bioterrorism attack using anthrax-laden letters merely highlighted the need to understand the biology of pathogens, develop better treatments and vaccines against them, and improve methods of diagnosis. The Bioterrorism Preparedness and Response Act of 2002 (BPRA) called for increased funding for bioterrorism-related basic research. The NIH funding targeted to bioterrorism increased sixfold from fiscal year 2002 to fiscal year 2003, to $1.75 billion [1]. In his 2003 State of the Union Address to Congress, President Bush proposed Project BioShield, a $5.6 billion initiative to expedite DHHS review and procurement of bioterrorism countermeasures [2]. The initiative provides financial incentives for pharmaceutical companies to develop vaccines, antibiotics, and antiviral agents to treat pathogens. If countermeasures are developed, the U.S. government promises to purchase them and maintain stockpiles for a possible bioterrorism event. Included in the proposed legislation is the power to permit the emergency use of countermeasures that lack FDA approval (see chapter 5). Congress appropriated close to $900 million in discretionary funding for bioterrorism research in the homeland security spending bill for fiscal year 2004, while the bill was still under consideration [3]. Both houses of Congress approved measures funding Project BioShield by summer 2004. The DHHS awarded $350 million to eight new biodefense study centers in fall 2003 [4]. Critics suggest that the targeted program will compromise other research programs, and that the development of new vaccines and other measures is not guaranteed (see chapter 2) [5]. In March 2005, a letter signed by over 750 microbiologists was sent to Elias Zerhouni, the NIH director, expressing concern that funding targeted at biodefense threatens public health by limiting funding for research on agents that present known public health risks, but have only limited potential as bioweapons [6].

In October 2005, the Senate Committee on Health, Education, Labor, and Pensions passed a bill on to the full Senate that includes a proposal to create the Biomedical Advanced Research and Development Agency (BARDA) to fund research into countermeasures for bioterror agents as well as natural agents such as the avian influenza [7]. BARDA is modeled on the DOD's Defense Advanced Research Projects Agency, which funds high-risk and high-payoff research (see Chapter 2). Scientific groups express concern about the duplication of effort already at the NIH and the CDC. Whether the proposal passes the full Senate is uncertain.

At the same time, concerns about the misapplication of information led to tighter regulations over the use, handling, and transfer of pathogens and the increased monitoring of research. The BPRA required all laboratories working with "select agents and toxins" to register with the federal government. Background checks of all researchers are mandated; researchers from countries on the federal "watch list" are barred from participating in sensitive work [8]. Imprisonment of up to five years and fines of $500,000 might be levied for noncompliance. Changes in immigration law also tightened rules for the granting of visas to students and workers from watch-list countries [9].

"Sensitive" Research and Calls for Regulation

A series of research papers published in leading journals in 2001 and 2002 heightened the dialogue over whether some kinds of research ought to be blocked. In February 2001, a group of Australian researchers reported the construction of a genetically engineered mousepox, a virus related to smallpox, that was capable of killing mice previously resistant to the infection [10]. The article described how an engineered virus might be made that was capable of overcoming natural resistance to a disease—making the virus more virulent. Two papers published in 2002 provided additional examples of information that could be misused by terrorists. The first described the complete chemical synthesis of poliovirus, suggesting that viruses might be generated independent of a biological system [11]. The paper received heavy media coverage, much of it sensational, claiming that the authors supplied the world with a recipe for a weapon. Poliovirus is an unlikely target for bioterrorism, however, given that much of the world's population is vaccinated against it [12]. Nevertheless, the paper demonstrated that the artificial creation of a virus is possible. The second paper identified the critical mutation in a gene in the avian influenza virus of 1997 that was responsible for its high level of lethality; of its eighteen human victims in Hong Kong, six died [13]. Critics suggested that this information might allow terrorists to construct a new lethal strain of influenza.

Scientists and policymakers called for increased discussion on how to deal with sensitive information [14, 15]. At issue is the balance between the free exchange of scientific information and national security. Scientists argue that the open exchange of information makes the world safer, rather than less secure. The U.S. government historically embraced this philosophy, and has been strongly supportive of unrestricted scientific inquiry, even during the height of the cold war. In his National Security Decision Directive 189, President Reagan stated in 1985 that fundamental research should remain unrestricted to the maximal extent possible—a position affirmed by President Clinton in 1995 [14]. The presidential order states that research should either be classified or freely available. Yet the Bush administration has expressed conflicting opinions, both supportive of open research and urging more control [9].

The American Society for Microbiology developed guidelines regarding the publication of microbiological information in its journals that were in line with the society's code of ethics [16]. It asked reviewers to express concerns to journal editors about the potential misuse of microbiology described in a manuscript [9]. Despite this, in summer 2002, a resolution introduced in the U.S. House of Representatives in response to the *Science* article on poliovirus called on the administration to examine its policies concerning the classification or publication of sensitive research [17]. The resolution also asked the publishers and editors of journals to establish ethical standards for published material to assure against aiding bioterrorists, and the scientific community to exercise restraint.

The National Research Council (NRC) convened a committee to review the current regulations on sensitive biotechnology research, assess their adequacy, and develop recommendations for changes to current practices, and released its report in October 2003 [18]. While embracing the idea that access to pathogens should be limited, the committee urged against overly restrictive legislation that might unintentionally hamper important research by legitimate researchers into the biology of dangerous pathogens. The committee recognized that most researchers have limited experience with issues of "dual-use" research, and have not considered the potential risks that their research might pose. Scientists have an "affirmative moral duty" to avoid contributing to biological warfare or bioterrorism [18].

The NRC report recommended a system of self-regulation by researchers and an expansion of oversight by the NIH, modeled after the Recombinant DNA Advisory Committee developed to oversee research on recombinant DNA in the 1970s (see "The Asilomar Conference" section) [18]. Each research institution would establish an institutional biosafety committee (IBC) that would review proposals, as is already done for research involving human or animal subjects. The NRC committee outlined seven types of research that might trigger review, including experiments that would demonstrate how to render a vaccine ineffective, confer resistance to antibiotics or antiviral drugs, enhance the virulence of a pathogen or other agent, increase the transmissibility of a pathogen, increase the number of hosts for the agent, allow for the evasion of diagnostic procedures, or allow for the weaponization of the agent [18]. Proposals approved by the IBC ordinarily would not require further review, but an IBC could request an NIH review to examine proposals for which the IBC had concerns. Further oversight would be provided by the National Science Advisory Board for Biodefense, announced by the DHHS in March 2004 and finally convened in June 2005.

The NRC committee further recommended that the scientific community practice self-governance regarding appropriate publication practices [18]. Government-mandated limits on publication might have a chilling effect on critical research on bioterrorism. While the committee recognized that the wide dissemination of results might offer opportunities to bioterrorists, it argued that the open publication of experimental results is necessary for the confirmation and challenge of data. Publication without a description of methods, for example, would make replication of the experiment impossible. Most important, the NRC committee recognized that no system of self-regulation would be effective without a broad level of international cooperation. It recommended that an international forum on biosecurity be developed to coordinate the responsible dissemination of information on a global scale. Such a meeting might be sponsored by the WHO, the United Nations Educational, Scientific, and Cultural Organization, or other nongovernmental organizations.

Critics suggested that the NRC report, while a first step, did not go far enough in guarding against the release of sensitive information. Since participation is voluntary, privately funded researchers in industry might choose not to cooperate [19].

Others worry that an open review process might inadvertently release information useful to terrorists [20]. In May 2005, the DHHS asked the *Proceedings of the National Academy of Science*, the journal of the National Academy of Science (NAS), not to publish an article that modeled a possible bioterror attack using botulinum toxin in the domestic milk supply [21]. After meetings with government officials, the NAS decided to proceed with publication in July, arguing that the article would not significantly aid a terrorist and instead provided useful information to combat a possible attack [22].

Lessons from a Changing Situation

The development of guidelines and regulations for sensitive research is still in progress. Most government response to date is directed toward preventing researchers from unfriendly countries from participating in research in the United States [9]. Recent DOD regulations, however, call for prior approval to publish or share information from certain DOD-funded projects, and seek to limit such sharing to researchers in North Atlantic Treaty Organization countries only [9, 23]. What remains clear is that controlling information on a national basis will be ineffective. Science is a global enterprise, and without the cooperation of scientists around the world, the spread of useful information to bioterrorists cannot be prevented.

References

1. Fauci, A. S. Bioterrorism: Defining a research agenda. *Food and Drug Law Journal* 57 (2002): 413–420.

2. Gottron, F. *Project BioShield*. RS21507. Washington, DC: Congressional Research Service, July 23, 2003.

3. Vastag, B. Project BioShield on track. *JAMA* 290, no. 14 (2003): 1844.

4. Malakoff, D. U.S. biodefense boom: Eight new study centers. *Science* 301 (September 12, 2003): 1450–1451.

5. Miller, J. D. Bioterrorism research: New money, new anxieties. *Scientist* 17, no. 7 (April 7, 2003): 52.

6. Altman, S., et al. An open letter to Elias Zerhouni. *Science* 307 (March 4, 2005): 1409.

7. Kaiser, J. Critics question proposed countermeasures agency. *Science* 310 (November 4, 2005): 755.

8. Bienstock, R. Anti-bioterrorism research: Laboratory security and safety under the bioterrorism preparedness and response act. *University of New Mexico*. July 2003. Available at <http://www.unm.edu/~ripls/summaries/1153.html> (accessed April 23, 2004).

9. Kellman, B. Regulation of biological research in the terrorism era. *Health Matrix* 13 (Winter 2003): 159–180.

10. Jackson, R. J., A. J. Ramsay, C. D. Christensen, S. Beaton, D. F. Hall, and I. A. Ramshaw. Expression of mouse interleukin-4 by a recombinant ectromelia virus suppresses cytolytic lymphocyte responses and overcomes genetic resistance to mousepox. *Journal of Virology* 75, no. 3 (February 2001): 1205–1210.

11. Cello, J., A. V. Paul, and E. Wimmer. Chemical synthesis of poliovirus DNA: Generation of infectious virus in the absence of natural template. *Science* 297 (August 9, 2002): 1016–1018.

12. Block, S. M. A not-so-cheap stunt. *Science* 297 (August 2, 2002): 769.

13. Seo, S. H., E. Hoffman, and R. G. Webster. Lethal H5N1 influenza viruses escape host anti-viral cytokine responses. *Nature Medicine* 8, no. 9 (September 2002): 950–954.

14. Atlas, R. M. National security and the biological research community. *Science* 298 (October 25, 2002): 753–754.

15. Freedom of information. *Nature Medicine* 8, no. 9 (September 2002): 899.

16. American Society for Microbiology. Policy guidelines of the publications board of the ASM in the handling of manuscripts dealing with microbiological sensitive issues. Available at <http://www.asm.org/misc/Pathogens_and_Toxins.shtml> (accessed April 23, 2004).

17. H. Res. 514, 107, 2nd sess., July 26, 2002. Available at <http://www.fas.org/sgp/congress/hres514.html> (assessed July 7, 2004).

18. National Research Council. *Biotechnology research in an age of terrorism*. Washington, DC: National Academy Press, 2004.

19. Miller, J. D. National Academy proposes scientists self-police. *Scientist* 4, no. 1 (October 9, 2003): 4.

20. Malakoff, D., and M. Enserink. Researchers await government response to self-regulation plea. *Science* 302 (October 17, 2003): 368–369.

21. Wein, M. L., and Y. Liu. Analyzing a bioterror attack on the food supply: The case of botulinum toxin in milk. *Proceedings of the National Academy of Sciences U.S.A.* 102, no. 28 (July 12, 2005): 9984–9989.

22. Alberts, B. Modeling attacks on the food supply. *Proceedings of the National Academy of Sciences U.S.A.* 102, no. 28 (July 12, 2005): 9737–9738.

23. Malakoff, D. Reports examine academe's role in keeping secrets. *Science* 304 (April 23, 2004): 500.

Further Reading

National Research Council. *Biotechnology research in an age of terrorism*. Washington, DC: National Academy Press, 2004.

• A discussion of the issue of sensitive research and recommendations for its oversight

10

Science Misunderstood: Genetically Modified Organisms and International Trade

International law and commerce is a complex and dynamic dance of national interests and global needs. Both producer and consumer nations must hammer out agreements that permit the effective and equitable exchange of goods and services. There are occasions where interests, beliefs, and values come into sharp conflict, however. A current example is the international marketing of genetically modified (GM) crops and foods. Many countries express reluctance to allow the United States to market crops and food products made from GM organisms within their borders. They cite undetermined risks to human health and the environment. Although no product can ever be determined to be without risk, considerable scientific evidence suggests that GM-generated products carry no unique risks compared to traditionally produced crops and foods. Most countries do not explicitly ban GM products, but have approval procedures that effectively block any imports. What is the justification for such actions?

Who Uses GM Crops?

By far the largest producer and consumer of GM crops is the United States. In 2002, the United States was responsible for 68 percent of the world's production, followed by Argentina (23 percent), Canada (6 percent), and China (4 percent) [1]. Forty percent of maize, 81 percent of soybeans, 65 percent of oilseed rape, and 73 percent of cotton grown in the United States are GM strains [2]. The FDA, the U.S. Department of Agriculture (USDA), and the EPA regulate the production of GM crops, which require extensive testing prior to the marketing of seed and some continued monitoring of crops after approval. Public acceptance of GM foods is generally high, although there have been calls to label all foods containing GM products.

Yet the opposition to GM crops is growing. Vermont enacted legislation in 2004 that requires labeling of all GM seeds [3]. Also in 2004, Mendocino County, a coastal wine-growing area in California, outlawed GM crops, even though none are currently grown there [4]. Concerns that growers might be unable to market wines in Europe led to the initiative.

Resistance to GM products is high in Europe, and in much of Asia and Africa. Public opinion remains strongly against the introduction of GM crops and the consumption of products, leading governments to move slowly on approving their use. The reluctance of other countries to accept GM products has led to a virtual trade war between the United States and its trading partners abroad.

The Biology of GM Crops

GM crops are agricultural plants (and animals) that have been modified using genes from other species to express a characteristic such as pest resistance, herbicide resistance, or an altered nutrient (including pharmaceuticals) content. Humans began to manipulate the traits of crops with the rise of agriculture ten thousand years ago. Crops such as maize (corn) and wheat were selectively bred for particular characteristics using laborious methods of cross-fertilization and selection. Seeds from the plants with the desired characteristics, such as increased yield or better taste, would be saved to plant the next crop. This process occurred without any knowledge of molecular genetics. Other "traditional" breeding approaches include the generation of new mutant strains by exposure to chemicals or radiation (induced mutation) or by forced hybridization between related species [5]. Modern "natural" crops bear little resemblance to the original unmodified plants. The same can be said of dogs, the many breeds of which have been created over time by selective and purposive manipulation.

What sets GM organisms apart from these traditionally bred crops is the use of recombinant DNA technology to selectively insert desired genes into the genome of the plant or animal (see Chapter 2 for more details) [5, 6]. Rather than taking the traditional scattershot approach to plant breeding, researchers select a particular gene that carries the desired trait and insert it into the DNA of the plant. Those plants that express the particular characteristic along with their normal traits may be used as parental stock for later generations. Since genes may be derived from other species, the traditional methods of crossbreeding are not able to generate such a transgenic organism. This characteristic leads critics to argue that the resulting product is "not natural."

GM organisms are not new species. They retain all the characteristics typical of the parental plant—GM corn is recognizably corn. Yet since scientists cannot control where the transgene is inserted into the host genome, it is possible that critical gene function might be altered by the introduction of the foreign gene. GM organisms that show significant alterations in the resulting plants are discarded after testing. Although a popular image in the media, new GM strains are not a bizarre melding of traits from different organisms.

Why Make GM Crops?

With the need to feed an ever-growing human population, crop losses represent a major problem. Crops may be lost to many pests, including insects, weeds, and pathogens. Although obtaining reliable data is difficult, estimates suggest that from 20 to 50 percent of crops are lost annually to pests [7]. While a variety of crop management approaches are used to control pests, pesticides and herbicides became the central strategy in the 1950s. Nevertheless, the use of chemical controls has many adverse consequences, including environmental pollution, loss of beneficial insect species and a reduction in biodiversity, human toxicity, and the development of resistance in pests. A shift from crop rotation to monoculture (planting a single strain of a food crop), associated with the move from small family farms to huge "factory farms," also increases crop losses, despite the increased use of chemical pesticides [7].

GM crops are one of several strategies aimed at increasing crop yields. Other approaches include the mixed cultivation of crops or crop rotation, a shift to "no-till" farming to reduce soil erosion, the use of biological control agents such as natural enemies of pests, and the more carefully timed application of new, less toxic pesticides [7]. In principle, the use of GM crops can help reduce the application of chemical pesticides by engineering pest resistance into the plants themselves. In addition, drought-tolerant strains of crops might be developed to grow in arid regions or where salt levels prevent the cultivation of traditional crops [5, 6].

Crops may also be modified to improve their nutritional content, thereby improving the health of those who consume them. Others propose to develop GM crops that produce important drugs, allowing less expensive manufacture and wider availability [5, 6].

Current GM Crops

Most crops are not directly used as human food but rather as animal feed, sources of oils, or components of processed food. Cotton was one of the first GM crops because it is a highly lucrative crop that is not eaten. The majority of GM crops currently grown are strains of maize, soybeans, cotton, and oilseed rape (canola) engineered to be resistant to either pests or herbicides [6].

Maize and cotton are particularly vulnerable to attack by insect borers. These plants are engineered to express a gene encoding one of the toxins (the Cry proteins) from a soil bacterium, *Bacillus thuringiensis* (Bt), that kills insect pests. Bt is a popular "natural" pesticide used by organic gardeners; it has been marketed for decades. By engineering plants to express the Bt toxin, less chemical pesticide is needed to control pests [5]. Recent data from India indicate that the cultivation of Bt cotton significantly reduces pesticide use while increasing yields by 80 percent [8].

Weeds can compete for soil nutrients and reduce crop yields. Chemical herbicides may be used to reduce weeds, but they must be selected so as not to damage the crops themselves. Different crops can tolerate different herbicides, so farmers must use several different chemicals if they cultivate more than one crop. Some herbicides persist in the soil, making crop rotation difficult. Some are toxic to animals, including humans. In order to reduce the need for repeated herbicide applications to control weeds, plants are engineered to be resistant to a common herbicide, glyphosate (Roundup), which is not toxic to animals [5]. Glyphosate-tolerant (Roundup Ready) strains of soybeans, cotton, oilseed rape, maize, and beets are in use. In some cases, a single treatment of herbicide is sufficient to control weeds for a season. Strains of crops engineered to tolerate other herbicides also are available [5].

Another approach is introducing virus and fungal resistance in crops such as potatoes, sweet potatoes, and papayas. Emergent plant viruses and fungi threaten to decimate these valuable crops. The development of a GM virus-resistant strain of papaya in 1998 is credited with saving the papaya industry in Hawaii, the state's second-largest fruit crop [9].

The most recent approach to GM crops is to modify the nutritional content of crops. Plant oils are a valuable commercial product, and crops are grown as a source of plant oils for both industrial and nutritional uses [5]. Plant oils contain a mixture of fatty acids, some of which can be toxic. For example, oilseed rape was not used as a food crop until traditional breeding methods produced a strain that had low levels of toxic fatty acids; this strain is now called canola. New GM strains of oilseed rape are now available that contain higher levels of desirable fatty acids; farmers can sell the produce at a premium price because of its desirability to end users [5]. New GM strains of soybeans with a healthier oil content also sell at higher prices.

Perhaps the most publicized example of a GM plant with an altered nutritional content is "golden rice," which contains higher levels of beta-carotene, the precursor to vitamin A [5]. Golden rice was developed in 2001 as a potential solution for vitamin A deficiency, a common malady in people who consume rice as their primary diet. Deficiencies can lead to blindness as well as an increased vulnerability to diarrheal and respiratory diseases. Golden rice is not yet marketed because the GM plant still needs to be crossed with commercial strains of rice to produce a strain that can be grown in the field. Developing a practical strain may take many more years of work [10].

Are GM Foods Safe to Eat?

Critics of GM foods suggest that they pose unique risks to consumers. No food, regardless of its derivation, is completely safe. Many commonly consumed traditional foods, such as soybeans and potatoes, contain compounds that are toxic or allergenic. Such compounds most likely became concentrated in food crops because they protected the plants against predation. Generations of breeding allowed for the selection of less toxic strains of crops, but traces still remain. Crossbreeding for desired characteristics may inadvertently increase toxin levels; GM crops also may express altered levels of plant constituents.

Some plants are known to trigger allergic reactions; these include nuts, wheat, and legumes such as soybeans and peanuts [11]. Other common allergens include cow's milk, fish, shellfish, and eggs. About 2 percent of adults and 4 to 6 percent of children are allergic to one or more foods [12]. Given the significant population of possibly susceptible individuals, does introducing foreign genes or altering endogenous levels of molecules increase the risk of allergic reaction? Critics of GM foods often cite a gene transfer from Brazil nuts into soybeans in order to increase the protein content as an example of the irresponsible manipulation of food. In this case, the soybeans containing the seed storage protein from Brazil nuts quickly were found to be allergenic, and the project was dropped [12].

What is less certain is whether Bt toxins or other introduced genes might prove to be allergenic. The evidence to date suggests that pesticide-engineered and herbicide-resistant crops do not pose an increased risk of allergy [11, 13]. For example, the Cry proteins of Bt toxin share some structural features with known allergens, raising the concern that consuming the protein from a GM crop could pose a hazard. In fall 2000, StarLink (Aventis Corp.), a GM strain of Bt corn approved for animal feed and industrial use only, was found in the human food supply [14]. The corn had inadvertently been commingled with other non-GM strains, and used in taco shells and other products. The event received heavy press coverage; the FDA received reports of fifty-one cases of possible allergic reactions. The CDC investigated the cases and tested individuals for evidence of antibodies to the Cry protein. No antibodies against Cry9c were found [14]. The CDC concluded that the allergic reactions in these individuals were not associated with the GM product.

Others suggest that the vectors used to transfer genes into plants might be dangerous themselves. In 1998, Arpad Pusztai, a researcher in the United Kingdom, announced to the media that rats fed a diet of raw GM potatoes expressing a lectin from snowdrops (a known toxin and natural pesticide) showed more changes in body organs than rats fed unmodified potatoes with added lectin. GM critics seized on his findings as evidence that the process of genetic modification itself posed a hazard [15]. But the rats fed GM potatoes also suffered from protein deficiency, since the modification reduced protein levels in the potatoes. Most scientists criticized Pusztai's conclusions, and his findings were discredited [16]. The NRC released a report in 2000 that concluded that the methods used to produce GM crops did not pose any novel health hazards [17].

The accumulated evidence suggests that while some GM products carry potential hazards, these are no different than the ones present in non-GM products. It is not possible to rule out problems that may occur after long-term consumption; the same may be said for non-GM foods.

Are GM Organisms Hazardous to the Environment?

Ecologists express concerns that GM organisms may pose a risk to the environment. Among their concerns are: GM plants engineered to express pesticides may have negative effects on beneficial species of insects; pesticide- and herbicide-resistance genes may "jump" to weedy relatives, producing "superweeds" that cannot be controlled by conventional methods; the widespread use of GM crops may select for resistant strains of pests, rendering the crops ineffective; GM organisms may become invasive and outcompete native plants; and the use of GM crops will reduce biodiversity [18]. How significant are these risks?

In 1999, a group of Cornell University researchers reported that in a laboratory experiment, pollen from Bt corn sprinkled on milkweed plants was highly toxic to monarch butterfly larvae [19]. Subsequent field experiments found that Bt corn pollen was carried no more than ten meters away from the field. While most larvae died after exposure to higher concentrations of pollen, more survived low-level exposure [20]. These findings demonstrate that Bt pollen can indeed have harmful effects on desirable insects. The results also suggest a solution: that "buffer zones" of unmodified corn might be planted along the fringes of GM crop fields to prevent the spread of GM pollen to adjacent areas.

Many heavily used crops have weedy relatives with which they might hybridize and transfer genes [18, 21]. Hybridization occurs naturally between species and is not a problem limited only to GM organisms. For example, hybridization can occur between two strains of crop, altering the oil content in undesired ways. Recent data suggest that hybridization between non-GM oil rapeseed and its weedy cousin *B. rapa* can occur under a variety of natural conditions, and that physical separation alone may be insufficient to prevent hybridization [22]. Strategies to limit hybridization include buffer zones, as suggested above, and the genetic manipulation of the GM crop to reduce gene flow via pollen, by rendering the male plants (which produce pollen) sterile [21]. The risk of transfer of herbicide or pest resistance from herbicide-resistant GM plants to undesired weedy plants is genuine yet difficult to assess. Traditional breeding methods may also pose a similar risk, however.

There is already evidence that resistance to common herbicides and pesticides is developing, and is not limited to GM crops. As a result of natural selection (not gene transfer), the development of resistance to herbicides and pesticides by weeds is inevitable [5]. What is remarkable about Round-up, a popular herbicide in wide use for nearly thirty years, is that resistance has emerged only recently [23]. Resistant strains of weeds have been found, mostly associated with GM plants, in several U.S. states. Although currently limited in area, the spread of herbicide-resistant weeds might limit the value and usefulness of GM crops engineered to tolerate Roundup. There is also evidence that insects are becoming resistant to Bt. Farmers try to slow the development of resistance by surrounding fields of Bt crops with non-Bt crops, to allow insects that might be selected for resistance to Bt to breed with those that have been feeding on Bt-free crops [24, 25]. Yet this strategy is not proving as effective as initially hoped. In addition, some insects may feed on more than one crop, making buffer zones ineffective [26]. The careful use of herbicides and pesticides may slow the development of resistance while new treatments are developed.

Another area of concern is whether the antibiotic resistance markers used in vectors transferring genes may escape from GM plants into other microorganisms, generating new populations of antibiotic-resistant bacteria. A common strategy when introducing genes into a host is to include a gene for resistance to an antibiotic such as neomycin. The resistance gene serves as a marker that the desired gene was introduced successfully into the new host organism. Bacteria mutate rapidly and readily exchange genetic information. Recent evidence suggests that such transfer is unlikely between GM soybeans and bacteria that reside in the human digestive tract [27]. Nevertheless, because the biocontainment of bacteria is difficult, the FDA recommends as a precaution that developers of GM organisms avoid using antibiotics employed for clinical purposes as resistance markers [11].

Whether GM crops have the ability to become invasive depends on the particular plant. An invasive species is one that when introduced in a new ecosystem, effectively outcompetes the native organisms present in the environment and alters the balance of other organisms within the ecosystem. Since most plant crops have been bred to grow under specific agricultural conditions, their escape into the environment is unlikely [18, 22]. Animals are another matter. For example, salmon and other fish are now commonly farm raised. These fish, whether GM or non-GM, are bred for more rapid growth and development; if they escape, they may outcompete wild populations. Containment is important to prevent the loss of natural populations [21].

The effects of GM organisms on biodiversity are currently under study. Field tests of GM oilseed rape and beets found a reduced diversity of weeds, insects, and seeds in GM fields compared with non-GM ones [28]. The GM plants were resistant to herbicides—and it was the herbicide treatment, not the GM plants, that reduced biodiversity. The reduction in biodiversity might also affect populations of birds and small animals that normally feed on weeds or insects found near crops [29]. Once again, the use of herbicides on non-GM crops poses the same hazards to biodiversity.

NRC committees conclude that while GM crops may present environmental risks, they are no different than those posed by genetic variations of crops generated by other methods [17, 30]. The transgenic process itself poses no new types of risk. Precautions are needed for both GM and conventional crops to limit the adverse environmental impacts, and new strains must be assessed on a case-by-case basis.

Trade Wars: What's Wrong with GM Products?

Like any other commodity, corporations wish to sell GM crops and the products derived from them to as broad a market as possible. While GM products are used heavily in the United States, other countries, especially in Europe, have resisted their introduction. What arguments have countries used to bar sale of GM crops?

European countries adhere more strongly to the precautionary principle (see also chapter 7) that argues that in the face of scientific uncertainty, the introduction of new technology should be done slowly [31]. These nations do not accept the U.S. premise that GM foods are "substantially equivalent" to non-GM foods, and contend that the scientific evidence does not rule out that the process of genetic engineering creates new risks for consumers and the environment [31, 32, 33].

Critics of GM technology often couch their concerns in scientific terms, but it is clear that their primary objections lie outside issues of biological safety. A central criticism is that GM organisms are not natural—that tinkering with the genome of our food is unethical and dangerous, and humans should not play God. Critics describe GM foods as Frankenfoods to suggest that they are horrors that might be unleashed by corporations on a helpless public. As mentioned earlier, though, the genetic manipulation of food is as old as agriculture, and the forced crossbreeding between species is nothing new [5, 15]. Nevertheless, the fear of biotechnology lies at the base of arguments that GM crops are not natural; those that espouse this position tend to disbelieve scientific information because of a lack of trust in scientists [15, 32, 34].

The development of GM organisms requires considerable expenditure in research and testing before a product makes it to the market. As a consequence, most GM organisms are made and marketed by large multinational corporations, among them the Monsanto Corporation. Distrust of the motives and ethics of large corporations is widespread; many believe that large corporations care little for the long-term effects of their products and instead only focus on profits. For example, Monsanto conducted studies on a "terminator" gene that would make it impossible for farmers to raise crops from seeds saved from the previous year's crop. Farmers would be forced to purchase new GM seeds each year, increasing Monsanto's profits at the expense of the farmers. Although the project was abandoned, probably influenced by the public outcry, critics still cite the terminator gene as emblematic of Monsanto's rapacious character [15]. In addition, Monsanto has aggressively pursued legal action against farmers who save seeds for future planting or obtain them without license, in direct violation of the seed purchase agreements and Monsanto's patent on the plants (see chapter 3) [35]. Developing countries resist the introduction of GM technology as a result of past misconduct by large corporations [2, 36].

The government resistance to GM crops is particularly emphatic in Africa, where the introduction of GM crops is strongly contested [37]. Only South Africa permits the growing of GM crops. Critics contend that corporations are seeking entrance into a lucrative market and express concerns that GM crops are not suited for African styles of agriculture. The governments of Angola, Malawi, Mozambique, Zambia, and Zimbabwe have refused to accept food aid if it contains GM seeds or products, despite severe famines in the countries [38, 39, 40].

Finally, the public's trust of its government officials is generally lower in countries other the United States [5, 32, 41]. In the aftermath of the outbreak of bovine spongiform encephalopathy (BSE, or mad cow disease) in the United Kingdom (see the following section), many members of the public simply do not believe government assurances that a GM product is safe [31, 33]. Non governmental organizations opposed to GM technology are much more active outside the United States, and further influence the public distrust of government [32, 37]. Examples of "failed" regulation in the United States, such as the release of StarLink maize into the human food supply as well as a second incident in which another unapproved transgenic Bt maize marketed by Syngenta was commingled with approved corn between 2001 and 2004, are cited as demonstrations that the public cannot rely on government for protection [5, 42].

The Regulation of GM Crops in the United States

The U.S. approach to GM crops is based on the policy that it is the end product, not the means of genetic modification, that is to be regulated; the U.S. government believes that GM crops are substantially equivalent to those derived by traditional agricultural methods. As noted above, the FDA, the USDA, and the EPA share oversight of GM crops and products. Each agency has a different responsibility: the FDA assures that the food is safe to eat; the USDA is responsible for ensuring that GM crops are safe to grow; and the EPA determines any risk to the environment [43]. The FDA uses the food safety provisions spelled out in the Federal Food, Drug, and Cosmetic Act (FFDCA, 21 USC §9) that governs food additives and adulteration.

If GM crops contain additives previously determined to be safe (or "generally recognized as safe"), then less testing is needed. In 2001, the FDA moved from a voluntary to mandatory premarket notification by manufacturers of GM organisms [43]. Manufacturers must provide evidence that GM plants are similar in chemical composition to unmodified crops, and show no evidence of potential toxicity or allergenicity. The FDA, however, does not spell out a methodology for testing food components [44].

The USDA's Animal and Plant Health Inspection Service (APHIS) regulates the movement, importation, and field testing of GM organisms. The Agricultural Risk Protection Act of 2000 mandates the protection of agricultural crops and extends the APHIS's authority to regulate GM plants [45]. Until approved for commercial release, GM crops are tightly regulated. Field trials must be approved by the APHIS to assure that the new GM crops are safe to grow, and will not escape into the environment and become plant pests [46].

The EPA monitors GM crops under the auspices of the Federal Insecticide, Fungicide, and Rodenticide Act (7 U.S.C. §136), and focuses on GM crops engineered for pesticide expression ("plant pesticides") [43]. The EPA sets limits on acceptable levels of pesticides and mandates that crops be grown in association with non-GM strains to limit potential environmental damage.

How effective is regulation? A GAO report in 2002 concluded that FDA oversight was generally acceptable, but recommended that the FDA spot-check industry raw data for accuracy and provide more detailed explanations for its decisions to the public [47]. An NRC study concluded that the approval process by the APHIS needs improvement, and recommended postcommercialization testing and monitoring [30]. The council also felt that the environmental monitoring of both GM and non-GM crops was inadequate, and suggested stronger oversight and field testing. A report by the Office of the Inspector General of the USDA in December 2005 was critical of the APHIS's oversight of field trials, arguing that adequate controls to prevent negative environmental impacts of GM plants were not in place [48].

Although there have been increasing calls by the public for the FDA to require the labeling of GM-containing food products, neither the FDA nor Congress has taken this action. The FDA contends that the mode of production of a new strain of food is not relevant—only the safety of the end product matters [49]. Labeling is required only if the product differs substantially from the traditional food.

Manufacturers may choose to avoid using GM components in order to claim that their food is "GM free." For example, baby food manufacturers may choose not to use any GM components in response to consumer demand. In 2001, the FDA issued guidelines warning manufacturers that labels must not be misleading—they may not imply that a product is superior because it was made without GM components, nor may they market a food as GM free when there is no GM counterpart product [49]. The exception to these guidelines are foods that may be certified as organic; USDA guidelines indicate that organic foods may not contain products that have been derived from bioengineered crops (National Organic Program Final Rule, 65 FR 80548) [49].

Whether the FDA's rules on labeling are deemed sufficient by the public remains unclear. Several bills requiring labeling have been introduced in Congress, but have failed to pass. Similar legislation failed in Oregon in 2002. The FDA acknowledges that its own focus groups indicate that the U.S. public desires that GM products be labeled to indicate that they were bioengineered [50].

The Regulation of GM Organisms in Europe
Unlike the United States, the regulation of GM products by the European Union (EU) is a cooperative activity between individual countries and more general EU agreements. The EU requires that each GM organism or product containing them be subject to a case-by-case assessment of the risks to human health and the environment [51]. Directives issued by the EU require a much more strict review of each GM organism prior to approval and mandatory postmarketing monitoring. Individual countries may block the approval of a GM product by invoking a safety clause contained within Directive 90/220/EEC (Article 16). From 1998 to mid-2004, the EU did not approve any new GM organisms or products by declaring a moratorium on reviews—a de facto ban on GM products. Under Directive 2001/18/EC, the deliberate release of GM organisms into the environment must be accompanied by clear evidence of their safety, and members argue that this has not been demonstrated [31]. In 2003, the EU passed stringent new labeling requirements to take effect in 2004; all GM food or animal feed must be labeled regardless of whether the GM material can still be detected [52, 53]. In 2004, however, the EU appeared poised to change its policies and approve the import of at least some GM products (see below). Approval of new GM products continues to lag, with over twenty applications unresolved as of spring 2006 [54, 55].

In the United Kingdom, oversight responsibility for GM organisms is divided between a number of advisory committees; each has fragmented authority over different aspects of GM organisms [32]. The United Kingdom moved slowly to approve GM crops, in the face of strong public opposition. In March 2004, the British government announced limited approval for the commercial planting of herbicide-resistant maize [56].

International Trade Policy and Other Agreements

A number of international trade agreements bear on the conflict between the United States and its trading partners over GM products. International trade agreements are intended to provide open markets while preserving the rights of individual nations. The World Trade Organization (WTO), founded in 1995, is a global organization dealing with the rules of trade between nations; it includes 147 member nations accounting for nearly 97 percent of all world trade [57]. The WTO is the successor to the General Agreement on Tariffs and Trade (GATT) organization, formulated after World War II. The policies it developed (known as the GATT agreements) are still in force, and serve as a framework for developing new trade agreements. The WTO also plays a role as arbiter to help resolve trade disputes between its members.

GATT requires members to give equal treatment to exports from all other members, and bars countries from discriminating between locally produced and imported products [58]. The WTO promulgated two additional agreements that have an impact on GM organisms: the Agreement on the Application of Sanitary and Phytosanitary Measures, which regulates steps to protect the health of humans, animals, and plants; and the Agreement on Technical Barriers to Trade, which covers packaging, marking, and labeling requirements [58]. The agreements state that members have a right to bar products only if objections are based on scientific principles and sound scientific evidence. Countries are not permitted to bar imports that are similar or identical to local products, or when steps taken represent a disguised restriction of international trade. Nevertheless, in the absence of clear scientific evidence of safety, countries may choose to protect health based on the precautionary principle.

Adding to the complexity is the recent enactment in 2003 of the Cartagena Protocol on Biosafety (the Biosafety Protocol), a codification of safety principles outlined in the Convention on Biological Diversity [58]. The convention, finalized in 1992, is aimed at developing a broad approach to the conservation of biodiversity, the sustainable use of natural resources, and benefit sharing [59]. The Biosafety Protocol provides that parties will ensure that the development, handling, transport, use, transfer, and release of GM organisms protect biological diversity and health. Risk assessment must be carried out prior to the release of GM organisms, and signatory nations may establish even more stringent guidelines. Exporters must have "advance informed agreements" with importers before any GM organisms intended for release into the environment may be shipped [58]. Yet the Biosafety Protocol also recognizes that prior international agreements (such as GATT and WTO agreements) may not be abrogated [58, 59].

The United States signed the Convention on Biological Diversity in 1992, but Congress failed to bring it to a ratification vote; this barred the United States from being a party to the Biosafety Protocol. A central objection was to the restrictions on intellectual property and patenting (see chapter 3) [60]. The Biosafety Protocol was signed by over 130 countries in 2000, and enacted by close to 100 by August 2003 [61]. It entered into force in September 2003.

The Biosafety Protocol does not deal with food safety, which is handled by a joint commission, the Codex Alimentarius Commission, of the Food and Agriculture Organization (FAO) and WHO of the United Nations. The commission includes a task force to study the issues surrounding foods derived from biotechnology. In May 2004, the commission concluded that GM foods are safe to eat and urged that efforts be made to develop crops that could feed hungry people in developing countries [62, 63]. This ruling weakens arguments against GM products used by many countries to justify their restrictive policies.

Trade Wars and Lawsuits
The EU has refused to allow GM imports by claiming that they are not "like products" as defined by GATT; countries argue that genetic modification means that the products are indeed different from unmodified organisms [31]. In addition, the public disapproval of GM products contributes to the perception that they are not like products.

In 2003, the United States and other GM producers initiated dispute settlement proceedings with the WTO against the EU and other trading parties for violation of GATT and WTO agreements [64]. The complaints cover three areas: the general moratorium on the consideration of applications for approval of GM products; the failure to consider specific applications under the EU's own legislation; and post-marketing bans imposed by EU member states. The WTO announced an interim ruling in favor of the United States in February 2006 [55]. Most analyses suggest that the United States and other GM producers will be successful, since the EU's scientific arguments against the safety of GM organisms are no longer tenable [52, 58, 64]. The EU's self-imposed moratorium on consideration of GM products, begun in 1998, expired in 2003. The EU approved its first GM product, canned GM corn, in May 2004, after its own review panel concluded that the product carried no health hazard [54]. Whether this marks the beginning of the end of the trade war with the United States will rest on whether the EU also approves GM produce in the future.

The Future of GM Organisms

Despite the likelihood of EU approval of more GM crops, corporations are restricting the range of GM products under development. Almost immediately after gaining approval in 2004 to market its Bt corn in the United Kingdom, the German company Bayer Crop Science decided not to move ahead in an uncertain market [65]. In May 2004, the Monsanto Corporation announced it was abandoning its program to develop herbicide-resistant wheat, citing consumer resistance and a limited market as the reasons [66]. Other corporations pulled back from developing other GM crops such as virus-resistant melons and pest-resistant potatoes, citing the high costs of regulation as well as consumer resistance [67]. Corporations already know that apparently consumer-oriented GM products, such as the Flavr-Savr tomato with its longer shelf life (the first GM product to reach the market in 1994), were commercial failures [67]. New product development will continue, but within a narrow range. China is working to develop its own transgenic rice to reduce pesticide use, while continuing to resist that developed by foreign companies [68].

At the same time, the Codex Alimentarius Commission of the FAO and the WHO urged the development of crops such as drought- and salt-tolerant strains of food crops that would benefit poor people in developing countries. It appears unlikely that corporations will be willing to engage in such research, because as with drugs for diseases that disproportionately affect developing world populations, it may not prove profitable. The strong resistance to GM crops in many developing countries is also a deterrent.

Although the use of GM crops is likely to spread internationally, the range of engineered organisms may remain small, and limited to those strains of maize, cotton, canola, and soybeans with the widest application. This suggests that the lofty promise of GM-based agriculture may never be met.

Lessons from GM Organisms

Given the lack of scientific evidence that GM organisms pose any unique, uncontrollable hazards, why have so many countries risked trade sanctions in order to block their introduction? Science is only one component of the political decision-making process. Science cannot serve as a "neutral arbiter" in making decisions involving risk, since decisions involving the assessment of acceptable risk always have nonscientific components [69, 70].

Governments find themselves in the awkward position of facing stiff and often violent opposition to GM products from their citizens, on the one side, and their trade obligations, on the other. Democratically elected administrations recognize that unpopular decisions may result in being voted out of office; they cannot afford to ignore public opinion. It is not surprising that their delaying strategy involves couching objections in vague scientific terms (the only acceptable arguments under GATT), citing undetermined risks to human health and the environment, and asking for additional research. Because it is never possible to conclude that any product is entirely risk free, the door remains open for continued claims of long-term risk of GM products. Delaying decisions by demanding more scientific research is a common political strategy, particularly when evidence is difficult to obtain. Similar delaying tactics have been used by the United States to block actions such as requiring reductions in fluorocarbon emissions to slow global warming or to obstruct the release of politically unpopular drugs such as the morning-after pill. Likewise, the failure of the United States to ratify the Convention on Biological Diversity was not by a simple up-or-down vote in the Senate but rather the failure to bring it to a vote before the full Senate. In principle, the Senate might bring the convention up for a vote at any time, although the current conservative makeup of the Senate makes this unlikely.

Is it possible to convince a skeptical public that GM products are safe? As a result of other public health events in the 1990s (see the following section), many citizens of the United Kingdom and other European countries no longer trust government and scientists' assurances about the safety of new products. No rational argument is likely to be persuasive unless attitudes toward authorities change. In the United States, where opposition is growing, GM products are already so entrenched that their continued use is likely despite objections. The FDA, however, may be pressured into developing some labeling systems for GM products.

The EU enacted stiff labeling regulations for GM products in 2004, and it remains to be determined whether labeling all GM-derived products as such will assuage consumer concerns. Corporations object strongly to such labeling on the grounds that it continues to suggest to the public that the products are different from unmodified ones; these objections fall on deaf ears. The next few years may determine whether the public will buy labeled GM products. The United States and other GM-producing nations may continue to file complaints with the WTO on the grounds of protectionism.

A real tragedy may unfold in Africa, where governments of famine-stricken countries are refusing to accept food aid if it contains GM seeds or, in some cases, any GM material at all. These decisions are made in order to express distaste for the United States and other developed nations, and for their global marketing policies. Sadly, such political posturing means that politicians are willing to allow their populaces to starve. While it may be possible to argue on political grounds against opening markets for GM products, the same cannot be said for playing politics with starvation. Developed countries with an abundance of food may have the luxury of debating over the merits of GM crops, but in famine-stricken countries, the need to feed people ought to trump any other positions.

References

1. James, C. Global review of commercialized transgenic crops. *ISAAA*. November 6, 2003. Available at <http://www.isaaa.org/Press_release/Briefs29-2003/GMReview_Nov6.htm> (accessed May 10, 2004).

2. Aldhous, P. Time to choose. *Nature* 425 (October 16, 2003): 655–659.

3. Pollack, A. National briefing science and health: Vermont requires labels on biotech seeds. *New York Times*, April 28, 2004, A13.

4. Holden, C. California county bans GM crops. *Science* 303 (March 12, 2004): 1608.

5. Halford, N. G. *Genetically modified crops*. London: Imperial College Press, 2003.

6. Human Genome Program. Genetically modified foods and organisms. *U.S. Department of Energy*. October 29, 2003. Available at <http://www.ornl.gov/sci/techresources/Human_Genome/elsi/gmfood.shtml> (accessed April 28, 2004).

7. Yudelman, M., A. Ratta, and D. Nygaard. *Pest management and food production: Looking to the future*. Discussion paper 25. Washington, DC: International Food Policy Research Institute, September 1998.

8. Qaim, M. and D. Zilberman. Yield effects of genetically modified crops in developing countries. *Science* 299 (February 7, 2003): 900–902.

9. Bren, L. Genetic engineering: The future of foods? *FDA Consumer* 37, no. 6 (November–December 2003): 603.

10. Nakanishi, N. Is golden rice the crop to prove GM's worth? *Reuters*. February 26, 2004. Available at <http://www.enn.com/news/2004-02-26/s_13486.asp> (accessed May 10, 2004).

11. Thompson, L. Are bioengineered foods safe? *FDA Consumer* 34, no. 1 (January–February 2000). Available at <http://www.fda.gov/fdac/features/2000/100-610.html> (accessed July 9, 2006).

12. Lehrer, S. B. Potential health risks of genetically modified organisms: How can allergens be assessed and minimized? In *Promethean science: Agricultural biotechnology, the environment, and the poor*, ed. I. Serageldin and G. J. Persely, 149–155. Washington, DC: CGIAR Secretariat, 2000.

13. Butler, D., and T. Reichardt. Long-term effect of GM crops serves up food for thought. *Nature* 398 (April 22, 1999): 651–653.

14. Centers for Disease Control. *Investigation of human health effects associated with potential exposure to genetically modified corn*. Washington, DC: Centers for Disease Control, June 11, 2001.

15. Allan, S. *Media, risk, and science*. Buckingham, UK: Open University Press, 2002.

16. Connor, S. Arpad Pusztai: The verdict; GM food: Safe or unsafe? *Independent* (London): February 19, 1999, 9.

17. National Research Council. *Genetically-modified pest-protected plants: Science and regulation*. Washington, DC: National Academy Press, 2000.

18. Wolfenbarger, L. L., and P. R. Pilfer. The ecological risks and benefits of genetically engineered plants. *Science* 290 (December 15, 2000): 2088–2093.

19. Losey, J. E., L. S. Raynor, and M. E. Carter. Transgenic pollen harms monarch larvae. *Nature* 399 (1999): 214–216.

20. Hansen Jesse, L. C., and J. J. Obrycki. Field deposition of Bt transgenic corn pollen: Lethal effects on the monarch butterfly. *Oecologia* 125 (2000): 241–248.

21. National Research Council. *Biological confinement of genetically engineered organisms*. Washington, DC: National Academy Press, 2004.

22. Wilkinson, M. J., L. J. Elliott, J. Allainguillame et al. Hybridization between Brassica napus and B. rapa on a national scale in the United Kingdom. *Science* 302 (October 17, 2003): 457–459.

23. Pollack, A. Widely used crop herbicide is losing weed resistance. *New York Times*, January 14, 2003, C1.

24. Liu, Y. B., B. E. Tabashnik, T. J. Dennehy, A. L. Patin, and A. C. Bartlett. Development time and resistance to Bt crops. *Nature* 400 (August 5, 1999): 519.

25. Clarke, T. Pest resistance feared as farmers flout rules. *Nature* 424 (July 10, 2003): 116.

26. Clarke, T. Corn could make cotton pests Bt resistant. December 4, 2002. Available at <http://www.nature.com/nsu/021202/021202-2.html> (accessed May 10, 2004).

27. Netherwood, T., S. M. Martin-Orúe, A. G. O'Donnell et al. Assessing the survival of transgenic plant DNA in the human gastrointestinal tract. *Nature Biotechnology* 22, no. 2 (February 2004): 204–209.

28. Giles, J. Biosafety trials darken outlook for transgenic crops in Europe. *Nature* 425 (October 23, 2003): 751.

29. Watkinson, A. R., R. P. Feckleton, R. A. Robinson, and W. J. Sutherland. Predictions of biodiversity response to genetically modified herbicide-tolerant crops. *Science* 289 (September 1, 2000): 1554–1556.

30. National Research Council. *Environmental effects of transgenic plants: The scope and adequacy of regulation.* Washington, DC: National Academy Press, 2002.

31. Wong, J. Are biotech crops and conventional crops like products? An analysis under Gatt. *Duke Law and Technology Review* 2003 (October 29, 2003): 27.

32. Hails, R., and J. Kinderlerer. The GM public debate: Context and communication strategies. *Nature Reviews Genetics* 4 (October 1, 2003): 819–825.

33. Compton, M. M. Applying world trade organization rules to the labeling of genetically modified foods. *Pace International Review* 15 (Fall 2003): 359–409.

34. Bond, C. Politics, misinformation, and biotechnology. *Science* 287 (February 18, 2000): 1201.

35. Simon, B. Monsanto wins patent case on plant genes. *New York Times*, May 22, 2004, C1.

36. Normile, D. Asia gets a taste of genetic food fights. *Science* 289 (August 25, 2000): 1279–1281.

37. Masood, E. A continent divided. *Nature* 426 (November 20, 2003): 224–226.

38. Cauvin, H. E. Between famine and politics, Zambians starve. *New York Times*, August 30, 2002, A6.

39. Lacey, M. Engineering food for Africans. *New York Times*, September 8, 2002, sec. 1, 16.

40. Wines, M. Angola's plan to turn away food imperils aid. *New York Times*, March 30, 2004, A3.

41. Freckleton, R. P., W. J. Sutherland, and A. R. Watkinson. Deciding the future of GM crops in Europe. *Science* 302 (November 7, 2003): 994–996.

42. Macilwain, C. US launches probe into sales of unapproved transgenic corn. *Nature* 434 (March 24, 2005): 423.

43. Marden, E. Risk and regulation: U.S. regulatory policy on genetically modified food and agriculture. *Boston College Law Review* 44 (May 2003): 733–787.

44. Krimsky, S., and N. K. Murphy. Epidemiology and science: Biotechnology at the dinner table: FDA's oversight of transgenic food. *Annals of the American Academy of Political and Social Science* 584 (November 2002): 80–95.

45. Agricultural Risk Protection Act of 2000. Pub. L. 106–224 (codified at 7 U.S.C. §§1501–1523, 7331, 7333).

46. Animal and Plant Health Inspection Service. Agricultural biotechnology. *USDA*. Available at <http://www.aphis.usda.gov> (accessed April 28, 2004).

47. U.S. General Accounting Office. *Genetically modified foods: Experts view regiment of safety tests as adequate, but FDA's evaluation process could be enhanced*. GAO–02–566. Washington, DC: General Accounting Office, May 2002.

48. Office of Inspector General. *Audit report: Animal and Plant Health Inspection Service controls over issuance of genetically engineered organism release permits*. Audit 50601–8–Te. Washington, DC: U.S. Department of Agriculture, December 8, 2005.

49. Center for Food Safety and Applied Nutrition. Voluntary labeling indicating whether foods have or have not been developed using bioengineering. *FDA*. June 22, 2001. Available at <http://vm.cfsan.fda.gov/~dms/biolabgu.html> (accessed May 24, 2004).

50. Center for Food Safety and Applied Nutrition. Report on consumer focus groups on Biotechnology. *FDA*. October 20, 2000. Available at <http://cfsan.fda.gov/~comm/biorpt.html> (accessed May 24, 2004).

51. Hedlund, E., B. Gminder, L. Mikkelsen, and C. Bunyan. Questions and answers on the regulation of GMOs in the EU. *European Union*. July 1, 2003. Available at <http://europe.eu.int/comm/dgs/health_consumer/library/press/press298_en.pdf> (accessed May 26, 2004).

52. Baumüller, H. *Domestic import regulations for genetically modified organisms and their compatibility with WTO rules: Some key issues*. Winnipeg: International Institute for Sustainable Development, August 2003.

53. Craddock, N. Flies in the soup: European GM labeling legislation. *Nature Biotechnology* 22, no. 4 (April 2004): 383–384.

54. Evans-Pritchard, A., and C. Clover. GM sweetcorn given the go-ahead as Europe bows to the US. *Daily Telegraph*, May 20, 2004, 6.

55. Alden, E., and J. Grant. WTO rule against Europe in GM food case. *Financial Times* (London, England) (February 8, 2006): 6.

56. Giles, J. Transgenic planting approved despite skepticism of UK public. *Nature* 428 (March 9, 2004): 107.

57. World Trade Organization. What is the WTO? *WTO*. Available at <http://www.wto.org> (accessed May 25, 2004).

58. Safrin, S. Treaties in collision? The Biosafety Protocol and the World Trade Organization agreements. *American Journal of International Law* 96 (July 2002): 606–628.

59. Secretariat of the Convention on Biological Diversity. *Cartagena Protocol on Biosafety to the Convention on Biological Diversity: Text and annexes*. Montreal: Secretariat of the Convention on Biological Diversity, 2000.

60. Fletcher, S. *Biological diversity: Issues related to the Convention on Biological Diversity*. 95–598. Washington, DC: Congressional Research Service, May 15, 1995.

61. Convention on Biological Diversity. Cartagena Protocol on Biosafety (Montreal, 29 January 2000): Status of ratification and entry into force. *Secretariat of the Convention on Biological Diversity*. April 24, 2004. Available at <http://www.biodiv.org/biosafety/signinglist.aspx?sts=rtf&ord=dt> (accessed April 28, 2004).

62. A call for a gene revolution. *New York Times*, May 24, 2004, A26.

63. Farley, M. Altered crops backed by the U.N.; genetically modified foods are safe, but staples should be stressed, agency says. *Los Angeles Times*, May 18, 2004, A3.

64. Institute of International Economic Law. GMOs in the WTO (overview). *Georgetown University Law Center*. 2003. Available at <http://www.law.georgetown.edu/iiel/current/gmos/gmos_wto.html> (accessed May 26, 2004).

65. Holden, C. GM corn dead for now in Britain. *Science* 304 (April 9, 2004): 203.

66. Pollack, A. Monsanto shelves plan for modified wheat. *New York Times*, May 11, 2004, C1.

67. Pollack, A. Narrow path for new biotech food crops. *New York Times*, May 20, 2004, C1.

68. Cyranoski, D. Pesticide results help China edge transgenic rice towards market. *Nature* 435 (May 5, 2005): 3.

69. Walker, V. R. The myth of science as a "neutral arbiter" for triggering precautions. *Boston College International and Comparative Law Review* 26 (Spring 2003): 197–228.

70. Lowrance, W. F. *Of acceptable risk: Science and the determination of safety*. Los Altos, CA: William Kaufman Inc., 1976.

Further Reading

Halford, N. G. *Genetically modified crops*. London: Imperial College Press, 2003.

• An excellent overview of the biology and politics of GM crops from a British perspective; this book generally takes a "pro-GM" stance

Mad Cow Disease, International Trade, and the Loss of Public Trust

The public response in Europe to GM crops might be very different if the outbreak of BSE, or mad cow disease, in the United Kingdom had not occurred in the 1980s and 1990s. Despite reassurances from British health officials that consuming British beef was safe, in 1996 the consumption of BSE-tainted beef was presumptively linked to a variant form of Creutzfeldt-Jakob disease (vCJD), a fatal neurodegenerative disease in humans. The beef industry in the United Kingdom was devastated; the appearance of BSE in cattle in other European countries further eroded the European public's trust that governments were able to assure the safety of food—a trust that had been already damaged by a series of food scandals in the 1980s. Single cases of BSE occurred in Canada in May 2003 and in the United States in December 2003, with similar severe economic impacts.

Prions and Disease

BSE is one of a group of illnesses called transmissible spongiform encephalopathies (TSEs), including scrapie in sheep, chronic wasting disease in deer and elk, and CJD and kuru in humans. TSEs can occur in other mammals, as a result of consuming infected meat or induced in the laboratory by researchers infecting research animals in order to study the disease [1]. The infectious agent appears to be a mutant misfolded protein, or prion, that triggers normal proteins to misfold, causing the disruption of the cellular function of brain cells. This leads to a characteristic degeneration of brain tissue, producing a spongy appearance to the brain. TSEs thus appear to be caused by an agent that contains no genetic material, contrary to most generally held views about infectious agents (i.e., viruses and bacteria). Prions are remarkably stable molecules; the normal cooking of meat does not destroy them [2]. While the prion hypothesis remains somewhat controversial, Stanley Prusiner, the researcher who first described prions in the 1980s, was awarded the Nobel Prize in Physiology or Medicine in 1997 for his discovery.

Diseases caused by prions are characterized by long latency periods between infection and the appearance of symptoms [1, 3]. The accumulation of misfolded proteins triggered by the infectious prions takes time to develop. Infected animals may show no symptoms for several years and develop severe symptoms over a period of six months to a year. Animals suffering from TSEs become nervous or aggressive, and develop gait and other motor problems. As the disease progresses, affected animals lose weight and the ability to walk. Most cows suffering from BSE begin to show symptoms after they are thirty months old.

Variant CJD differs from the sporadic form of CJD: it attacks younger people and causes different changes in the brain. Like other TSEs, the first symptoms are nervousness or other behavioral changes, followed by motor problems leading to death.

Scrapie, a TSE affecting sheep, was described in the eighteenth century [1]. BSE was not described until 1986. Where did it come from? The mostly likely sources were either prions from scrapie-infected sheep that "jumped species" to infect cows or a spontaneous mutation in prions in cattle. The opportunity for infecting other animals arose from the practice of feeding rendered inedible meat by-products—meat and bonemeal (MBM)—from sheep and cattle to livestock and other animals to supplement their diets; the material presumably contained portions of BSE-infected nervous systems. This practice was common everywhere. Why, then, did BSE appear first only in the United Kingdom?

In the 1980s the United Kingdom changed its meat by-product processing methodology in ways that decreased exposure to heat and solvents, and also pooled large numbers of carcasses for processing [1]. These alterations might have allowed prions to remain infective in MBM and thus increased the likelihood that large batches of MBM were contaminated. As carcasses from infected cattle were processed into MBM, the prion level rose to a level of transmissibility to cows that consumed the feed. Some calves of BSE-infected cows might have been infected in utero. The new appearance of TSEs in felines (both zoo and domestic), zoo antelopes, and primates also is linked to MBM consumption [1]. BSE, like other TSEs, can be confirmed only by the testing of brain tissue after death. There are no blood tests or other screens that reveal the presence of BSE prions in an otherwise-healthy cow.

How were humans exposed to BSE prions, leading to vCJD? The working hypothesis is from the consumption of beef that contained fragments of brain or spinal cord from BSE cattle [1]. Most solid cuts of meat are unlikely to contain prions since little is found in muscle, although some cuts of steak that contain parts of the spinal column might contain prions. The most likely source is processed meat that contains "mechanically recovered meat" (MRM), material from a crude mechanical process used to recover as much meat as possible from a carcass. The consumption of tainted MRM containing fragments of spinal cord in hamburgers, hotdogs, and other processed beef provides an exposure route; the higher incidence of vCJD in children and young adults may reflect a greater consumption of these products.

BSE and vCJD in the United Kingdom

A veterinary commission announced in 1986 that a new TSE appeared in British cattle. The number of cases of infected cattle increased rapidly, peaking in 1992–93, when nearly 1,000 new cases occurred weekly; about 50 percent of dairy herds were affected [4]. Nearly 180,000 cows were confirmed to have BSE by 2001, and millions of other cattle were slaughtered to prevent further spread [5]. New cases declined through the 1990s because of the institution of a 1988 ban on feeding MBM to cattle. New cases still appeared, though, because of the poor enforcement of the ban and the long latency period for the disease [6]. The government offered to compensate farmers for the loss of revenue by paying them for culling herds of cows suspected of having BSE instead of sending them to market. The payment was about half a cow's market price, providing a strong incentive for farmers to sell even suspect cattle to meatpackers [6].

The British government continued to maintain throughout the late 1980s and early 1990s that British beef was safe to eat. The government nonetheless instituted measures in 1995 to protect meat for human consumption from contamination with bovine brain and spinal cord [4]. This may have been too late, since in March 1996, the government announced that ten young Britons had been diagnosed with a new variant of CJD, and that it was linked to the possible consumption of BSE beef. The damage to public confidence was severe, despite actions by the British government to tighten regulations and assure meat safety. A total of 144 cases of vCJD occurred in the United Kingdom by 2000—far below a feared epidemic, but still the majority of the 154 total cases reported worldwide [1].

The Ministry of Agriculture, Fisheries, and Food along with the Ministry of Health were targets of scathing criticism for their failure to handle the BSE epidemic—not only did they fail to protect public health in an effort to support the beef industry but they also misled and misinformed the public throughout the crisis [7]. One motivation may have been to protect the cattle industry.

Much of the outrage following the 1996 announcement of a link between BSE and vCJD stemmed not from the direct fear of developing vCJD but rather from revealing the more gruesome details of the meatpacking industry. Many Britons were revolted by the idea that cows were fed material from other cows; the media provided graphic discussions of "cannibal cows." Remarkably, in view of the public outcry, the consumption of beef in the United Kingdom dropped only by a third, mostly reflecting a decreased demand for processed meat products [8]. The net result of the BSE crisis, coming after a decade of other food scares, was a British consumer population hypersensitive to issues of food safety. Demands for organic food doubled in the late 1990s, and activist and environmental groups became increasingly vocal about the dangers of food industrialization [9].

Political and Economic Fallout in the EU

While the BSE epidemic raged in the United Kingdom, other countries responded in a variety of ways. The United States banned imports of all British cattle beginning in 1989, and fresh beef and beef by-products (such as MBM) in 1991 [10]. Britain's trading partners in Europe, members of the EU, on the other hand, acted in the interest of the beef industry, and only limited the export of cattle born before the MBM ban of 1988 and the offspring of affected animals [4]. It was not until after the announcement of a link between BSE and vCJD that the EU instituted a worldwide ban on British beef exports [6, 11]. The United Kingdom argued that this response represented an overreaction to a limited problem; the British beef industry suffered billions of dollars of losses. Meanwhile, beef consumption in European countries dropped by close to 50 percent after the announcement of the possible link between BSE and vCJD in 1996 [6].

Political jockeying began as the United Kingdom attempted to restore public trust while EU members sought to protect their own beef industries [6]. The EU accused the United Kingdom of failing to enforce its ban on MBM feed, and allowing contaminated feed and cattle to be exported [5]. The EU demanded that the United Kingdom trace all cattle born after 1989 to their original herds and destroy the herds before the EU would consider lifting the export ban. The United Kingdom responded by threatening to block central and eastern countries from EU member-ship unless the plan for the wholesale slaughter of British cattle was reformed [6]. In 1999, the EU began to relax export restrictions on British beef, although France did not remove its ban until 2002 [12].

Protecting against the Spread of BSE

BSE-infected cattle appeared in several European countries beginning in 1997 and in Japan in 2001. In response, the United States and other countries banned imports of cattle and beef from BSE-affected countries, and slaughtered animals already imported [13]. The EU instituted a systematic screening of the brains of cattle for a BSE in the late 1990s; all obviously sick animals or those unable to walk ("downer" cows) over twenty-four months old as well as all cows over thirty months old intended for human consumption are now tested. Annually, 10.4 million animals are tested, at a cost of over $300 million [14, 15]. Japan tests all cattle slaughtered for human consumption, plus at-risk animals on farms, totaling about 1.2 million cattle at a cost of $30 to $40 million [14, 15]. Most countries also instituted meas-ures banning the inclusion of central nervous system tissue in meat products, although enforcement is limited.

The United States instituted its own ban on feeding MBM to cattle in 1997; as in the United Kingdom, however, enforcement of the ban is weak [10]. In 2002, the GAO issued a report critical of the current practices, and suggested that the United States remained at risk for a BSE outbreak unless feed regulations, inspections, and surveillance were tightened, and the screening of animals expanded. The U.S. screen-ing program is more limited than those in countries with active BSE. In 2001, fewer than five thousand animals were tested for BSE. In response to GAO recommenda-tions, the number was increased. About twenty thousand animals in 2002 and 2003 were tested from a population of apparently sick or downer cows brought to slaugh-terhouses [10, 14].

BSE Reaches North America

In May 2003, a single cow in Canada tested positive for BSE; its carcass was not used for human food [16]. The cow quickly was traced and its source herd destroyed, but not before the United States and other countries added Canada to the list of those barred from exporting cattle or meat products. The United States gradually lifted restrictions on meat during 2003, but most live animals are still barred. Two new cases appeared in December 2004 and January 2005. The Canadian beef industry, a close to $6 billion annual enterprise, has collapsed; over 50 percent of beef production normally was for export, with 70 percent sent to the United States [17, 18].

In December 2003, a single cow in the state of Washington was found to be suffering from BSE, but not before its carcass was processed for human consumption [15]. Within days, the cow was traced to a Canadian herd [19]. The United States, whose beef export industry had boomed in the aftermath of the Canadian BSE scare, now found its exports cut off; thirty countries barred the import of U.S. cattle or beef products [20, 21, 22]. Cattle prices plummeted.

The event caused beef industry critics to call for the testing of all downer cattle and more surveillance testing of healthy animals. In late December 2003, the USDA issued new rules that banned the use of downer cows in the human food supply, increased testing (but not of all animals), and required better tracking of animals [22]. Also, the USDA ordered changes in meat recovery to reduce the possibility of spinal cord tissue entering the food supply. The reaction appeared more directed at restoring consumer confidence than protecting human health. Some beef industry spokespersons criticized the government response as an expensive and impractical overreaction, and an attempt to appease foreign markets, particularly Japan, the U.S. beef industry's largest foreign market. Increased testing of over 484,000 cattle revealed only a single cow confirmed to have BSE in June 2005 [23], and a third cow with BSE in March 2006; all the animals were born prior to the feed ban of 1997. In October 2005, the FDA announced new guidelines regarding the use of bovine material in animal feed; brains and spinal cord from older cattle are banned in all animal feed [23].

Japan announced in 2004 it would continue to block U.S. beef imports unless all animals were tested for BSE. In February 2004, a beef producer in Kansas asked for permission to test all its animals, in order to resume sales of its prime beef to Japan [24]. The USDA denied the request on the grounds that such blanket testing might imply that U.S. beef is not safe. The department added that Japan's demand is irrational because it requires testing of young animals that have never been shown to harbor BSE. The beef producer argued that it was losing upward of $50,000 a day in revenue. Nevertheless, the USDA continues to block blanket testing.

As of June 2004, only one country had relaxed its ban to allow some imports of U.S. beef products [21]. In December 2005, the United States and Japan announced a tentative agreement to resume limited beef trading; U.S. cattle must be under twenty months of age [25]. Whether trading partners lift the bans on U.S. beef remains to be determined. The ban is beneficial to other beef exporting nations seeking to replace the U.S. market share.

Does BSE Pose a Real Hazard to Humans?

When BSE first was linked to vCJD, many feared that an epidemic of thousands of fatal TSEs would occur in the United Kingdom and abroad. This has not happened; the development of vCJD by the consumption of prions has yet to be unequivocally confirmed, and it appears that infection rates are low [1]. The overall risk of developing vCJD from tainted beef is low, and is likely to decline further as more stringent control of beef production is instituted.

As in many areas of risk perception, the public's view may be influenced by misinformation and perceived government secrecy. In the United States, trust in the government's ability to assure food safety generally is high. Although the U.S. public expressed concern, beef consumption did not decline markedly in the weeks after the BSE scare in 2003, in marked contrast to Europe in 1996 [26]. In the United Kingdom and Europe, the mishandling of the BSE crisis provided additional confirmation to citizens that they could not trust their governments to keep their food safe. The repercussions of BSE are now reflected in trade barriers against GM foods and other biotechnology.

Even a hint of BSE in a country shuts down beef exports. In the complex trade battle globally, one country's misfortune is another's economic boon. Countries will continue to jockey for market share in the competitive meat industry, and will use whatever opportunities arise to gain an advantage.

References

1. Brown, P. Mad-cow disease in cattle and human beings. *American Scientist* 92, no. 4 (July–August 2004): 334–341.

2. Schwartz, M. *How the cows turned mad.* Berkeley: University of California Press, 2003.

3. Prusiner, S. Prion diseases and the BSE crisis. *Science* 278 (October 10, 1997): 245–251.

4. Demko, V. Appendix 1: A chronology of BSE and CJD developments. In *The mad cow crisis: Health and the public good*, ed. S. Ratzan, 209–216. New York: New York University Press, 1998.

5. Freeman, H. B. Trade epidemic: The impact of the mad cow crisis on EU-U.S. relations. *Boston College International and Comparative Law Review* 25 (Spring 2002): 343–371.

6. Goethals, C., S. Ratzan, and V. Demko. The politics of BSE: Negotiating the public's health. In *The mad cow crisis: Health and the public good*, ed. S. Ratzan, 95–100. New York: New York University Press, 1998.

7. Lang, T. BSE and CJD: Recent developments. In *The mad cow crisis: Health and the public good*, ed. S. Ratzan, 65–85. New York: New York University Press, 1998.

8. How mad cow disease hit the beef industry. *BBC News*. June 10, 1998. Available at <http://news.bbc.co.uk/1/hi/health/background_briefings/bse/110131.stm> (accessed June 16, 2006).

9. Brooks, K. History, change, and policy: Factors leading to current opposition to food biotechnology. *Georgetown Public Policy Review* 5 (Spring 2000): 153–164.

10. U.S. General Accounting Office. *Mad cow disease: Improvements in the animal feed ban and other regulatory areas would strengthen U.S. prevention efforts.* GAO–02–183. Washington, DC: General Accounting Office, January 2002.

11. EU issues worldwide ban on British beef exports. *CNN*. March 26, 1996. Available at <http://www0cgi.cnn.com/WORLD/9603/mad_cow/26/eu_beef_ban/> (accessed June 30, 2004).

12. France lifts ban on British beef after six-year dispute. *Guardian Newpapers Limited*. October 2, 2002. Available at <http://www.buzzle.com/editorials/text10-2-2002-27447.asp> (accessed June 30, 2004).

13. Enserink, M. Is the U.S. doing enough to prevent mad cow disease? *Science* 292 (June 1, 2001): 1639–1641.

14. Normile, D. First U.S. case of mad cow sharpens debate over testing. *Science* 303 (January 9, 2004): 156–157.

15. McNeil, J., and G. Donald. Mad cow disease in the United States: The overview; mad cow disease leads government to consider greater meat testing. *New York Times*, December 26, 2003, A1.

16. Importation of processed Canadian beef products regulatory timeline. *APHIA/USDA*. June 10, 2004. Available at <http://www.aphis.usda.gov/lpa/issues/bse/bsechronjune10.pdf> (accessed June 24, 2004).

17. Goddard, E., and J. Unterschultz. BSE leaves Canada's beef industry mired in uncertainty. *University of Alberta*. January 23, 2004. Available at <http://www.expressnws.ualberta.ca/expressnews/articles/ideas.cfm?> (accessed July 2, 2004).

18. Economic Research Service. Background statistics on U.S. beef and cattle industry. *USDA*. June 1, 2004. Available at <http://www.ers.usda.gov/news/BSEcoverage.htm> (accessed July 2, 2004).

19. Clemetson, L. U.S. officials say ill cow is linked to Alberta herd. *New York Times*, December 28, 2003, sec. 1, 1.

20. Moss, M., R. A. Oppel Jr., and S. Romero. Mad cow forces beef industry to change course. *New York Times*, January 5, 2004, A1.

21. Animal and Plant Health Inspection Service. BSE trade ban status as of 6–21–04. *USDA.* June 21, 2004. Available at <http://www.aphis.usda.gov/lpa/issues/bse/trade/ bse_trade_ban_status.html> (accessed June 24, 2004).

22. Grady, D. U.S. issues safety rules to protect food against mad cow disease. *New York Times*, December 31, 2003, A1.

23. McNeil, J., and G. Donald. To prevent mad cow disease, F.D.A. proposes new restrictions on food for animals. *New York Times*, October 5, 2005, A18.

24. McNeil, J., and G. Donald. U.S. won't let company test all its cattle for mad cow. *New York Times*, April 10, 2004, A8.

25. Kaufman, M. U.S. and Japan to resume beef trade. *Washington Post*, December 13, 2005, A3.

26. Day, S. A time for finesse: Marketing beef after a mad cow discovery. *New York Times*, January 1, 2004, C1.

Further Reading

Brown, P. Mad-cow disease in cattle and human beings. *American Scientist* 92, no. 4 (July–August 2004): 334–341.

• An excellent overview of the science and history of BSE.

11

Dangers in the Environment: Air Pollution Policy

Few areas of science policymaking pose more challenges than environmental issues. Scientific data are often limited or contradictory, and policymakers find themselves arguing the relative merits of different interpretations. The competing interests of public health and welfare, economic needs, and environmental preservation also inform the debate. This part explores the challenges of environmental policymaking by focusing on the regulation of air pollution, and particularly those pollutants regulated by the National Ambient Air Quality Standards (NAAQS).

A Brief History of Air Pollution

Air pollution, the release of undesirable chemicals into the atmosphere, is not a new problem; erupting volcanoes emitted pollutants into the air long before human history. The first human-generated pollution occurred when fire became a tool [1]. Air pollution became a significant problem with the Industrial Revolution, as the increased use of coal and other fossil fuels released hazardous gases and smoke into the air. Urban areas in particular became cloaked in smog (combinations of smoke and fog).

It is a common misconception that no one worried about air pollution prior to the mid-twentieth century [2]. Yet, concerns tended to be local in nature. In Great Britain, most people in cities were aware of the negative effects of air pollution on health by the early seventeenth century. Attempts to mitigate the effects of increased coal burning were unsuccessful. In the United States, the availability of wood for burning delayed the switch to coal; the heavy use of coal did not begin until the early nineteenth century [2]. U.S. cities on the East Coast primarily used anthracite coal, which burned cleaner than the soft bituminous coal. Midwest areas, however, relied on locally available bituminous coal. The degradation in air quality appeared quickly; industrialized areas located in river valleys, such as Pittsburgh, developed serious pollution problems in the nineteenth century. Local politicians expressed concern that the increasing smoke might make residents ill and slow further economic development.

By the early twentieth century, dense smoke triggered traffic accidents, illnesses, and deaths as well as damage to forests and buildings. By 1912, most of the large cities (greater than two hundred thousand inhabitants) had smoke abatement ordinances, but they were ineffective [2]. Local citizens demanded that something be done to remedy pollution problems. New methods of coal burning and furnace design were developed to mitigate pollution, but most either did little to solve the problem or were deemed too expensive to institute. The sole exception was Saint Louis, Missouri, which in the 1930s instituted strong measures to deeply limit the burning of bituminous coal, over the strong objections of the coal industry. Saint Louis's air was noticeably cleaner by the early 1940s [2].

The Science of Air Pollution

Earth's atmosphere consists of a small number of gases maintained in stable proportions (fixed gases) with small amounts of additional gases that vary with changing conditions (variable gases). Nitrogen gas makes up about 78 percent of the atmosphere, with oxygen gas making up close to 21 percent; small amounts of the inert gases argon, neon, helium, xenon, and krypton make up the remaining 1 percent of fixed gases [3, 4]. Variable gases include water vapor, carbon dioxide, carbon monoxide (CO), ozone (O_3), sulfur- and nitrogen-containing gases, chlorofluorocarbons (refrigerants and aerosol propellants containing fluorine and chlorine), and a number of volatile organic compounds (VOCs) including benzene, perchloroethylene (cleaning fluid), formaldehyde, and other chemicals. These gases come from a number of sources, as described below. These pollutants generally are broken into two categories for regulatory purposes: criteria pollutants, including CO, nitrogen dioxide (NO_2), sulfur dioxide (SO_2), O_3, lead, and particulate matter (PM); and hazardous air pollutants (HAPs), those pollutants known to cause human cancer or other significant health problems. The atmosphere also contains clusters of fine particles known as aerosols. PM includes carbon particles (soot), nitrogen and sulfur compounds, soil dust particles, heavy metals such as lead and mercury, and biological material such as pollen or fungal spores [3, 4].

The chemistry of the atmosphere is complicated [3, 4, 5]. Secondary pollutants may be formed as a result of interactions between pollutants emitted directly from sources. Many atmospheric components are chemically active; reactions between many variable gases or particulates and water vapor produce sulfuric or nitric acids that are toxic to living organisms. VOCs are precursors to ozone. Sunlight also contributes to the production of ozone and other toxic gases.

Where Do Pollutants Come From?
The primary source of air pollution is the burning of coal, oil, and other fossil fuels to generate electricity, power automobiles, provide heat, and support industrial activities ranging from the smelting of metal ore to plastics or textiles manufacture. The combustion of fossil fuels releases large amounts of carbon and carbon-containing compounds into the atmosphere. In addition, coal and oil contain varying amounts of sulfur and nitrogen that are released into the atmosphere; low-grade soft bituminous coal ("dirty coal") produces much more soot and sulfur than high-grade hard anthracite coal. Other sources from combustion include the burning of biomass such as trees and grasslands to clear areas for agriculture along with the burning of trash and other wastes [3, 4].

Pollution may arise from noncombustion sources as well. VOCs are emitted in part by combustion sources, but also in large part because of evaporative sources—such as the storage of fuels and solvents, and the application of paints, coatings, and inks [3, 4, 5]. Sources of PM emissions include not only combustion but also brake lining and tire wear, fugitive dust, and secondary formation from condensable VOCs. Deforestation and land-use changes can affect biogenic emissions. There are natural causes of flux of some pollutants into the environment (e.g., sea spray, volcanic eruptions, lightning, and unmanaged fires).

Pollution sources generally are broken into two categories: stationary sources, such as power plants and heavy industry, which are further subdivided by their range of effects as point, line, and area sources; and mobile sources, which include automobiles and trucks.

The term smog was coined in 1905 to describe the combination of smoke and fog that commonly settled over cities in Great Britain [3]. Smog develops because climatic conditions promote temperature inversions that trap pollutants for extended periods. Industrial activity involving the burning of coal and use of a variety of chemicals in the late nineteenth and early twentieth centuries produced thick smog events that killed hundreds of people. In London, where lingering fog is common, dense black smog developed on many occasions in the late nineteenth and twentieth centuries; in December 1952, between twenty-eight hundred and four thousand people died over a weeklong period during a temperature inversion that trapped industrial smoke [3, 6]. Similar "killer fogs" occurred in other industrialized regions. In October 1948, a smog event took place in Donora, Pennsylvania, a small town in a valley not far from Pittsburgh that had a zinc smelter plant for the steel industry. Five days of heavy smog killed twenty people directly and caused respiratory illness in nearly half the town's population of fourteen thousand [3, 6]. These events helped raise awareness of the hazards of air pollution and triggered efforts to reduce air pollution to protect public health.

A second form of smog is responsible for persistent air pollution problems in areas that have extended sunny periods, such as Los Angeles, Mexico City, and Tokyo. This form of pollution, photochemical smog, develops when sunlight helps produce ozone from nitrogen-containing pollutants and oxygen [3, 4]. Automobile exhaust is a major source of pollutants in photochemical smog [7]. Rather than the thick black conditions of London-style smog, photochemical smog presents as a light haze because of less PM present in the atmosphere.

The Health Effects of Air Pollution

Many air pollutants may irritate or damage the respiratory system; these include sulfuric and nitric acid, ozone and many organic solvents [8]. Individuals with compromised respiratory systems or those suffering from heart disease may be particularly vulnerable to the effects of air pollution [7, 9]. Organic solvents may also be toxic to many organ systems; some are carcinogenic. Heavy metals such as lead and mercury damage the nervous system and other organ systems. In contrast, CO prevents the uptake of oxygen by red blood cells by binding to hemoglobin, the oxygen-carrying protein in blood. Low-level exposure causes headaches; higher exposure produces unconsciousness leading to death.

As part of its efforts to control pollution, the federal government sets limits on the acceptable levels of some pollutants present in the air. These levels, usually presented as parts per million or micrograms per cubic meter of air ($\mu g/m^3$), represent a best-guess limit above which health effects occur. But accumulating evidence suggests that there are no safe levels for HAPs, particularly for individuals with compromised respiratory systems [10]. The regulation of HAPs is not based on any acceptable threshold level [5]. The lack of certainty as to what constitutes "safe" levels of pollutants represents a major problem for policymakers charged with protecting public health.

The Environmental Effects of Air Pollution

Pollutants have broad effects on the environment. In areas downwind of pollution sources, acid rain, which is rainwater containing sulfuric and nitric acids, contributes to environmental degradation by acidifying soils and bodies of water [11]. Acid rain leaches important nutrients from the soil while raising the levels of sulfur, nitrogen, and aluminum (which is toxic to plants). Acid rain kills trees by affecting their ability to take up nutrients from the soil, and makes them less tolerant of cold winters by damaging their leaves or needles. Increasing acid levels in lakes and ponds kill many organisms in the complex food webs found in aquatic ecosystems. Acid rain has a particular impact on mountain forests in the eastern United States that lie downwind of the major industrial areas in the Midwest. Similar effects may be seen in Europe, where the fabled Black Forest in eastern Europe is dying from the effects of acid rain produced by industrial activity in western Europe [3, 12].

Air pollution impacts visibility, and even areas remote from industrialization show increased haze from drifting particulates. In addition to indicating low-level pollution, the loss of visibility affects tourism by reducing the views in national parks. Both Yosemite in California and the Grand Canyon in Arizona are now plagued with frequent haze. One strategy to improve the air quality in these parks is to reduce auto traffic within them. For example, tourists in Yosemite are encouraged to leave their cars in designated parking areas, and use free transportation to scenic overlooks and sites; only limited parking is available at sites within the park [13].

Air pollution also contributes to global warming, the increase in global temperatures, by increasing the level of CO_2 and other greenhouse gases in the atmosphere. Estimating the effects of global warming is a challenging and contentious issue, but will not be discussed in this book.

Air Pollution Policy in the United States

Events such as the Donora killer fog in 1948 raised public awareness to the health risks of air pollution, but the federal government was slow to become involved in air pollution control except to intervene in cases in which pollution crossed the border with neighboring Canada. In 1955, Congress passed the Air Pollution Control Act [14]. The act provided limited funds to the states to support their own research into air pollution, but left the resolution to local communities and states. The federal Clean Air Act (CAA) of 1963 was the first to establish a permanent federal program with funding to conduct research, provide assistance to the states, and help local communities initiate pollution abatement programs. The law also called on the Department of Health, Education, and Welfare (now the DHHS) to develop air quality criteria with the goal of protecting public health. The law was limited in scope, however, and mandated no steps to control pollution on a national level. The Air Quality Act of 1967 called for uniform emissions standards for industrial pollutants, controls based on "regional airsheds" for pollutants that crossed state boundaries, assistance for states to develop programs for motor vehicle inspections, the registration of fuel additives, and further research [2]. Not surprisingly, industries fought hard against increased federal involvement in pollution control. The resulting compromise bill was weak, with many built-in delays and poor coordination of programs. The public quickly criticized the law as toothless and inadequate [2].

Auto Emissions

Most early efforts to control air pollution focused on stationary sources such as heavy industry plants and power-generating stations. By the mid-1950s, however, increasing evidence pointed to automobile exhaust as a major cause of air pollution. Los Angeles, with its growing problem of photochemical smog, led the way in developing regulations to limit auto emissions. Ironically, Los Angeles had one of the best light-rail public transportation systems in the country from the late nineteenth century to the 1950s. The rail system failed as a result of overcrowding, high fares, limited routes in an ever-expanding city, and some claim, efforts by General Motors and other automobile manufacturers to limit the growth of public transportation [15, 16, 17].

In the late 1940s, Arie Haagen-Smit, a biochemist at the California Institute of Technology in Pasadena, determined the chemistry of photochemical smog. His data pointed to gasoline as a central source of the reactants necessary for this kind of smog. The population of Southern California grew explosively in the 1950s, and ever-increasing numbers of cars crowded the expanding highway system. The resulting smog worsened throughout the 1950s, despite local efforts to control pollution from stationary sources. Efforts to expand pollution control were met with strong opposition from industry; the auto manufacturers denied that cars were responsible for smog *and* contended that smog was not a health hazard [2]. Local governments appealed to auto manufacturers to develop devices to reduce the emission of polluting gases and aerosols. Although the auto companies made some effort to develop such devices, they continued to argue against further pollution controls on both health and economic grounds.

Frustrated by the lack of voluntary response by auto manufacturers, Los Angeles activists persuaded the California State Legislature in 1960 to pass the first laws requiring emission controls [2]. The Motor Vehicle Control Act mandated that vehicles sold in California after January 1, 1961, must have emission control devices. The law's enactment was delayed until at least two emission control devices were available, to avoid establishing a monopoly. The auto manufacturers successfully stalled the enactment of the law until 1964, when under increasing federal as well as state pressure, they agreed to begin to install devices on all model vehicles beginning in 1967 [2]. California continues to maintain the strictest emission standards in the United States.

The federal government weighed in on auto emissions in 1965 with the passage of amendments to the 1963 CAA that mandated the adoption of California's emission standards nationwide [2, 14]. California had already determined that the levels were too lax to resolve the state's pollution problems. The 1967 federal Air Quality Act further mandated auto emission controls, but failed to provide strong federal involvement in assuring that the standards would be met [2].

The CAA of 1970

The CAA of 1970 (42 U.S.C. § 7401 et seq.) is credited as the first to genuinely involve the federal government in air pollution control. Strictly speaking, the law is a set of amendments to the original legislation of 1963. The 1970 CAA provided the legal framework to promote cleaner air in order to improve the public's health and welfare. It set five goals: to mitigate possible harmful concentrations of six pollutants (criteria pollutants)—CO, NO_2, SO_2, O_3, lead, and PM; to limit exposure to HAPs; to protect and improve wilderness area and national parks; to reduce emissions that cause acid rain; and to curb the use of chemicals that may deplete the atmospheric ozone layer [10].

The CAA authorized the newly established EPA to set maximum allowable atmospheric concentrations for the six criteria pollutants (lead was added in 1976) by establishing NAAQS; these limits are set to provide a "margin of safety" from the adverse effects of pollution [14, 18]. These standards were to be uniform across the country, but states were allowed to adopt stricter standards [14]. States were directed to develop implementation plans applicable to their particular pollution sources and submit them to the EPA for approval [18]. Mandated reductions in pollutants from stationary sources were to be achieved by 1975, with the possibility of one extension for two years to 1977 [19]. The strictest limits were placed on industrial plants under construction; new plants were required to install pollution control devices meeting the best available control technology standard to limit emissions [14]. Emissions from motor vehicles would be reduced by 90 percent from 1970 levels beginning with the 1975 model year.

The CAA was amended repeatedly during the 1970s, mostly to allow delays in implementation of its goals. Auto emissions standards were extended three times, driven by the gasoline shortage and rising gas prices; auto manufacturers pointed out that installing emission control devices would decrease fuel economy [19]. Steps to eliminate leaded gasoline were initiated in 1976, as it became clear that lead inactivated catalytic converters that were an effective means to control tailpipe emissions.

Amendments in 1977 codified the EPA's classification of three regional levels of priority: class I areas included national parks, which were to be protected from any further deterioration in air quality, with class II and III areas, already impacted by varying levels of pollution, subject to less regulation until the overall air quality goals were reached, with the expectation that these goals would be attained by 1982 [20]. The major change was the introduction of strict emission guidelines for any new construction in order to preserve the air quality of class I areas. The amendments allowed older pollution sources such as power plants to delay the installation of pollution control devices to reach "new source standards" until they underwent significant renovation and modification; routine maintenance and other activities were exempt from the new standards. This process of "new source review" proved to be among the most contentious. The EPA filed suit against some polluters who attempted to skirt the requirement for retrofitting old plants.

The CAA did recognize that HAPs posed a hazard at all exposure levels, and hence set regulations in place to first reduce exposure as much as current technology might permit and then determine what additional controls were needed to meet a risk-based standard [19, 20].

Criticisms of the CAA

Implicit in the CAA is the assumption that there are safe levels of criteria pollutant exposure that do not pose a health risk to the public. Increasing evidence suggests that this assumption is incorrect—even low levels of pollutants pose a hazard to at least some individuals. Since there were no clear data on the health effects of low-level pollution, the EPA made assumptions based on hypothetical models, leaving itself open to criticism that the limits set for pollutants were arbitrary [14, 21]. Since it is not possible to prevent the production of all pollutants without shutting down all use of fossil fuels and modern chemicals, the standards are criticized as both impractical and inadequate. Critics also suggest that the levels should be set to protect the health of the general public, not the relatively small number of highly sensitive individuals [19].

A second major criticism of many of the EPA standards is that they were established without any consideration of the economic costs of attaining them; no cost assessment was included in the enabling legislation [1, 14, 21]. If industries were required to sharply limit emissions, the cost of such mitigation might raise the price of products for all citizens. Critics argued that the cost of new pollution control equipment must be balanced against the overall benefits of cleaner air, and this is the case for new construction. They also suggested that the process of obtaining permits for new plant construction was so slow that it damaged the country's economic growth. Auto manufacturers maintained that emissions standards raised the cost of automobiles and trucks, and reduced overall gas mileage [19].

Through the 1980s, the antiregulatory stance of the Reagan administration led to the weakening of the CAA by the simple measure of removing its authorization and funding [19]. Although Congress passed appropriation bills providing funding for the EPA, enforcement of standards proved impossible. EPA administrator Ann Gorsuch made it clear that the Reagan administration intended to work with industry groups to avoid aggressive enforcement measures [19]. The sole exception was the decision by the EPA in 1985 to reduce the lead content of gasoline by 90 percent (see the "Environmental Poisoning" section below).

Amendments of 1990

With the election of George H. W. Bush as president in 1988, the regulatory climate changed; bipartisan support of new clean air legislation led to significant revisions to the CAA, despite continued industry opposition [19]. The changes were motivated by a general perception that the CAA and its 1977 amendments were inadequate to improve ambient air quality nationwide. The 1990 amendments to the CAA added new categories of polluted areas; the most polluted would have more time to attain air quality standards. Standards were established for additional chemicals, including those believed to contribute to O_3 loss from the stratosphere. Controls were tightened for most emission sources, with specific goals mandated. Auto emissions were tightened, with mandated "clean fuels" in the most polluted areas.

The amendments also explicitly addressed controls to sharply reduce acid emissions contributing to environmental degradation. The problem of pollution transport from one region to another was dealt with by calling for the establishment of multi-state air quality management regions.

Economic Incentives for Reducing Pollution

How should states proceed to implement CAA provisions and its amendments? Regulation is more than the passage of law—manufacturers cannot be forced to produce cleaner cars or plants if the technology does not exist to do so. Nonetheless, some policies have been developed for "technology forcing," including California's mandate in the 1980s that 2 percent of automobiles sold in the state by 1998 had to be zero emission vehicles [22]. Various manufacturers attempted to bring electric cars to market. These requirements were loosened by California amendments in 1996 and 1998, but the coincident development of hybrid cars, involving engines driving generators that power electric motors for the vehicle's operation, has been successful.

It may be easier for some polluters to meet existing limits than others. Providing economic incentives to polluters to reward them for a response is now a central strategy [19, 20]. The goal is to achieve cleaner air at a lower cost overall. As early as 1974, the EPA allowed states and municipalities to establish a system of "emissions trading," whereby companies may earn "credit" for reducing pollution in some areas of the company in exchange for more emissions in others in the same region [20]. Pollution permits, a form of "pollution tax," also allow companies to purchase the right to emit a certain level of a given pollutant; those who do not purchase permits may not emit the pollutant. As the total number of available permits is reduced, overall levels of the given pollutant decline. If industries produce less pollution, the number of permits they need to purchase declines, saving the companies money. Creating a market for pollution would also enable environmental groups to purchase permits, thereby helping to reduce discharge in an area.

The 1990 amendments expanded emission trading to allow midwestern polluters to trade credits for the emission of SO_2, the major source of acid rain. Each coal-fired power plant was allotted a certain number of credits that might be traded, bought, or sold. A 50 percent reduction in SO_2 emissions was mandated by 2000 [19].

A possible downside of emissions trading is that those who live in proximity to a dirty plant, or downwind of a coal-fired power plant, may find themselves breathing dirtier air than if all sources were held to similar standards. No hard data are available to support this claim. Although the cost of pollution control by industry may have dropped, overall air quality may not have improved [20]. Arguably, permits should be priced to reflect social environmental costs as well as be high enough to motivate firms in the long term to cut pollution in lieu of paying the tax. Critics also suggest that any regulatory system that permits delays in air quality improvement sends a message that pollution is acceptable as long as companies can afford to pay for it [19].

The Clear Skies Initiative of 2003

A significant expansion of emissions trading is included in the so-called Clear Skies Initiative proposed in President Bush's State of the Union Address in 2003. When his generally antiregulatory administration took office in 2001, the EPA was asked to develop changes to clean air regulations to give industries more flexibility in controlling emissions. Two significant changes were included: the relaxation of the new source review requirements for retrofitting old plants, and new guidelines for emissions trading that limited the costs incurred by polluters. An expanded "cap and trade" emission program for nitrogen oxides, SO_2 and mercury was proposed, with provisions to limit the costs to industry if the caps cannot be met without excessive expense. For example, the emissions cap for mercury was acknowledged to be too expensive for industry, and therefore it would not be met unless new technology was developed [23]. The proposed caps would reduce emissions by roughly 70 percent over fifteen years [23].

Critics of the new initiative argue that regulations are a thinly disguised move to weaken provisions of the CAA and its amendments. The proposed caps are higher than those in the current CAA requirements and delay the deadlines for reaching lower emission goals [24, 25, 26]. The proposed regulation for new source review for older plants would assure that only limited retrofitting would take place by setting a cost limit of 20 percent of the replacement cost of the entire plant as the upper limit for exemption [27].

The EPA issued the final rule of its new source review criteria in August 2003 that would go into effect on November 2003. The state attorneys general for twelve states, including all northeast states, and several cities filed suit in September 2003 to block the enforcement of new EPA regulations for new source review for coal-burning power plants; in December 2003, the U.S. Circuit Court of Appeals blocked the enforcement of the new regulations until the suit is heard [28].

Congress has yet to move on the Clear Skies Act. In October 2003, the GAO released a report to Congress that expressed concern that the proposed new source review revisions might impact on current enforcement cases and make it more difficult for the public to obtain information concerning current emissions from polluters [29]. The U.S. Department of Energy issued a report requested by Congress that compared the Clear Skies Act and other proposed bills to the current CAA regulations, and concluded that the Clear Skies Act would be more expensive than the current CAA regulations [24, 30]. The bill failed to make it out of committee in the Senate as of March 2005.

Has the Clean Air Act Been Successful?

There is no question that air quality is markedly better in the United States than it was in 1970. The total emissions of the six principal air pollutants declined by over 50 percent between 1970 and 2003 [31]. During the same period, the gross domestic product increased by 176 percent, the number of miles driven by vehicles grew by 155 percent, energy consumption increased by 45 percent, and the U.S. population grew by 39 percent [31]. Not everyone agrees that the passage of the CAA and its amendments should be credited with the progress made to date, however. Because of actions taken by the states prior to 1970, air quality improved throughout the 1970s, even before the regulations promulgated by the CAA were in effect. Some argue that air quality would have continued to improve even without the CAA's passage because the states would take measures controlling emissions [1].

Critics also argue that the cost of the federal regulation of air quality is an inefficient use of federal funds. The direct costs of implementing the CAA are estimated to be from $20 to $30 billion per year [10]. Costs are "justified" by the benefits they confer. The EPA asserts that the benefits of air pollution control far outweigh their costs; the cost-benefit ratio is around four to one [14]. For air quality management, most calculations are based on the improvements in public health. As part of the 1990 amendments, the EPA was asked to generate a cost-benefit analysis of clean air regulations. In 1997, the EPA released estimates based on epidemiological studies linking air pollution to adverse health effects. For example, the EPA calculated there were 184,000 fewer premature deaths in 1990 because of reductions in particulates; the cost of each saved life was estimated to be $4.8 million [14]. Critics contend that such estimates are based on assumptions that pollution would not have declined anyway, and hence are meaningless [1, 21].

One problem with the EPA's estimates is that they do not address the economic costs of the environmental impacts of pollutants. Although the 1977 and 1990 amendments to the CAA include explicit mandates to reduce emissions associated with environmental degradation, little effort was placed on obtaining the data needed to assess the economic impacts of environmental damage.

The American Lung Association released a report in 2004 indicating that despite major improvements in air quality, almost half of the U.S. population lives in areas with unhealthful O_3 levels, and close to 25 percent lives in areas that have chronically high levels of particulate pollution [32]. After decades of improving air quality, conditions began to decline in Los Angeles beginning in 2001 [33]. Considerably more effort is needed to provide clean air for the entire population [10]. For environmental effects the story is similar: SO_2 emissions were reduced by regulations introduced by the 1990 amendments, but the situation is far from remedied. An additional 80 percent reduction in SO_2 emissions is needed to allow damaged areas to recover from the effects of acid rain [11].

Air Pollution Regulation in the Future

Despite significant improvements in overall air quality, the reduction rate of emissions declined in recent years, as the U.S. population continued to increase and new methods for further controls on emissions became more expensive. Finding a cost-effective and less cumbersome way to monitor and remediate problems remains elusive. Recent NRC reports argued that the EPA needs better data on the health and environmental effects of pollution in order to make recommendations [10, 34]. In addition, the NRC expressed concern that the impact of air pollution is felt most heavily by poor and minority populations who live in areas near heavy industry or other polluters; the questions of "environmental justice" have not been adequately addressed [10].

Industry continues to argue that further reductions in emissions would be so expensive that the country's economy will suffer. It claims that current regulations are so cumbersome that they deter efforts to reduce emissions. Some also suggest that emissions will continue to decline as the U.S. economy shifts from heavy industry to a more service-oriented structure; will more controls be necessary [14]?

There seem some obvious targets for further reductions. Switching to less polluting fuels may mitigate some pollution problems. Diesel-powered vehicles and heavy trucks have been exempt from many emissions regulations on the grounds that available technology to limit emissions is both expensive and ineffective. Still, the net reductions in pollutants may justify the costs. Encouraging Americans to reduce automobile use would significantly reduce overall emissions; however, this requires greater commitment on the government's part toward public transportation systems.

Lessons from Air Pollution

The recognition that pollutants posed hazards to human health came early in the process of developing policy. For a time, local and state efforts to control pollution were at least minimally successful. Pollution, though, does not respect state boundaries—the federal government needed to become a party in pollution control. The challenge is to develop policy that recognizes that pollution differs regionally; California's photochemical smog is different from the emissions that produce acid rain in the Northeast. While early federal legislation adopted a one-size-fits-all approach to emissions control, the CAA amendments reflected a growing understanding of the science of the problem. Nevertheless, the regional approach to pollution control requires a measure of interstate cooperation that is difficult to achieve. Although sulfur emissions from midwestern power plants have declined, states in the Northeast still suffer from the consequences of being downwind.

The enforcement of clean air regulations varies with the politics of each new presidency and Congress. Those who support economic growth and industry flexibility seek to weaken the enforcement of regulations or to modify them in industry-friendly ways. Those with a stronger proenvironmental stance will attempt to tighten regulations. It is not surprising that the EPA finds itself caught between its own regulations and the administration's desire for particular actions.

How does science fit into the mix? There is no question that the analysis of the effects of air pollution is incomplete. Particularly problematic are projections of the amount of pollution in future years. Because of the lack of hard data, nearly any position may be "supported" using these same data. A lack of certainty in projections may be taken to mean that: there is no reason to jeopardize economic growth because air quality may not decline or may improve without further steps; or aggressive measures must be taken to assure that air quality does not decline.

Air pollution remains a significant problem in the United States and globally. Policymakers continue to grapple with uncertain data, conflicting economic needs, and an ever-growing population. Finding the effective strategy to protect public and environmental health while allowing economic growth will not be easy.

References

1. Goklany, I. *Clearing the air: The real story of the war on air pollution.* Washington, DC: Cato Institute, 1999.

2. Dewey, S. H. *Don't breathe the air: Air pollution and U.S. environmental politics, 1945–1970.* College Station: Texas A&M University Press, 2000.

3. Jacobson, M. Z. *Atmospheric pollution: History, science, and regulation.* Cambridge: Cambridge University Press, 2002.

4. Seinfeld, J. H., and S. N. Pandis. *Atmospheric chemistry and physics: From air pollution to climate change.* New York: John Wiley and Sons, Inc., 1998.

5. U.S. Environmental Protection Agency. *Taking toxics out of the air.* EPA–452/K–00–002. Washington, DC: Environmental Protection Agency, Office of Air Quality, August 2000.

6. Davis, D. *When smoke ran like water.* New York: Basic Books, 2002.

7. U.S. Environmental Protection Agency. *Smog: Who does it hurt?* EPA–452/K–99–001. Washington, DC: Environmental Protection Agency, July 1999.

8. Christiani, D. C., and M. A. Woodin. Urban and transboundary air pollution. In *Life support: The environment and human health,* ed. M. McCally, 15–37. Cambridge, MA: MIT Press, 2002.

9. Brook, R., B. Franklin, and W. Cascio et al. Air pollution and cardiovascular disease: A statement for healthcare professionals from the expert panel on population and prevention of the American Heart Association. *Circulation* 109 (June 1, 2004):2655–2671.

10. National Research Council. *Air quality management in the United States.* Washington, DC: National Academy Press, 2004.

11. Driscoll, C., G. Lawrence, and A. Bulger et al. *Acid rain revisited: Advances in scientific understanding since the passage of the 1970 and 1990 Clean Air Act Amendments.* Hanover, NH: Hubbard Brook Research Foundation, 2001.

12. Francis, B. M. *Toxic substances in the environment.* New York: John Wiley and Sons, 1994.

13. U.S. National Park Service. Yosemite. *U.S. Department of the Interior.* December 2004. Available at <http://www.nps.gov/yose/index.htm> (accessed December 15, 2004).

14. Portney, P. R. Air pollution policy. In *Public policies for environmental protection,* ed. P. R. Portney and R. N. Stavins, 77–123. Washington, DC: Resources for the Future, 2000.

15. Bottles, S. L. *Los Angeles and the automobile.* Berkeley: University of California Press, 1987.

16. Foster, M. S. The role of the automobile in shaping a unique city: Another look. In *The car and the city: The automobile, the built environment, and daily urban life,* ed. M. Wachs and M. Crawford, 325. Ann Arbor: University of Michigan Press, 1992. 186–193.

17. Snell, B. The streetcar conspiracy: How General Motors deliberately destroyed public transit. *TomPaine.com.* Fall 1995. Available at <http://www.lovearth.net/gmdeliberatelydestroyed.htm> (accessed April 10, 2005).

18. Clean Air Act. Vol. 42 U.S.C. § 7401 et seq., 1970.

19. Bryner, G. C. *Blue skies, green politics: The Clean Air Act of 1990 and its implementation*. 2nd ed. Washington, DC: CQ Press, 1995.

20. Smith, Z. A. *The environmental policy paradox*. 3rd ed. Upper Saddle River, NJ: Prentice Hall, 2000.

21. Stroup, R. L. Air toxics policy: Liabilities from thin air. In *Cutting green tape: Toxic pollutants, environmental regulation, and the law*, ed. R. L. Stroup and R. E. Meiners, 59–82. New Brunswick, NJ: Transaction Publishers, 2000.

22. California Energy Commission. *ABCs of AFVs: A guide to alternative fuel vehicles*. 5th ed. P500–99–103. Sacramento, CA: California Energy Commission, November 1999.

23. U.S. Environmental Protection Agency. Clear Skies Act of 2003. *EPA*. July 2003. Available at <http://www.epa.gov/air/clearskies/fact2003.,html> (accessed December 7, 2004).

24. Janofsky, M. Study ranks Bush plan to curb air pollution as weakest of 3. *New York Times*, June 10, 2004, A16.

25. Lee, J. E.P.A. plans to expand pollution markets. *New York Times*, December 15, 2003, A24.

26. Mund, N. Clean air: Facts about the Bush administration's plan to weaken the Clean Air Act. *Sierra Club*. 2003. Available at <http://www.sierraclub.org/cleanair/clear_skies.asp> (accessed December 23, 2004).

27. Johnson, J. Clean air revamp: Industry applauds, states sue over revised EPA regulation. *Chemical and Engineering News* 81, no. 35 (September 1, 2003): 7.

28. Associated Press. Appeals court blocks Bush clean air changes. *CNN.com*. December 24, 2003. Available at <http://cnn.law.printthis.clickability.com/pt/> (accessed October 18, 2004).

29. U.S. General Accounting Office. *Clean Air Act: New source review revisions could affect utility enforcement cases and public access to emissions data*. GAO–04–58. Washington, DC: General Accounting Office, October 2003.

30. Energy Information Administration. *Analysis of S. 485, the Clear Skies Act of 2003, and S. 843, the Clean Air Plannng Act of 2003*. SR/OIAF/2003–03. Washington, DC: U.S. Department of Energy, September 2003.

31. U.S. Environmental Protection Agency. Air emissions trends: Continued progress through 2003. *EPA*. September 27, 2004. Available at <http://www.epa.gov/airtrends/econ-emissions.html> (accessed December 7, 2004).

32. American Lung Association. State of the air, 2004. April 29, 2004. Available at <http://lungaction.org/reports/stateoftheair2004.html> (accessed December 24, 2004).

33. South Coast Air Quality Management District. South Coast air basin smog trend. *AQMD*. May 5, 2004. Available at <http://www.aqmd.gov/smog/o3trend.html> (accessed December 28, 2004).

34. National Research Council. *Estimating the public health benefits of proposed air pollution regulations*. Washington, DC: National Academy Press, 2002.

Further Reading

Davis, D. When smoke ran like water. New York: Basic Books, 2002.

• A critical and personal account of air pollution politics in the United States

Dewey, S. H. Don't breathe the air: Air pollution and U.S. environmental politics, 1945–1970. College Station: Texas A&M University Press, 2000.

• A lively account of efforts to control air pollution prior to the Clean Air Act of 1970

Environmental Poisoning: Mitigating Lead Exposure

The public health risks of lead exposure, particularly for children, have been known for decades. The federal government began to take steps in the 1970s to reduce lead exposure; these efforts have been successful, and may serve as a model for how government intervention can reduce an environmental health hazard. Nevertheless, in 2001 over four hundred thousand children in the United States had blood levels of lead higher than the recommended ten micrograms per deciliter of blood (μg/dL) [1]. Why does lead continue to be a problem?

Lead (Pb) is a heavy metal, widely used in the manufacture of batteries, ammunition, plumbing supplies, and solder, and formerly widely used in gasoline and paint. The health consequences of lead exposure were known as early as Roman times [2]. Lead damages nearly all organ systems, but the nervous system is particularly vulnerable. Lead poisoning has adverse effects on adults and humans, but children, whose bodies are still growing, are more severely affected. Permanent development damage occurs in fetuses exposed to lead in utero, or in children under the age of six who inhale or ingest lead particles. Children with even low levels of lead exposure may develop learning disabilities, behavioral problems, hearing damage, and stunted growth [3]. Acceptable levels of lead exposure dropped steadily as more evidence became available; it appears likely that even the current standard of ten μg/dL is too high.

In 1988, the CDC identified six environmental sources of lead: leaded paint, leaded gasoline, stationary sources such as smelters, contaminated soils and dust, food (through the use of lead-based solder in tin cans), and water (via lead pipes) [4, 5]. Once lead is deposited it may persist for years, creating long-term problems. Unlike other toxic chemicals, lead is not easily broken down; long after the source of pollution is removed, lead is still present.

Lead paint became popular early in the twentieth century; paint replaced wall-paper as the preferred home decorating tool. Early paints contained as much as 50 to 70 percent lead in a linseed oil and kerosene base. This mixture produced a white paint that could then be colored with added chemicals [2]. Because the lead paint tended to flake off as a fine dust, the surface could be washed repeatedly to restore the original color. Yet this tendency to produce fine dust particles contributes to the problem of lead poisoning. The growing recognition of the hazards of lead paint led to calls to regulate lead levels by midcentury. Paint manufacturers lowered the lead content of paint to 1 to 2 percent in the 1950s in an effort to stave off federal action [2]. In 1971, Congress passed the Lead-Based Paint Poisoning Prevention Act, banning the use of lead in interior paints, and barred all residential use in 1978 [6]. Further amendments sought to develop strategies to deal with the large number of homes built prior to 1970 that contained lead paint. The presence of lead paint in older homes remains the greatest current source for lead poisoning [1, 7].

Lead was added to gasoline beginning in 1921 to improve the efficiency and power of gasoline engines; by the mid-1930s, leaded gasoline had claimed 90 percent of the market [5]. Prior to the phase out of leaded gasoline beginning in 1976, the largest source of lead exposure came from the burning of leaded gasoline in auto-mobiles. Ironically, the central motivation for the removal of lead from gasoline was that it interfered with the action of the catalytic converters intended to reduce auto emissions. The EPA mandated further reductions in leaded gasoline in 1985, and leaded gasoline was fully phased out in 1990. Aerosol particles containing lead continue to contaminate soils, however, contributing to persistent lead toxicity, especially in children who play in contaminated areas near roads or in urban areas [7, 8]. Although leaded gasoline is no longer in use in the United States, it is still in use in many other countries [9].

A broad range of law seeks to protect the public from exposure to lead. For example, portions of the Clean Water Act ban the use of lead solder and lead pipes in plumbing fixtures; the CAA includes measures to reduce lead emissions from smelting plants and manufacturers using lead; and the Consumer Product Safety Commission issues regulations to protect children from contact with lead-containing toys and other products [10].

The Lead-Based Paint Hazard Reduction Act, Title X of the Housing and Community Development Act of 1992, called for interventions to prevent lead poisoning in children [11]. These included requiring that known lead hazards must be reported at the time of the sale or lease of properties built before 1978 (when the residential use of lead paint ended), providing information to the public of the hazards of lead-contaminated dust, and establishing standards for cleanup and lead remediation. Title X mandated that the Department of Housing and Urban Development, along with the EPA and the CDC, take primary responsibility for reducing the risk of childhood exposure to lead paint and dust. These agencies have been aggressive in enforcing the Title X requirements, bringing suit against realtors and property owners who violate its provisions [7].

General recommendations to the public for preventing lead poisoning involve maintaining old paint and avoiding household dust that might contain lead, having household water checked for lead, avoiding using leaded crystal or ceramics for food, covering bare soil with mulch to reduce exposure to lead-contaminated dirt, and having children wash their hands before eating [7]. These measures, however, do not get at the root problem: the persistence of lead paint in millions of homes nationwide.

The measures taken beginning in the 1970s, particularly the removal of lead from gasoline and food containers, had a significant impact on blood lead levels in children [12]. Between 1976 and 1991, the mean blood levels of lead dropped 72 to 77 percent in children aged five and under. Higher blood lead levels occurred in Hispanic and low-income children. Public health advocates were encouraged that further reduction in lead poisoning in children was an achievable goal.

Nevertheless, the goal set in 1990 by the DHHS to reduce blood levels of lead in all children under five years old to less than twenty-five μg/dL was not met, even in the face of the recognition that levels of ten μg/dL are more desirable [1]. Of around two million children tested in 2000, almost nine thousand had blood lead levels over twenty-five μg/dL. The DHHS set a new goal of eliminating blood lead levels of over ten μg/dL to be achieved by 2010. Although the total number of children with elevated lead levels declined steadily, the persistence of large numbers of affected children in a small number of urban areas indicates that the current measures to reduce lead exposure are inadequate. While lead poisoning does not respect socioeconomic boundaries, the children of low-income families who live in urban areas with large numbers of homes built prior to 1950 are at greatest risk. In 1998, the American Academy of Pediatrics recommended that pediatricians should inform parents of potential sources of lead poisoning and screen all children at risk [13]. Government-sponsored lead screening programs should be targeted to high-risk areas and coupled with the appropriate public health measures to respond to cases of elevated blood levels.

Why does childhood lead poisoning persist, in view of a long-term government commitment and strong legislation? The main limiting factor is the availability of funds to remediate lead paint in older homes. As long as deteriorating lead paint produces dust, children will be exposed to significant levels of lead through inhalation and ingestion. Many of these same children may play in vacant lots or along roadsides where they are exposed to lead deposited years ago from gasoline emissions. Low-income families, even if they are aware of the risk of lead exposure, are unlikely to be able to afford to take steps to remove or cover lead sources. Although laws state that property owners are responsible for providing safe housing for tenants, the removal of lead paint is a hazardous and expensive process. Lead remediation in federally owned public housing may cost hundreds of millions of dollars alone [14]. Community service organizations provide services to reduce the risks of lead paint, but the number of homes reached is still limited [15]. Until significantly more monetary support is provided to reduce lead exposure, children will continue to suffer from lead poisoning.

References

1. Meyer, P. A., T. Pivetz, T. A. Dignam, D. M. Homa, J. Schoonover, and D. Brody. Surveillance for elevated blood lead levels among children—United States, 1997–2001. *Morbidity and Mortality Weekly Report* 52, no. SS–10 (September 12, 2003): 1–21.

2. Rosner, D., and G. Markowitz. The politics of pollution: Lead poisoning and the industrial age. *Columbia University*. 2002. Available at <http://ci.columbia.edu> (accessed October 26, 2004).

3. Needleman, H. L., and D. Bellinger. The health effects of low level exposure to lead. *Annual Review of Public Health* 12 (1991): 111–140.

4. Agency for Toxic Substances and Disease Registry. *The nature and extent of lead poisoning in children in the United States: A report to Congress.* Atlanta: Centers for Disease Control, July 1988.

5. Jacobson, M. Z. *Atmospheric pollution: History, science, and regulation.* Cambridge: Cambridge University Press, 2002.

6. U.S. Department of Housing and Urban Development. History of lead-based paint legislation. *HUD.* June 18, 2004. Available at <http://www.hud.gov/offices/cpd/affordablehousing/training/leadsafe/ruleoverview/legislationhistory.cfm> (accessed December 27, 2004).

7. National Safety Council. Lead poisoning. March 18, 2004. Available at <http://www.nsc.org/library/facts/lead.htm> (accessed December 13, 2004).

8. Xintaras, C. Analysis paper: Impact of lead-contaminated soil on public health. *ATSDR–CDC.* May 1992. Available at <http://www.atsdr.cdc.gov/cxlead.html> (accessed December 13, 2004).

9. Hu, H. Human health and heavy metal exposure. In *Life support: The environment and human health*, ed. M. McCally, 65–81. Cambridge, MA: MIT Press, 2002.

10. Hazardous Wastes & Toxics Reduction Program. Federal lead regulations and guidance. *Washington State Department of Ecology.* August 5, 2003. Available at <http://www .ecy.wa.gov/programs/hwtr/demodebeis/pages 2/leadregsfed.html> (accessed June 17, 2006).

11. U.S. Environmental Protection Agency. Lead in paint, dust, and soil: Rules and regulations. *EPA.* November 30, 2004. Available at <http://www.epa.gov/lead/pubs/regulation .htm> (accessed June 17, 2006).

12. Pirkle, J., D. Brody, E. Gunter et al. The decline in blood lead levels in the United States: The national health and nutrition examination surveys. *JAMA* 272, no. 4 (July 27, 1994): 284–291.

13. American Academy of Pediatrics. Screening for elevated blood lead levels. *Pediatrics* 101, no. 6 (June 1998): 1072–1078.

14. U.S. General Accounting Office. *Department of Housing and Urban Development: Requirements for notification, evaluation, and reduction of lead-based paint hazards in federally owned residential properties and housing receiving federal assistance, OGC–99–67, September 29, 1999.* B–283667. Washington, DC: General Accounting Office, September 29, 1999.

15. ClearCorps. Protecting children from lead poisoning. October 12, 2004. Available at <http://www.clearcorps.org/> (accessed December 28, 2004).

Further Reading

Needleman, H. L., and D. Bellinger. The health effects of low level exposure to lead. *Annual Review of Public Health* 12 (1991): 111–140.

• A landmark publication in the recognition that even low levels of lead pose a significant health hazard to children

Can We Ever Be Safe? Risk and Risk Assessment

The term risk appears in the discussion of many policy decisions, but remains broadly misunderstood. Developing effective methods to estimate risk as well as compare the likelihood of harm with the costs and benefits of avoiding it are central to policy decisions. Obtaining hard data is often difficult, particularly in cases of environmental hazards. In addition, the public's perception of risk is often at odds with its "scientific" measure. How should policymakers proceed?

In its narrow definition, risk is a mathematical measure—the probability that a substance or situation will produce harm under specific conditions [1]. Two factors are important: the probability that an adverse event will occur, and the consequences of the event. In other words, if an adverse event is likely, but has only mild effects, the risk might be considered low; an unlikely event that has potential catastrophic effects might be viewed as more serious. A third, often-neglected component is the range of exposure—if the hazard is present widely, its relative risk is greater than if it is confined to a small area or subpopulation [2]. Obtaining data that provide the underpinnings for these calculations is often difficult. The thresholds for adverse health or environmental effects may be unknown, requiring scientists to use mathematical models or estimates to calculate risk. Another problematic method is extrapolation, taking data from a small sample and using them to estimate what the broad effect might be [3]. This approach may produce large errors in calculating the possible risk.

A long-standing approach to risk assessment is the consideration of each potential hazard in isolation. This tendency is enhanced by the fragmentation of regulatory organizations charged with the responsibility of protecting health and the environment [4]. While it may be easier to make estimates of the potential risk using this approach, it neglects the fairly obvious observation that risks do not happen in isolation—managing one risk may inadvertently increase another [4, 5].

For example, efforts to reduce the health effects of pollution from automobiles led to calls in the 1980s to develop "clean fuels," gasoline with additives that resulted in fewer emissions. One popular additive, MTBE, proved to be toxic to humans, however, and readily leached into the water table from leaky underground gasoline storage tanks. The resulting water pollution and risk of serious health effects triggered calls for the phase out of MTBE beginning in 1999 [6]. Sometimes trade-offs between risks are inevitable; one risk must be accepted in order to generate a greater benefit from managing another [4]. There is a growing recognition that risk assessment and management must take a more integrated approach [1, 5].

Probabilities alone cannot drive policymaking; the relative costs and benefits of action are critical. In general, the benefits of an action to reduce risk ought to justify its cost. Many federal policies, though, are criticized for their excessive expense. Critics of environmental regulation argue that in the absence of clear-cut benefits, the high costs of regulation are not justified (see chapter 11). In order to be responsive to such criticism, policymakers have turned to cost-benefit analysis [6]. This approach aims to determine the economic benefits of an action relative to its costs. But when good data are hard to obtain, such analysis is fraught with potential error [7]. Putting a price tag on benefits is also problematic—how much is a human life worth? Cost-benefit analysis requires that a monetary value be placed on human life and health as well as noneconomic factors such as environmental quality or biodiversity. Not surprisingly, critics of cost-benefit analyses view such attempts with skepticism [7].

Cost-benefit analysis may also be used as a political tool to delay the implementation of policy or block it entirely. The Reagan administration instituted rules that required detailed cost-benefit analysis *before* new regulations could be placed in effect; this strategy delayed the implementation of regulations for years [6]. By simply manipulating the assumptions in a cost-benefit analysis, an economic "justification" for taking no action against a given risk may be obtained [7]. Among the more ludicrous conclusions drawn from such assumptions is an analysis that "undervalued" children and yielded data that suggested that the costs of reducing lead poisoning in children were not justified by their economic benefits [7].

The largest challenge in the rational management of risk is that the public usually perceives risks very differently from what the scientists may calculate. This perceived risk and the public's response, if poorly handled, can trigger bad policy decisions that neither address a problem effectively nor make good use of limited funds. Yet most policy analysts agree that fears and perceptions cannot be dismissed merely as irrational and subjective, and must be part of the process of risk management [1, 5, 8, 9, 10].

Why does the public's perception of risk differ so widely from that calculated by scientists? Psychologists define a number of factors that contribute to risk perception:

• *Trust*   The more people trust the source of information, the less fear is felt
• *Dread*   A hazard that might kill in a particularly horrible way is viewed as more risky than one that kills in a less dramatic manner
• *Control*   Having a sense of control over the risk reduces the perceived magnitude
• *Natural or human-made*   Human-caused risks are viewed as worse than natural ones
• *Voluntariness*   Taking a risk by choice lessens its impact
• *Children*   Any risks to children are less acceptable
• *Uncertainty*   The more uncertain the risk, the worse it is
• *Novelty*   Newly emerging risks are viewed as less acceptable than more familiar ones
• *Awareness*   The greater the awareness of risk, the greater the fear
• *Proximity*   A hazard that might affect an individual directly is more feared
• *Risk-benefit trade-off*   If taking a risk yields an identified benefit, it is more acceptable
• *Catastrophic or chronic*   A hazard that might kill many people at once is perceived as more risky than one that kills over a long period of time [9].

What is remarkable about these characteristics is that they appear to be shared worldwide; risk perception is similar across cultural and socioeconomic boundaries [2, 8, 11].

Public attitudes toward air and automobile travel are an excellent example of how the perceived risk differs from the calculated one. Particularly after a plane crash, many individuals avoid air travel and use their cars for long trips instead. Motor vehicle accidents, however, killed nearly forty-two thousand people in the United States in 2000; only ninety-two people died in plane crashes [2, 12]. The perceived risk of air travel is distorted by the catastrophic nature of air crashes and the (false) sense of control people have when in their cars.

How should policymakers deal with perceived risk? Excluding the public from deliberations only reduces trust. Simply providing the public with statistical information is also not effective because of the limited understanding of the basis of such calculations. Providing more science education, while desirable, will not necessarily give citizens the ability to weigh conflicting information from other sources. The public needs to be part of the deliberations when risk assessment and management decisions are made [1, 9, 13]. When the rationale for decisions is clear and the public's views are heard, the chance that a new policy will be supported is enhanced.

Science's contribution to risk assessment and management is central, but it is only one part of the decision-making process. Many scientists express frustration that the scientific data supporting certain actions may be ignored or distorted by public perceptions [8, 14]. The uncertainty in many calculations also places scientists in difficult positions—policymakers demand clear evidence that may not be available, despite all efforts. Despite such challenges, scientific data on risk continue to inform the policy process in critical ways. Improving communication to the public will only strengthen the role of science in the deliberative process.

References

1. The Presidential/Congressional Commission on Risk Assessment and Risk Management. *Framework for environmental health risk management.* Final report, Vol. 1. Washington, DC: U.S. Government Printing Office, 1997.

2. Ropeik, D., and G. Gray. *Risk: A practical guide for deciding what's really safe and what's really dangerous in the world around you.* Boston: Houghton Mifflin Co., 2002.

3. Calabrese, E. J. Hormesis: A revolution in toxicology, risk assessment and medicine. *EMBO Reports* 5 (2004): S37–S40.

4. Wiener, J. B. Managing the iatrogenic risks of risk management. *Risk: Health, Safety, and Environment* 9 (Winter 1998): 39–82.

5. Renn, O., and A. Klinke. Systemic risks: A new challenge for risk management. *EMBO Reports* 5 (2004): S41–S46.

6. Sunstein, C. R. *Risk and reason: Safety, law, and the environment.* Cambridge: Cambridge University Press, 2002.

7. Heinzerling, L., and F. Ackerman. *Pricing the priceless: Cost-benefit analysis of environmental protection.* Washington, DC: Georgetown Environmental Law and Policy Institute, 2002.

8. Weingart, P. Science in a political environment. *EMBO Reports* 5 (2004): S52–S55.

9. Ropeik, D. The consequences of fear. *EMBO Reports* 5 (2004): S56–S60.

10. Stern, P. C., and H. V. Fineberg, eds. *Understanding risk: Informing decisions in a democratic society.* Washington, DC: National Academy Press, 1996.

11. Slovic, P. The perception of risk. *Science* 236 (1987): 280–285.

12. National Transportation Safety Board. *U.S. air carrier operations, calendar year 2000: Annual review of aircraft accident data.* ARC–04/01. Washington, DC: N.T.S.B. June 17, 2004.

13. McComas, K. When even the "best-laid" plans go wrong: Strategic risk communication for new and emerging risks. *EMBO Reports* 5 (2004): S61–S65.

14. Gago, J. M. Science policy for risk governance. *EMBO Reports* 5 (2004): S4–S6.

Further Reading

Fischhoff, B., S. Lichtenstein, P. Slovic, S. L. Derby, and R. L. Keeney. *Acceptable Risk.* Cambridge: Cambridge University Press, 1981.

• An early examination of the issues and proposals for working through the problems

Ropeik, D., and G. Gray. *Risk: A practical guide for deciding what's really safe and what's really dangerous in the world around you.* Boston: Houghton Mifflin Co., 2002.

• A lively and informative tabulation of common risks and how their perceived risk does not match reality

Who Lives and Who Dies? Organ Transplantation

Scientists may be asked to weigh in on questions that do not have a scientific basis. One example already explored is human embryos and the beginning of life. Many of these "nonscientific" issues concern the definitions of life and death, or the quality of life. This chapter will examine scientific input into nonscientific questions by looking at organ transplantation and the ever-growing shortage of organs for transplant. Who will determine who should receive an organ? What steps may be taken to assure the equitable distribution of organs? What can be done to alleviate the organ shortage?

A Donation Scenario

A twenty-three-year-old man arrives at a hospital emergency room after being shot in the head in a drive-by shooting. The patient is placed on mechanical life support (respirators and other machines) and stabilized in order to assess the extent of injury. The hospital notes that the patient has indicated on his driver's license that he wishes to donate his organs. The man's family arrives at the hospital and is informed of the seriousness of his injuries. The physicians assure the family that all steps are being taken to save his life, but that the prognosis is grave. It becomes clear that the patient's injuries are so severe that it is unlikely that he can survive. Brain function is assessed and found to be nonexistent; tests include the ability to breathe without mechanical support and the assessment of brain electric activity.

The hospital contacts the local Organ Procurement Organization (OPO; see below) about a potential organ donor. The OPO sends a transplant coordinator, a trained nurse, to speak with the family members. The nurse explains the concept of brain death and asks if they are willing to consider the donation of his organs. After some thought, the family members, although devastated by the loss of their loved one, agree to donate the patient's organs. Blood and tissue typing is carried out, and tested for HIV and other communicable diseases. Tests to determine brain death are repeated after another twelve hours, and the patient is declared brain-dead. The respirator is left on to keep blood flowing to the man's organs, however, and physicians determine that his organs will be suitable for transplant. The OPO identifies potential organ recipients on its waiting lists and notifies the transplant centers that organs are available. The transplant centers notify recipients that they will receive the organs for which they have been waiting.

A transplant surgical team opens the body and removes the organs; the heart, liver, lungs, pancreas, kidneys, and part of the intestine are recovered. The organs are packed in ice and sent to several transplant centers within a few hours away by car or air. Mechanical support is discontinued when the surgeon clamps the blood vessels to the heart prior to its removal. The incisions are neatly closed. The donor's heart is transplanted into one recipient, and his liver, lungs, pancreas, intestine, and kidneys into six others. Seven lives are extended by the donation of the young man's organs.

The Biology of Organ Transplantation

Organ transplantation is possible because of advances in two areas: surgery and immunology. In order to successfully transplant organs, techniques had to be developed that permitted the connection of severed blood vessels. Alexis Carrel developed methods for vascular anastomosis, the joining of blood vessels, in the early 1900s [1]. He was awarded the Nobel Prize in Medicine in 1912. Carrel also conducted experiments in kidney transplantation in animals. When he attempted to transplant kidneys between two dogs (allotransplant), the kidneys quickly failed. In 1914, Carrel published an article that concluded that the organs failed because of "biological incompatibility"—a phenomenon now recognized as rejection. The first successful solid organ transplant between humans occurred in 1954, when Joseph Murray and John Harrison removed a kidney from a living donor and transplanted it into his ailing identical twin [2]. In 1967, Christiaan Bernard became an international celebrity when he successfully carried out the first heart transplant [3]. The first successful liver transplant also occurred in 1967. Most patients, however, only survived a short time before rejecting the organs, and transplantation was viewed as more of a desperate intervention than a medical procedure [4].

Immune System Function and Rejection

Organ rejection is the result of immune system function. The immune system is a defense against foreign pathogens such as viruses and bacteria; immune system cells recognize such pathogens as different from the cells of the body. Control of the immune system function is critical to the survival of the organism. If the immune system begins to attack its host, serious medical conditions such as rheumatoid arthritis or lupus can result. It is thus critical that immune system cells be able to distinguish between "self" and "nonself." This recognition is achieved using cell surface proteins.

All living organisms carry proteins on their cell surfaces. These proteins play many functional roles for cells. A subset of cell surface proteins are involved in cellular recognition, either for immediate cell-to-cell interactions or more distant ones. These proteins commonly have sugar molecules attached to them and are called glycoproteins. The first glycoprotein system characterized was the A, B, and O blood system of red blood cells. Karl Landsteiner determined in 1907 that all red blood cells from a single individual expressed the type A glycoprotein, the type B glycoprotein, both (type AB), or neither (type O). When blood of different types was mixed, the cells clumped together. Landsteiner reasoned that a mismatch of blood cell types explained why some blood transfusions, which had been attempted as early as the seventeenth century, were successful and some were not. Landsteiner's research opened the door for the safe transfusion of blood between humans, and he was awarded the Nobel Prize in 1930 for his work.

Even after the characterization of the A, B, and O blood groups along with the recognition of the need to match blood types, attempts to transplant kidneys failed. In the 1940s, Peter Medawar studied mechanisms of skin graft rejection in mice and determined that the rejection of grafts was mediated by the immune system. Medawar was awarded the Nobel Prize in 1960. Further research determined that the immune system recognized a particular class of cell surface proteins, the major histocompatibility loci. In humans, these proteins are called human leukocyte (white blood cell) antigens (HLA). HLA proteins are encoded by a highly diverse gene complex; only identical twins share identical HLA proteins [5]. Yet HLA proteins may be grouped into several classes that are somewhat similar. The greater the similarity in these classes of HLA proteins between a donor and a recipient, the less the immune system is activated. When organ transplantation is attempted between unrelated people, the immune system recognizes the HLA proteins on the transplanted organ as foreign and mounts a response against them. Depending on the degree of mismatch, the organ may be rejected in days or weeks (acute rejection), or after months or years (chronic rejection). In both cases, differences in HLA cause T-cells, the immune system cells closely involved in tissue rejection, to attack the transplanted organs [2].

A dramatic form of rejection occurs when organs are especially mismatched. The transplanted organ begins to blacken and fail within minutes of transplant. The hyperacute reaction occurs because the immune system recognizes glycoproteins that are present on the endothelial cells that line the blood vessels, in addition to HLA proteins. In all mammals but primates, a complex sugar, galactose-alpha,1,3,-galactose (GAL), is expressed on the endothelial cells. If an organ from a pig is transplanted into a human, circulating antibodies immediately recognize the GAL as foreign and mount a response against it [2, 6]. The hyperacute response poses a serious barrier to successful xenotransplantation, the use of animal organs in human recipients (see "Animals as Organ Factories" section below). Hyperacute rejection also may occur in human-human transplantation if a person has been sensitized to foreign antigens by a previous blood transfusion.

Preventing Rejection

After the basis of rejection was determined, scientists worked to develop methods to suppress the immune response. Screening for HLA antigens to achieve the closest possible match is one approach, but even the closest match differs from identity, and the organ will be rejected. Clinicians also use chemicals that block parts of the immune response.

Early immunosuppressant drugs included corticosteroids, a group of anti-inflammatory drugs that inhibit the activation of T-cells [7]. Corticosteroids, such as prednisone, appear to function nonspecifically to reduce rejection. A second approach is to prevent the proliferation of activated T-cells. Azathioprine is one of the first antimitotic drugs to be developed; it blocks DNA synthesis and prevents cell division. Another approach to blocking cell proliferation is to treat bone marrow with radiation; irradiation kills dividing cells by damaging their DNA. Early transplant patients were treated with corticosteroids and either azathioprine or whole body irradiation. These treatments could not fully suppress the immune response at nontoxic doses, though [8]. Organ transplants continued to have limited success.

The organ transplantation world changed in 1979, when cyclosporine, purified from a soil fungus, was developed as an immunosuppressant [8]. Cyclosporine effectively blocks T-cell activation and successfully prevents transplant rejection. Related drugs include FK-506 (tacrolimus) and rapamycin (sirolimus). When cyclosporine or its relatives is taken with azathioprine and corticosteroids, the long-term blockade of rejection is achieved. Since the development of cyclosporine, organ transplantation has become an almost routine procedure, although the recipients need to take immunosuppressant drugs for the rest of their lives.

The irony of successful immunosuppression is that it places the transplant recipient at increased risk for infection and cancer because other immune functions are also blocked. Patients are immunocompromised; they are less able to ward off bacterial or viral infection, and their immune systems are unable to recognize and kill cells that may become cancerous. Immunosuppressed transplant recipients exhibit many of the same problems as individuals suffering from AIDS (see Chapter 5), including infections and unusual cancers. Immunosuppressive drugs may also be toxic to organs; cyclosporine is notorious for its damage to kidneys. Because of this, efforts are made to keep doses as low as possible, but the patient's maintenance becomes a delicate balancing act between drug toxicity, infection, and transplant rejection. Nevertheless, close management of transplant recipients by physicians has allowed many patients to live for over ten years with their new organs—providing not only life but also a high quality of life.

Advances in medicine make possible the replacement of damaged organs with new ones. Patients whose failing kidneys, livers, pancreas, intestines, lungs, or hearts would otherwise cause their deaths can have their health restored by the transplantation of a new organ. The development of immunosuppressive drugs in the late 1960s and 1970s permitted successful transplants from unrelated donors for the first time. Since then, thousands of individuals have benefited from organ transplant. Almost 27,000 transplants were performed in the United States in 2004, from over 14,000 donors [9]. Unfortunately, the number of patients who might benefit from transplants far exceeds the number of available organs. As of December 2005, the waiting list for organs exceeded 90,700; in recent years, around 7,000 people died each year while on the waiting list [9].

The severe shortage of available organs leads to difficult ethical and policy challenges. Most organs will not remain viable for more than a few hours after blood stops flowing through them. Clinicians try to minimize the time between the organ removal from a donor and the transplant into a recipient. Donors may be kept on mechanical support in order to complete arrangements for transplant (see below). The narrow window of time for successful transplants plays a major role in determining who might be the recipient for an organ, and also contributes to the limited number of actual donors.

Obtaining Organs: Developing Policies

Although death is inevitable, its definition is difficult. Most agree that death is a process, not an event, and so the moment of death is not definable in scientific terms. Before medical devices such as mechanical ventilators and cardiac resuscitators, people died when they irreversibly stopped breathing and their heart stopped beating. Defining death was a simple matter; death was determined by a mirror under the nose and fingers to the artery in the neck. The majority of people still die when respiration and heartbeat cease (cardiovascular death) [3], and no one questions their deaths. But advances in medicine complicate what it means to die. Medical science has now enabled the lengthy maintenance on respirators of individuals who would otherwise die.

Unfortunately, without intervention, most organs from cadavers are unusable because of damage from a lack of oxygen after the patient's death. Unless organs are recovered within minutes after death, this hypoxic damage renders organs unusable. Most people who die are not good candidates for organ donation because their organs are damaged as their cardiovascular function fails. The need for organs led to the development of a new definition of death—brain death. Shifting to a different definition would allow clinicians to maintain body organs for recovery using life support even if the brain dies.

When Is a Person Dead?
In 1968, a committee at Harvard Medical School proposed that persons whose brains had irreversibly ceased functioning should be considered dead, regardless of their cardiovascular status. Irreversible brain damage might result from interruptions in blood flow (ischemia) or oxygen (hypoxia), or massive trauma. By the 1970s, countries around the world began to discuss a new definition of death based on brain function. A notable exception is Japan, where organ transplantation from dead donors remains exceedingly rare [4]. Guidelines for the medical tests required to declare brain death were codified by panels of physicians associated with medical schools [10]. Brain death is defined as irreversible whole brain damage so extensive that there is no potential for recovery and the body can no longer maintain normal homeostasis (functions including respiration, heartbeat, and temperature regula-

tion). Tests for brain death include the absence of spontaneous respiration when the ventilator is turned off, no electric brain activity, the absence of reflexes other than those from the spinal cord, the persistence of the condition (usually for twelve hours), and other measures to assure that the person has not taken drugs that might depress function or is suffering from hypothermia (low body temperature).

During the 1970s, most states moved to a brain death definition; death now had a *legal* criterion. Most people who die, however, do not formally meet the criteria for brain death; they die because of the cessation of cardiovascular function. In 1980, the National Conference of Commissioners on Uniform State Laws approved the Uniform Determination of Death Act that provides for *two* definitions of death. These are the "(1) irreversible cessation of circulatory and respiratory functions, or (2) irreversible cessation of all functions of the entire brain, including the brain stem" [11]. The act does not spell out specific criteria for the determination of death, and instead requires that accepted medical practice be followed. With the increased emphasis on brain death, many people today do not realize that there are two ways to legally define death. This confusion has led to controversy about the status of potential organ donors, as described below.

Despite the general acceptance of brain death, considerable misunderstanding still exists among the general public. If a patient is determined to be brain-dead, mechanical support may be discontinued without approval from the family. Some family members may still persist in demanding that the mechanical support be maintained for their loved one because they do not understand brain death. It can be confusing to see a body that appears alive and be told that the person is dead [12].

Where Do Organs Come From?

The success of organ transplantation through the use of immunosuppressive drugs creates a new problem: where do physicians obtain organs for transplant?

Living Donors
In the 1950s, the majority of kidneys made available for transplant came from living donors. The first successful transplant occurred between identical twins. The use of living donors limited transplantation to those organs that are paired, such as kidneys, and organs that can be split into sections, such as livers and bone marrow. Single whole organs, such as hearts and pancreases, are obviously not available, since medical ethics do not permit the removal of organs that lead to the donor's

death. Most individuals are also unwilling to give up an organ to a stranger. Nevertheless, the number of transplants involving organs from living donors has increased dramatically since 1992 [13]. At some transplant centers, about half of the kidneys transplanted come from living donors [9, 14]. Recipients of organs from living donors tend to have a better outcome because they tend to receive organs when they are less sick. Most living donors give organs to family members or friends, allowing them to avoid waiting lists.

Kidneys are the most common organs donated by living donors; portions of livers and lungs also may be donated. A donation entails considerable medical risk to the donor, from the surgery and the complications of surgery. Professional societies and ethicists express concern whether potential donors are fully aware of the risks associated with procedures, and whether they agree to donate without coercion [14, 15]. The American Medical Association (AMA) recommends that donors must be fully informed of the risks of donation, and be made aware of the possible outcomes for the recipients (including negative ones). They must be given the opportunity to decline to be donors, even to family members, and must be assessed for their psychological competency for making the decision to be a donor. Donors must not bear any costs associated with donation; however, they may not be paid directly for a donated organ, as stipulated by law (see below). Despite this admonition, donors do risk economic losses from missed work time and the like [16].

Sometimes family members are not suitable donors because of a mismatch of blood types. For kidneys, a rare but novel approach is to arrange for kidney "swapping"; two (or more) individuals agree to a direct donation to a stranger in exchange for a kidney for a family member [17]. Kidney swapping again allows individuals to jump the line on waiting lists, and reduces the number of organs for individuals on waiting lists who do not have a family member who can donate an organ.

Cadaveric Donation
In 1968, the Uniform Anatomical Gift Act (UAGA) [18] allowed individuals over eighteen to stipulate in advance that they wished to donate their organs on their deaths. Donors might complete a donor card when obtaining a driver's license or other documents to indicate their wishes. At the time of death, family members are asked if they wish to donate the organs of their loved one. Family members may also approve the donation of the deceased's organs in the absence of a written directive. An important element of the law is that the physician determining the patient's death may not participate in removing or transplanting organs. This stipulation

responds to concerns that physicians will not make every effort to treat a patient if he/she indicated a willingness to be an organ donor. Yet a Gallup Poll taken in 1985 [19] revealed that although 75 percent of Americans polled approved of the concept of organ donation, only 27 percent indicated that they were likely to donate their organs, and only 17 percent had signed a donor card. Half had not discussed organ donation with family members.

In 1987, the UAGA was revised to make it easier for an individual to express a willingness to donate organs. It also dropped the requirement of obtaining approval by the next of kin [19] at the time of death and permitted the limited recovery of organs in the absence of a paper trail. Although all states adopted the earlier version of the UAGA, only twenty-five states adopted the revised version [20, 21]. Other states found the loosened criteria for donation troublesome [21].

The Patient Self-Determination Act of 1990 [22] made clear a patient's right to refuse treatment such as life support and express these wishes in an advance directive (a "Living Will," or Durable Power of Attorney for Health Care). Patients might indicate their wish to donate organs on their death in such a directive. Despite the new laws, however, hospitals continue to request permission from family members to recover organs [23, 24, 25, 26]. Over 40 percent of families refuse to allow organ donation, even if the deceased indicated a wish to donate. Still, families are more likely to agree to the donation if they were aware of the deceased's wishes. Physicians appear unwilling to risk the legal consequences if they proceed to harvest organs without family approval, and some states do not regard living wills as legal documents.

The Dead Donor Rule

A central tenet of organ transplantation is that organs may only be harvested from dead donors (the exception being voluntary donations from living donors); the removal of organs for transplant may not cause the individual's death [12, 27]. The shortage of organs, however, has led many to suggest that the dead donor rule is outmoded and should be reconsidered. If waiting for death only assures that organs will not be usable, why not loosen the restrictions on harvesting organs? [27, 28]. In particular, the extreme shortage of organs for pediatric patients, especially infants, has led to calls to permit donations from anencephalic babies prior to their death.

Anencephaly, the absence of cerebral cortices of the brain and overlying structures, results from the failure of the neural tube to close during early embryonic development. Anencephalic babies are permanently unconscious, but retain some brain stem function; they can breathe and exhibit reflex behavior reminiscent of normal newborns. These babies have no chance for survival and typically die with a few days of birth [2]. Many parents of anencephalic babies wish to allow their children to be organ donors, in order to save the lives of other infants. Yet if anencephalic babies are maintained until the brain stem function ceases, their organs are often too damaged to be useful for transplantation.

Since these children have no future, it is suggested that the dead donor rule should be waived to allow for the recovery of organs before the babies are legally dead. In 1995, the AMA presented a report that supported permitting parents to donate organs from anencephalic babies prior to their death [29]. The AMA argued that the possibility of saving the lives of other children represented a significant ethical benefit. The fundamental requirement to protect the interests of the organ donor does not operate in the case of anencephalic infants, who cannot experience consciousness and hence do not have interests in remaining alive. Appropriate safeguards would be established to prevent other severely disabled children from being used as organ donors prior to their death. Concerns that waiving the dead donor rule might open the door to abuse of other disabled populations are answered by contending that anencephalic babies are a unique case. Nevertheless, a few months later the AMA rescinded its own recommendation in the wake of strong public disapproval [2]. No laws have been passed that permit the waiver of the dead donor rule to allow for the recovery of organs from anencephalic babies prior to their death, even if their parents request it.

A second population that has been proposed as potential donors is individuals in persistent vegetative states (PVS) (see "Is There a Right to Die?" section below). In PVS, the cerebral cortices die as a result of ischemia or hypoxia. The brain stem remains alive, though, allowing the individual to breathe without a respirator. Individuals with PVS go through periods of apparent wakefulness and sleep, but have no conscious awareness. With proper nutritive and medical support, PVS patients may live for years. Some suggest that PVS patients, although they do not meet the criteria for brain death, are not truly alive because they have no consciousness. Since the rest of the body is generally healthy, such patients might be suitable organ donors [28]. But many ethicists find this proposal troubling. PVS is difficult to diagnose [30], and in order to ease concerns about abuse of the severely disabled, a reliable diagnosis is critical. Critics assert that allowing PVS patients to be organ donors sets a disturbing precedent for using other populations, such as the severely mentally retarded, as well.

Non-Heart-Beating (NHB) Donors

In 1996, *60 Minutes* ran a story that suggested that the Cleveland Clinic, a highly respected hospital in Ohio, was collecting organs from patients who were not dead, violating the dead donor rule [2]. According to the report, the Cleveland Clinic used NHB criteria to determine death: the patient would be taken off life support, and their organs were removed within minutes of the cessation of the heartbeat and respiration. Patients are not brain-dead using this protocol, but it is necessary to declare them dead using the cardiac and respiratory criteria spelled out in the Uniform Definition of Death Act. Concerned individuals went to the Cleveland district attorney, who launched an investigation. No evidence of wrongdoing was found. Apparently, someone made public a preliminary version of an NHB protocol that had never been adopted by the hospital.

Persistent concerns about the abuse of dying patients led the DHHS to ask the Institute of Medicine (IOM) to explore the medical and ethical issues associated with obtaining organs from NHB donors. The IOM report, released in 1997 [31], found considerable variability in the criteria for procuring organs from NHB donors, and made a series of recommendations for standards and protocols. Among the more controversial recommendations was setting a five-minute standard for heartbeat cessation before declaring a patient dead. What is not certain is whether a patient might be resuscitated after this point or whether some brain function remains [3, 32]. Those who support NHB donation suggest that removing life support from a non-brain-dead patient is legal [3, 28], given a patient's right to refuse medical treatment and sign advance directives under the Patient Self-Determination Act [22]. The controversy arises because organs are to be recovered and the timing of the declaration of death is therefore critical [32]. Under other circumstances, it is unlikely that anyone would challenge a physician's pronouncement of death.

Proponents of NHB donation point out that the number of potential donors who fulfill brain death criteria is limited. Using the cardiovascular definition of death expands the potential pool of donors while adhering to the principles of the dead donor rule. An advantage of NHB donation is that the timing of death resulting from the withdrawal of mechanical support can be coordinated to maximize the likelihood that healthy organs may be obtained, although these are limited to kidneys and livers for technical reasons. The IOM published a set of guidelines and protocols for NHB donation in 2000 [33]. The report directs OPOs to work to develop protocols for donation.

NHB donation is limited to a small number of transplant centers; only 2 percent of kidney transplants from cadavers were conducted using organs obtained in this manner in 2000 [34]. NHB donation remains controversial, despite its potential to increase the number of organ donors.

Who Gets Organs?

The success of transplantation creates a high demand for donor organs. Waiting lists for organs, particularly for kidneys and livers, increase every year. As of December 2005, over 90,700 individuals are on waiting lists for organs; almost 65,000 are waiting for kidneys [9]. The shortage of organs for transplant creates intense competition. How can they be distributed fairly?

The general premise underlying waiting lists for organs is medical necessity. Patients are added to waiting lists when it is determined that they will not survive without an organ transplant and are otherwise healthy enough to benefit from the transplant. A patient who may die within days is higher priority than one who will live for months, regardless of how long each has been on the waiting list. Put simply, sicker patients may "cut in line"; one's position on a waiting list is determined by the severity of the illness, not the time on the waiting list. The other central criterion is the ability to pay for the transplant; patients without medical insurance may not be placed on waiting lists for some organs unless they can demonstrate a means to pay for the procedure [2]. Poor patients are effectively denied access to transplants because they cannot get referrals to transplant centers for medical evaluation.

While there is a national database of potential organ recipients, waiting lists are local or regional. Geographic disparities in population and the number of transplant centers mean that waiting times are variable. Tragically, over seven thousand people died in 2004 waiting for organs [9].

Rapid advances in the success of organ transplantation in the early 1980s made clear both the promise of transplantation and the growing shortage of available organs. The National Organ Transplant Act (NOTA) [35], enacted in 1984, provided funds to establish the Organ Procurement and Transplantation Network (OPTN) to coordinate organ sharing between OPOs. The OPTN was designed to help OPOs allocate organs fairly by standard criteria both locally and nationally. The goal was to assure that the limited number of available organs would all be used in appropriate ways. The United Network for Organ Sharing (UNOS), a nonprofit organization established earlier as a central registry for kidney donation, was designated as the OPTN coordinator.

NOTA also established a task force to study issues related to the shortage of organs and develop recommendations to encourage organ donation [36]. A recommendation to require hospitals to request donations from family members ("required request") became law in 1986 [37]. Hospitals also must comply with all UNOS regulations concerning organs allocation in order to receive Medicare funds. Despite the new regulations, the organ demand continued to grow, as supply lagged far behind.

The Final Rule

NOTA was amended in 1988 and 1990, clarifying the OPTN's role in determining equitable organ sharing. The 1990 amendments made clear the intent that OPTN should act as a national network. In 1998, the DHHS recommended a series of significant changes to OPTN operations that would expand OPOs to cover much larger regions and eliminate the regional disparities in access to organs. Organs should be given to the sickest patients, regardless of geographic location. Uniform criteria for placement on waiting lists and the determination of medical status should be developed [38, 39].

Transplant centers and patients voiced strong objections to the proposed rule. They argued that organs would go only to the large centers and the small transplant centers would be forced to close. Organ donations would decline because people preferred their organs to be used locally. By always choosing the sickest patients, the percentage of successful transplants would decline and this would also discourage donations, exacerbating the organ shortages [2, 40]. The state of Louisiana filed suit against the DHHS to block the implementation of the final rule; the courts delayed implementation for one year. Congress mandated that the IOM convene a panel to review the implications of the DHHS final rule. The IOM focused on liver transplants, the organs in shortest supply with a high waiting list mortality. The IOM released its report in 1999 [41]; it recommended that organ allocation areas serving populations of at least nine million people be established for liver donations, unless the time needed to transport the organs would exceed their survival time. The IOM concluded that the median time on a waiting list was not a good measure of access to organs and suggested that the number of waiting list deaths was a better measure. The IOM also concluded that there was no evidence that the broader sharing of organs would reduce donations.

The DHHS revised the final rule in response to the IOM report. After additional wrangling, the OPTN final rule was enacted in 2000. Despite the rule, considerable disparities in waiting times between regional areas persist.

Why Do So Few People Donate Organs?

Because OPOs continue to request family members' permission to retrieve organs, the refusal by about half of the families asked is a major contributor to organ shortages. Numerous polls indicate that the majority of Americans support the idea of organ donation [23, 26]. Why, then, do so few complete donor cards or advance directives?

Many individuals do not sign a donor card for fear that physicians will not work as hard to save their lives. Others are opposed to the donation for religious reasons. Many families believe that if organs are removed, it is not possible to have on open casket at a funeral. Some remain highly suspicious of the fairness of organ distribution; this suspicion is often voiced by family members who refuse permission to donate a loved one's organs [23, 24, 42]. Finally, families frequently are approached when they are suffering the emotional shock of unexpectedly losing a loved one, and thus are unwilling or unable to consider a request to donate organs. Studies suggest that both the person making the request and the mode of request can significantly affect whether families give permission. Some studies indicate that hospitals, in order to reduce the costs of maintaining bodies on mechanical support, prevent OPOs from approaching families; instead, patients are removed from life support and allowed to die before they can be declared brain-dead.

How Can the Supply of Organs Be Increased?

Despite new initiatives and legislation, the gap between organ donation and the demand for organs continues to grow [43]. Medical advances loosened the criteria for potential donors, thereby allowing for older donors with more medical problems. The central problem, however, appears to be the failure to obtain organs from potential donors—even those who expressed their willingness. Only a small fraction of potential cadaveric donors provide organs [24, 26], and efforts to increase the fraction are not succeeding.

The growing shortage has led to a variety of suggestions to increase donations. The first is a call for hospitals to adhere to the 1987 UAGA and follow any advance directives left by dying patients, even if it means overriding family objections [44]. The second suggestion is to require all individuals to complete a donor card, which then becomes part of a national registry [40, 45]. This "mandated decision" would mean that no potential donor would be rejected because of the absence of information. Some view mandated donation as coercive; they claim that such a process would actually decrease the number of individuals who agree to be donors by making them feel pressured to make a decision [45, 46].

Another option is "presumed consent." All adults are presumed to be donors unless they opt out of the donor pool by registering their objection. Presumed consent is used in a number of European countries, including Belgium, Austria, Spain, and France [3, 40, 47]. The number of people who opt out varies from country to country, but in general, the rates of donation are higher. A similar system is unlikely to be instituted in the United States, where individual decision making is held to be paramount. A possibly more palatable version for the United States might be an approach that informs a potential donor's family that organs will be harvested unless the family objects.

A final approach to increase organ donation is the use of financial incentives. To date, organ donation has been based on altruistic assumptions; NOTA includes a ban on payments for organs [35]. The "gift of life" approach requires potential donors to accept the premise that it is generous and ethical to give up one's organs on death. The reward to the donor or the donor's family is the feeling that the organs will preserve the life of others. Organs become a public good, and those who would give their organs if something happened to them may appreciate the benefit they or their family and friends might realize by the similar actions of others. In contrast, taking a purely economic approach to organ donation is suggested as an option to alleviate much of the current shortage [46]. What is necessary is a determination of what payments will constitute an adequate incentive and compensation for donation. The payment may come in several forms: as a direct payment for the organs, as tax credits or refunds, or as an indirect payment for funeral expenses or other costs. There are no data to support the contention that payment will increase the donation rate, though.

Pennsylvania enacted legislation in 1994 (Act 102) to increase organ and tissue donations by means of education and public awareness programs. The law also allowed for the payment of up to $300 to funeral homes to help cover the costs of a donor's funeral. The latter part of the act has not been placed in practice because it may violate NOTA's ban on the exchange of organs for anything of monetary value. In 2002, the AMA's Council on Ethical and Judicial Affairs recommended that pilot programs be established to determine whether financial incentives will increase donation rates [48]. Several bills have been introduced in Congress to allow the payment of the "expenses" of living organ donors, but none have passed to date [49].

Payment for organs is strongly opposed by both professional groups and ethicists [49, 50, 51]. Payment for donation could exploit the poor, who already have limited access to transplants. It provides an incentive to be dishonest about medical conditions that might otherwise eliminate a potential donor, and might cause family members to prematurely withdraw care in order to receive payment for organs. Payment for organs commodifies the human body. Finally, there is little direct evidence that some form of payment would increase the likelihood of organ donation. Still, even opponents suggest that token payments for funerals, paid medical leave for living donors, and insurance for donors might be acceptable [49].

Abuses of living donors are documented in other countries, such as India and China, where an active black market in organs exists [52]. Wealthy people already travel to foreign countries to obtain organ transplants [49]. There is also concern that living donors in the United States may be given informal financial incentives to donate an organ to an unrelated recipient. In 1998, two Chinese nationals were arrested in the United States for trafficking in organs allegedly collected from executed prisoners in China [53]. In the aftermath, professional organizations discussed the possibility of organ donation by prisoners, either at the time of execution or as a means to reduce sentences. Most found that the approach was clearly coercive [54].

Ironically, even if every potential donor contributed organs for transplant, there will not be enough organs to meet the demand [43]. The number of appropriate brain-dead donors has not increased in recent years, and most increases in transplants result from living donors. Unless the donation criteria are changed or there is an increased use of NHB donors, the gap between the supply and the demand will continue to grow.

Lessons from Organ Transplantation

Organ transplantation has enabled thousands of gravely ill individuals to gain a new lease on a high quality of life. Transplant recipients have lived over thirty years with their donated organs. Continued advances in immunosuppression allow patients to live longer with their new organs with a reduced risk of rejection. The more progress is made in preventing rejection, however, the worse the organ shortage becomes. Improvements in technology also result in making more people acceptable candidates for transplants, further swelling the waiting lists.

As long as thousands remain on waiting lists, a "medical" choice may mean life for one person and death for another. The desperate state of many patients has led to increased discussion of moving from an altruism-based approach to organ donation to an incentive-based one [55].

Organ transplantation produced a new legal definition of death, based on medical criteria that are difficult to attain in many cases. The shortage of organs also leads to calls to reevaluate this definition of death in order to perhaps identify other populations of potential donors (see "Is There a Right to Die" section). Can science provide a firm definition of death to aid policymakers?

The organ shortage also motivates researchers to look for other sources of organs. While artificial hearts have been studied, they seem most effective as stopgap measures, not permanent solutions. Kidney dialysis can keep patients alive for years, but with a relatively poor quality of life. Xenotransplantation, the use of animal organs for transplant, is under increasing study (see "Animals as Organ Factories" section). This approach carries with it a new set of medical and ethical challenges. The potential of embryonic stem cells to generate replacement organs and tissues (see chapter 4) also raises nonscientific questions because of the cell source.

References

1. Papalois, V. E., N. S. Hakim, and J. Najarian. The history of kidney transplantation. In *History of organ and cell transplantation*, ed. V. E. Papalois and N. S. Hakim, 76–99. London: Imperial College Press, 2003.

2. Munson, R. *Raising the dead: Organ transplantation, ethics, and society.* Oxford: Oxford University Press, 2002.

3. Price, D. *Legal and ethical aspects of organ transplantation.* Cambridge: Cambridge University Press, 2000.

4. Lock, M. *Twice dead: Organ transplants and the reinvention of death.* Berkeley: University of California Press, 2002.

5. Duquesnoy, R. J. Histocompatibility testing in organ transplantation. Available at <http://tpis.upmc.edu/tpis/immuno/wwwHLAtyping.htm> (accessed November 10, 2003).

6. Cooper, D. K. C., and R. P. Lanza. *XENO: The promise of transplanting animal organs into humans.* Oxford: Oxford University Press, 2000.

7. Helderman, J. H., W. M. Bennett, D. M. Cibrik, D. B. Kaufman, A. Klein, and S. K. Takemoto. Immunosuppression: practice and trends. *American Journal of Transplantation* 3, supp. 4 (2003): 41–52.

8. Kamps, M. The history of immunosuppressive drugs. In *History of organ and cell transplantation*, ed. N. S. Hakim and V. E. Papalois, 335–346. London: Imperial College Press, 2003.

9. United Network for Organ Sharing. U.S. transplantation data. *UNOS.* December 27, 2005. Available at <http://www.unos.org/data/> (accessed December 27, 2005).

10. Plum, F., and T. B. Posner. *The diagnosis of stupor and coma.* 3rd ed. Philadelphia: F. A. Davis Company, 1980.

11. National Conference of Commissioners on Uniform State Laws. Uniform Determination of Death Act. *NCCUSL.* Available at <http://www.law.upenn.edu/bll/ulc/fnact99/1980s/udda80.htm> (accessed March 10, 2005).

12. Koppelman, E. R. The dead donor rule and the concept of death: Severing the ties that bind them. *American Journal of Bioethics* 3, no. 1 (2003): 1–9.

13. Organ Procurement and Transplant Network. OPTN/SRTR annual report 2003. Available at <http://www.optn.org/AR2003/default.asp> (accessed July 16, 2004).

14. The Authors for the Live Organ Donor Consensus Group. Consensus statement on the live organ donor. *JAMA* 284, no. 22 (2000): 2919–2926.

15. Kallich, J., and J. F. Merz. The transplant imperative: Protecting living donors from the pressure to donate. *Iowa Journal of Corporation Law* 20 (1994): 139–154.

16. Jacobs, C., and C. Thomas. Financial considerations in living organ donation. *Progress in Transplant.* 13 (2003): 130–136.

17. Menikoff, J. Organ swapping. *Hastings Center Report* 29, no. 6 (1999): 28–33.

18. National Conference of Commissioners on Uniform State Laws. Uniform Anatomical Gift Act. *NCCUSL.* Available at <http://www2.sunysuffolk.edu/pecorip/ SCCCWEB/ETEXTS/DeathandDying_TEXT/Uniform%20Anatomical%20Gift%20Act.pdf> (accessed November 7, 2003).

19. National Conference of Commissioners on Uniform State Laws. Uniform Anatomical Gift Act (1987). *NCCUSL.* Available at <http://www.law.upenn/bill/ulc/fnact99/uaga87.htm> (accessed March 10, 2005).

20. National Conference of Commissioners on Uniform State Laws. Anatomical Gift Act (1987). *NCCUSL.* Available at <http://www.nccusl.org/Update/uniformact_factsheets/uniformacts-fs-aga87.asp> (accessed March 10, 2005).

21. Keller, K. A. The bed of life: A discussion of organ donation, its legal and scientific history, and a recommended "opt-out" solution to organ scarcity. *Stetson Law Review* 32 (2003): 855–895.

22. Patient Self-Determination Act of 1990. Pub. L. 101-508 (codified at 42 U.S.C. §§ 1395, 1396).

23. Siminoff, L. A., N. Gordon, J. Hewlett, and R. M. Arnold. Factors influencing families' consent for donation of solid organs for transplantation. *Journal of the American Medical Association* 286, no. 1 (2001): 71–77.

24. Langone, A. J., and J. H. Helderman. Disparity between solid-organ supply and demand. *New England Journal of Medicine* 349, no. 7 (2003): 704–706.

25. Wendler, D., and N. Dickert. The consent process for cadaveric organ procurement: How does it work? How can it be improved? *JAMA* 285, no. 3 (2001): 329–333.

26. Siminoff, L. A., R. M. Arnold, A. L. Caplan, B. A. Virnig, and D. L. Seltzer. Public policy governing organ and tissue procurement: Results from the national organ and tissue transplant procurement study. *Annals of Internal Medicine* 123, no. 1 (1995): 10–17.

27. Robertson, J. A. The dead donor rule. *Hastings Center Report* 29. no. 6 (1999): 6–14.

28. Truog, R. D. Is it time to abandon brain death? *Hastings Center Report* 27, no. 1 (1997): 29–37.

29. American Medical Association, Council on Ethical and Judicial Affairs. The use of anencephalic neonates and organ donors. *Journal of the American Medical Association* 273, no. 20 (1995): 1614–1618.

30. American Medical Association. Persistent vegetative state and the decision to withdraw or withhold life support. *JAMA* 263, no. 3 (1990): 426–430.

31. Institute of Medicine. *Non-heart-beating organ transplantation: Medical and ethical issues in procurement.* Washington, DC: National Academy Press, 1997.

32. Menikoff, J. Different viewpoints: The importance of being dead: Non-heart-beating organ donation. *Issues in Law and Medicine* 18 (2002): 3–19.

33. Institute of Medicine, Committee on Non-Heart-Beating Transplantation. II: The scientific and ethical basis for practice and protocols. In *Non-heart-beating organ transplantation: Practice and protocols.* Washington, DC: National Academy Press, 2000.

34. Cecka, J. M. Donors without a heartbeat. *New England Journal of Medicine* 347, no. 4 (2002): 281–293.

35. National Organ Transplant Act of 1984. Pub. L. 98-507 (codified at 42 U.S.C. §§ 273, 274).

36. National Attorneys' Committee for Transplant Awareness. Organ and tissue donation and transplantation: A legal perspective. *NACTA.* Available at <http://www.transweb.org/reference/articles/donation/nacta.html> (accessed November 19, 2003).

37. Omnibus Budget Reconciliation Act of 1986. Pub. L. 99-509 (codified in pertinent part at Vol. 42, U.S.C., sec. 1320b-8).

38. U.S. Department of Health and Human Services, Health Resources and Service Administration. *Organ procurement and transplantation network: Final rule.* Vol. 42, C.F.R. part 121, 1999.

39. U.S. Department of Health and Human Services. HHS rule calls for organ allocation based on medical criteria, not geography. *HRSA.* Available at <http://www.hhs.gov/news/press/1998press/980326a.html> (accessed November 7, 2003).

40. Davis, R. M. Meeting the demand for donor organs in the US. *British Medical Journal* 319 (1999): 1382–1383.

41. Institute of Medicine, Committee on Organ Procurement and Transplantation Policy. *Organ procurement and transplantation: Assessing current policies and the potential impact of the DHHS final rule*. Washington, DC: National Academy Press, 1999.

42. Peters, T. G., D. S. Kittur, L. J. McGaw, M. R. First, and E. W. Nelson. Organ donors and nondonors: An American dilemma. *Archives of Internal Medicine* 156, no. 21 (1996): 2419–2424.

43. Sheehy, E., S. L. Conrad, L. E. Brigham et al. Estimating the number of potential organ donors in the United States. *New England Journal of Medicine* 349, no. 7 (2003): 667–674.

44. Capron, A. M. Reexamining organ transplantation. *JAMA* 285, no. 3 (2001): 334–336.

45. Klassen, A. C., and D. K. Klassen. Who are the donors in organ donation? The family's perspective in mandated choice. *Annals of Internal Medicine* 125, no. 1 (1996): 70–73.

46. Kaserman, D. L., and A. H. Barnett. *The U.S. organ procurement system: A prescription for reform*. Washington, DC: AEI Press, 2002.

47. Johnson, E. J., and D. Golstein. Do defaults save lives? *Science* 302 (2003): 1338–1339.

48. Taub, S., A. H. Maixner, K. Morin, and R. M. Sade. Cadaveric organ donation: Encouraging the study of motivation. *Transplantation* 76, no. 4 (2003): 748–751.

49. Delmonico, F. L., R. M. Arnold, N. Scheper-Hughes, L. A. Siminoff, J. Kahn, and S. J. Youngner. Ethical incentives—not payment—for organ donation. *New England Journal of Medicine* 346, no. 25 (2002): 2002–2005.

50. Caplan, A. L. *Am I my brother's keeper? The ethical frontiers of biomedicine*. Bloomington: Indiana University Press, 1997.

51. Arnold, R. M., S. Bartlett, J. Bernat et al. Financial incentives for cadaver organ donation: An ethical reappraisal. *Transplantation* 73, no. 8 (2002): 1361–1367.

52. Cohen, L. Where it hurts: Indian material for an ethics of organ transplantation. *Daedalus* 128, no. 4 (1999): 135–165.

53. Josefson, D. Two arrested in US for selling organs for transplantation. *British Medical Journal* 316 (1998): 723.

54. Gianelli, D. M. Ethics forum debates prisoners as donors. *American Medical Association*. Available at <http://ama-assn.org/amdenews/1998/pick_98/inta1221.htm> (accessed November 26, 2003).

55. Joralemon, D., and P. Cox. Body values: The case against compensating for transplant organs. *Hastings Center Report* 33, no. 1 (2003): 27–33.

Further Reading

Fox, R. C., and J. P. Swazey. *Spare parts: Organ replacement in American Society*. New York: Oxford University Press, 1992.

• A classic text dealing with the challenges of organ transplantation

United Network for Organ Sharing Web site. Available at <http://www.unos.org/>.

• Daily updates of statistics on organ transplantation in the United States and a rich source of background information

Munson, R. *Raising the dead: Organ transplantation, ethics, and society.* Oxford: Oxford University Press, 2002.

• A readable and thoughtful account of the dilemmas of organ transplantation, using largely fictitious scenarios to promote questions

Animals as Organ Factories: Xenotransplantation

Given the growing shortage of human organs for transplant, can animal organs be used instead? Harvesting organs from animals might not be subject to the same restrictions as harvesting them from humans. Animals might be bred especially as organ donors, instead of for food or other purposes. Since mammals share a similar physiology, it seems logical that animal organs might function well in a human host. The number of organs is, in principle, unlimited, and thus might solve the organ shortage.

Xenotransplantation, the transplantation of organs between species, has been attempted in the past and has a long history of failure. Blood transfusions between species occurred in the seventeenth century; while the immune system undoubtedly destroyed the foreign blood cells, the recipients apparently survived [1]. The first serious attempts at xenotransplantation occurred in the 1960s. Keith Reemtsma, a surgeon at Tulane University, transplanted chimpanzee kidneys into human recipients. Most were unsuccessful, but one patient lived for nine months after receiving the transplant. Tom Starzl, who pioneered liver transplantation in humans, attempted to transplant baboon kidneys into humans. His patients survived for at most sixty days [1, 2]. Attempts to use pig hearts in humans resulted in the dramatic hyperacute rejection of the organs within minutes of the transplant.

Despite these failures, transplant physicians continued to discuss the possibility of using animal organs as "bridges" to maintain a patient until a suitable human organ became available. In 1984, xenotransplantation gained international prominence when Leonard Bailey, who led a neonatal transplant unit at Loma Linda Hospital in California, transplanted a baboon heart into a newborn baby who was born with a severely malformed heart. The organ was the correct size for an infant. "Baby Fae" rejected the organ and died twenty days after the transplant; no human heart became available. One possible reason for the rapid rejection (despite immunosuppressive drugs) was a mismatch of blood types. Baby Fae was type O; baboons are only types A, B, or AB. The poor success rate dampened enthusiasm for xenotransplantation, but research continues. A primary stimulus remains the dire shortage of human organs.

Candidate Animals

The major limiting factor for xenotransplantation is the rapid rejection of organs in cross-species transplants. Organs from other species differ not only in HLA antigens but also may express other cell surface molecules that trigger an immune response. In principle, the rejection of foreign organs might be overcome using immunosuppressive drugs, provided that the donor animal is not too different from the human recipient. Our closest genetic relatives are the great apes—bonobos, chimpanzees, and gorillas. The best "success" in xenotransplantation so far is with primate organs. All three species are endangered, however, and also breed slowly. For simple pragmatic reasons, these animals are not good candidates for the mass production of organs. In addition, the idea of using our close relatives as "organ factories" is, for most people, morally and ethically repugnant. Other primates, such as monkeys and baboons, are more distantly related to humans, and they breed well in captivity. Yet research suggests that rejection will be hard to control. In addition, these animals are too small to be good donors for adult humans.

The most popular candidate animal is the pig. Pig organs function similarly to human organs, and their size is a good match for humans. Pigs are inexpensive, reproduce rapidly, and have large litters. Because many people consume pork as part of their diet, the notion of using pigs as organ donors is somewhat less ethically problematic. Unfortunately, pig organs are immediately subjected to hyperacute rejection because of the presence of GAL on the surface of the cells lining blood vessels. Researchers are focusing on reducing the reaction to GAL, either by blocking the antibodies that trigger the response or by genetically engineering pigs that lack GAL [1, 3, 4]. Progress has been made in this area, although a fully compatible animal has yet to be generated. The human immune system recognizes other porcine proteins as foreign, and controlling rejection remains a major challenge. It is necessary to breed pigs that have tissues that are more compatible with humans ("humanized" pigs). Another focus is to find ways to generate "tolerance" in humans, in which the recipient's immune system accepts the foreign organ without mounting a response against it.

Is Xenotransplantation Safe?

In addition to the currently intractable problem of rejection, xenotransplantation carries the risk of the transmission of animal microorganisms and viruses to humans [5]. Scientists recognize that many pathogens are capable of jumping species and creating new illnesses in new hosts. Well-documented examples are the AIDS virus (see chapter 5), which is believed to have jumped from a primate host to humans; the influenza virus, which can have both avian and porcine hosts; and SARS, whose animal host is not yet determined (see chapter 9). Organisms that do not cause illness in their animal host also may cause illness in humans. The greatest risk of xenozoonoses, animal microorganisms or viruses that become active in human hosts, is believed to be associated with primates because of their close genetic relationship to humans. This provides another reason not to use them as potential organ donors.

Can pigs be raised in a germ-free environment to eliminate the risk of transmission of a new disease? Pigs and humans have coexisted for centuries with a low occurrence of cross-species infection (with the notable exception of the influenza pandemic of 1918). Nevertheless, the contact between human and pig is much more intimate in organ transplantation than in animal husbandry. Microorganisms and viruses will have direct entry into the recipient's bloodstream. In 1997, Robin Weiss and others reported that a pig virus called PERV was stably integrated into the genome of pigs and therefore could not be removed using the methods of raising germ-free animals. PERV is able to infect cultured human cells [2], though it apparently does not cause disease. These findings suggest that it may be impossible to assure that cryptic or unknown viruses will not spread into human recipients of pig organs.

Policy Decisions

In the mid-1990s, increasing concerns about the risks of xenozoonoses led to calls for the U.S. government to develop policies to regulate the research and clinical applications of xenotransplantation [6, 7, 8]. Some argued that a moratorium should be placed on clinical applications until the safety of the procedures could be determined. The IOM released a report in 1996 that recommended the continued study of xenotransplantation under strict control [9]. The Public Health Service responded by drafting guidelines for xenotransplantation that permitted transplant trials and invited public comment. Some continued to contend that trials were premature and that the unknown public health risk outweighed the potential individual benefits [10, 11]. The Public Health Service finalized its guidelines in 2001 [12]. The guidelines permit testing under tight regulation and call for the maintenance of a national registry of all recipients in order to track possible new diseases. Organ recipients and close family members are barred from donating blood or other tissues to reduce the risk of spreading unknown infections. In 2003, the FDA also released guidelines for appropriate animal husbandry as well as the screening of animals used as sources of organs and tissues [5]. The FDA guidelines also include detailed procedures for preclinical and clinical testing. The FDA claims jurisdiction over any transplant of foreign tissues or cells as investigational new drugs under the Federal Food, Drug, and Cosmetic Act (21 U.S.C. 321 et seq.).

Xenotransplantation is being studied in a number of foreign countries too; China in particular is actively conducting research. Regulations vary from country to country—we do not know whether new diseases will emerge as a result of studies conducted in countries with less rigorous controls.

Ethical Issues

In addition to the undetermined risk of new diseases, xenotransplantation is criticized as an inappropriate exploitation of animals. Most criticism is directed against using primates as donors, but some also feel that raising pigs as a source of "parts" for humans is unethical [13]. Others suggest that in the absence of alternatives, xenotransplantation should be pursued if it can be shown to be safe [14, 15].

Widespread xenotransplantation remains a distant possibility. Yet progress is being made in engineering pig organs and tissues. Whether this approach to solving the organ shortage will prove successful, or whether new diseases will emerge, can only be determined in the future.

References

1. Cooper, D. K. C., and R. P. Lanza. *Xeno: The promise of transplanting animal organs into humans.* Oxford: Oxford University Press, 2000.

2. Holzknecht, R. A., and J. L. Platt. The history of xenotransplantation. In *History of organ and cell transplantation*, ed. N. S. Hakim and V. E. Papalois, 320–334. London: Imperial College Press. 2003.

3. Cooper, D. K. C. Clinical xenotransplantation: How close are we? *Lancet* 362 (August 16, 2003): 557–559.

4. Phelps, C. J., C. Koike, T. D. Vaught et al. Production of α1,3-galactosyltransferase-deficient pigs. *Science* 299 (2003): 411–414.

5. U.S. Department of Health and Human Services, Food and Drug Administration. Guidance for industry: Source animal, product, preclinical, and clinical issues concerning the use of xenotransplantation products in humans. *CBER*. April 2003. Available at <http://www.fda.gov/cber/gdlns/clinxeno.pdf> (accessed December 1, 2003).

6. Chapman, L. E., D. R. Salomon, and A. P. Patterson. Xenotransplantation and xenogenic infections. *New England Journal of Medicine* 333, no. 22 (1995): 1498–1501.

7. Michler, R. E. Xenotransplantation: Risks, clinical potential, and future prospects. *Emerging Infectious Diseases* 2, no. 1 (1996): 1–9.

8. Murphy, F. A. The public health risk of animal organ and tissue transplantation into humans. *Science* 273 (1996): 746–747.

9. Institute of Medicine. *Xenotransplantation: Science, ethics, and public policy*. Washington, DC: National Academy Press, 1996.

10. Bach, F. H., J. A. Fishman, N. Daniels et al. Uncertainty in xenotransplantation: Individual benefit versus collective risk. *Nature Medicine* 4, no. 2 (1998): 141–144.

11. Butler, D. Last chance to stop and think on risks of xenotransplants. *Nature* 391 (1998): 320–324.

12. U.S. Department of Health and Human Services, Public Health Service. PHS guideline on infectious disease issues in xenotransplantation. *DHHS*. January 19, 2001. Available at <http://www4.od.nih.gov/oba/sacx/xenoguide01.pdf> (accessed December 1, 2003).

13. Fox, M., and J. McHale. Xenotransplantation: The ethical and legal ramifications. *Medical Law Review* 6 (1998): 42–61.

14. Caplan, A. *Am I my brother's keeper? The ethical frontiers of biomedicine*. Bloomington: Indiana University Press, 1998.

15. Munson, R. *Raising the dead: Organ transplantation, ethics, and society*. Oxford: Oxford University Press, 2002.

Further Reading

Cooper, D. K. C., and R. P. Lanza. *Xeno: The promise of transplanting animal organs into humans*. Oxford: Oxford University Press, 2000.

• A readable and positive view of xenotransplantation

Munson, R. *Raising the dead: Organ transplantation, ethics, and society*. Oxford: Oxford University Press, 2002.

• A broad-based book on organ transplantation with a thoughtful section on xenotransplantation

Is There a Right to Die?

In fall 2003, the national media was flooded with stories about Terri Schiavo, then a thirty-nine-year-old woman in Florida who had been in a PVS for thirteen years. Her husband wanted to withdraw food and water, thereby allowing her to die, as he maintained she would have wished. Her parents, the Schindlers, objected, and insisted that she was conscious and could recover. After the courts ruled repeatedly that Schiavo's husband could order the withdrawal of nutrition and hydration, the Florida legislature passed "Terri's law," which explicitly allowed Governor Jeb Bush to set aside the court ruling and order the reinsertion of her feeding tube, which he did. Mr. Schiavo sued to invalidate the law [1]. In May 2004, the U.S. District Court granted a summary judgment for Mr. Shiavo, and this decision was affirmed on appeal [2]. The U.S. Supreme Court declined to hear the case. The Florida trial court again ordered that Ms. Schiavo's feeding tube be withdrawn on March 18, 2005. Having exhausted their appeals, Ms. Schiavo's parents turned to Congress for help.

Congress quickly passed a law ordering a review of the case in the federal courts. President Bush, making an emergency trip back to Washington for the occasion, signed it into law on March 21 [3]. Although passage of so-called private laws— those intended to aid a single individual or corporation on matters of immigration or tax relief—is not unusual, the bill was unprecedented in that it created federal jurisdiction over a matter typically confined to a state court. The intention of the law was that a federal court would order the reinsertion of Ms. Schiavo's feeding tube while the case was reviewed further in federal courts. But Judge Whittemore of the U.S. District Court for the Middle District of Florida denied the Schindlers' request to reinsert the feeding tube, determining that they were unlikely to prevail in court based on their claims that Ms. Schiavo was denied "due process" [4]. The Schindlers appealed the ruling unsuccessfully to the U.S. Court of Appeals [5], and the U.S. Supreme Court again declined to hear the case. Florida governor Bush filed additional suits in an effort to aid Ms. Schiavo's parents. The governor also ordered the Florida Department of Children and Families to enter Ms. Schiavo's hospice and take her body to another medical facility, but the trial court prevented the department from doing so. After additional legal wrangling, Ms. Schiavo died on March 31, 2005, thirteen days after her feeding tube was withdrawn [6].

The case again received heavy media coverage, and attracted demonstrators who camped outside Ms. Schiavo's hospice to demand that her feeding tube be replaced. Some conservative members of Congress criticized both the federal and state courts for failing to follow the intent of Congress, and appeared opposed to the constitutional separation of powers. Whether new laws are made at the state or federal level in the aftermath of this highly publicized case remains to be determined.

Patient Autonomy and the Right to Die

The development of modern medical technology complicated the process of dying (see chapter 12). For many, a lingering death prolonged by the use of machines is highly distasteful. Most discussions center around issues of individual autonomy—should people be allowed to control how and when they die?

The right to refuse medical treatment is well established in law and medicine, and physicians have a duty to obtain consent before carrying out medical procedures. In a seminal decision in New York in 1914, Justice Benjamin Cardozo wrote, "Every human being of adult years and sound body has a right to determine what shall be done with his own body" [7]. This ruling reflected a general recognition that a physician may not touch a patient without his/her consent, except in emergencies. This common law rule has been extended to acknowledge that competent patients may refuse any medical treatment, even those that would sustain life [8].

A complicating factor is that in many cases, patients may not be able to indicate their wishes at the time of a medical emergency. In such cases, the physician's duty to act in the best interest of a patient creates a legal and ethical privilege for the physician to intervene. When a patient regains consciousness, however, the patient once again has the right to demand that unwanted treatment be withdrawn. In practice, this often proves to be difficult for ethical and moral reasons. Is the withdrawal of treatment the cause of the patient's death, or is the cause the patient's underlying disease or condition [9]?

The issues become more complex when the dying patient no longer is able to participate in medical decisions. While still competent, patients may indicate their desires, both verbally and in writing, for refusing treatment once a certain stage of deterioration is reached (such as a "do not resuscitate" order), and in general physicians should follow their patients' directives. When patients are comatose or judged incompetent, who should make decisions for them? Should family members be permitted to speak for their loved one and decide on withholding life-sustaining treatment? What about close friends? How do we decide which person to turn to?

When there is disagreement between the patients or their surrogates and the physicians offering care, resolution may occur in the courts [10]. In 1976, the New Jersey Supreme Court ruled in *In re Quinlan* [11] that the parents of a comatose woman, Karen Ann Quinlan, could have her ventilator turned off to allow her to die, consistent with her stated wishes and inherent right to privacy. Rather than dying, Quinlan began breathing on her own and remained comatose in a PVS until her death in 1985. The Quinlan decision was seminal since it suggested that proxies were able to make medical decisions for a comatose patient that included the withdrawal of life-sustaining treatments. In *Cruzan v. Director, Missouri Dept of Health*, the U.S. Supreme Court ruled in 1990 that the state of Missouri could require "clear and convincing" evidence about the patient's preferences about withholding care before allowing the withdrawal of a feeding tube [12]. The ruling also reaffirmed the right to refuse medical treatment. In this case, the parents of Nancy Cruzan requested permission to withdraw nutrition and hydration to allow their daughter, who was in a PVS, to die. The state denied the request because Ms. Cruzan had not made her wishes clear. Notably, the state of Missouri also prevented the parents from transporting Ms. Cruzan to a health care facility in another state. After the Supreme Court ruling, her parents obtained additional information supporting their claim and Ms. Cruzan was allowed to die.

The Patient Self-Determination Act of 1990 (PSDA) was a federal law enacted in the aftermath of the Cruzan case, having the goal of getting patients to complete advance directives by requiring hospitals to tell people they have the right to complete such documents. The PSDA allows patients to indicate, in accordance with state law, what kinds of treatments they do and do not want in the event of a medical emergency in the future, and also to name individuals to speak for them. Such advance directives allow patients to refuse life support if recovery to a good quality of life is impossible, indicate a willingness to donate organs, and designate what types of medical procedures may be conducted. A "Durable Power of Attorney for Health Care" names individuals who will act as proxies should the patients be unable to speak for themselves. The law allows patients to refuse treatment that would otherwise keep them alive. Advance directives are not always followed, though; they may be unavailable, or family members may object to their contents. Many individuals do not complete forms prior to a medical emergency and also may never discuss their wishes with family members.

Physician-Assisted Suicide and Euthanasia

Euthanasia (meaning "good death") can take several forms. It involves interventions that hasten the patient's death, such as an injection of lethal drugs. Another form of active intervention is termed physician-assisted suicide (PAS), in which a doctor supplies a patient with the means to commit suicide.

Although polls suggest that as much as 65 percent of the U.S. population favors PAS or euthanasia under some conditions, such practices are illegal except in Oregon (see below) [13]. The PSDA does not give patients the right to take measures that might accelerate their death, except under limited circumstances. In other words, patients may not ask for medical procedures that deliberately hasten their deaths. One acceptable treatment is "terminal sedation," in which a terminally ill patient is rendered unconscious using medication, thereby relieving all pain until the patient's death. Such sedation may depress respiration and hasten the patient's death [14]. Morally, this is permissible under the principle of "double effect." This principle, which is typically attributed to Saint Thomas Aquinas, holds that when an act can have both good and bad effects, the act is acceptable as long as the actor's intention is to accomplish the good and not the bad. In end-of-life decisions, the intent of physicians is not to hasten death but to make dying patients comfortable, knowing that their treatment of pain is likely to be terminal [15].

For some patients, the prospect of a death associated with the loss of personal autonomy and dignity, or a death with intractable pain, is disturbing enough to lead them to request aid in dying. These requests are not rare; studies estimate that from 18 to 50 percent of physicians, depending on their clinical specialty, have been asked by a patient for assisted suicide [13, 16]. While more than half of physicians do not view PAS or euthanasia as ethical, a small but significant proportion of physicians (and nurses) have performed either PAS or euthanasia on request [13]. Because PAS is illegal (except in Oregon), these arrangements are made surreptitiously [17].

Discussions about euthanasia began after World War I; the first society supporting euthanasia was established in the United States in 1938 [17]. Open meetings were sponsored by the Hemlock Society, which published how-to manuals for committing suicide. In 1991, Timothy Quill, a physician in Rochester, New York, reported that he had provided a long-standing patient suffering from terminal cancer with sleeping pills to use to commit suicide [18]. His letter in the *New England Journal of Medicine* provoked considerable debate about euthanasia. Quill was not charged with any crime in New York [19]. Also in the 1990s, Jack Kevorkian achieved considerable notoriety by assisting in the suicides of over one hundred patients, using a homebuilt machine mounted in the back of a van. In 1998, Kevorkian administered a lethal injection to a man suffering from amyotropic lateral sclerosis (ALS, or Lou Gehrig's disease); the euthanasia was televised on *60 Minutes*. Kevorkian was convicted of second-degree murder and sentenced to ten to twenty-five years in prison. His actions helped stimulate the passage of several state laws making PAS illegal.

In two states passing laws criminalizing PAS, Washington and New York, lawsuits resulted in the overturning of the bans at the state court level. Both cases were appealed to the U.S. Supreme Court. In 1997, the court ruled on *Washington v. Glucksburg* [20] and *Vacco v. Quill* [21]. The plaintiffs in both cases argued that patients had an inherent right to assisted suicide based on the rights to privacy and equal protection under the law. The Supreme Court rejected these contentions, and ruled that there was no inherent right to assisted suicide and that states may make laws concerning it. The court affirmed that patients retain the right to refuse life-sustaining treatments. It also suggested that patients are allowed access to medication to relieve pain and suffering, even if the medication may hasten their deaths [14, 22].

In contrast to the trend of banning PAS by a number of states, the state of Oregon narrowly passed a citizens' initiative in 1994, the Death with Dignity Act [23], legalizing PAS under certain conditions. Initially blocked by a court order, the act was reaffirmed by the voters in 1997 [23]. Oregon remains the only state permitting PAS. Patients may request prescriptions for lethal drugs if the following criteria are met: the patient is over eighteen years old, a resident of Oregon, capable of making the decision, and diagnosed with a terminal illness (defined as death expected within six months). Patients must make two oral requests to a physician, at least two weeks apart, and submit a written request. A second physician must confirm the diagnosis, and both must ascertain that the patient is mentally competent and not impaired by a psychiatric or psychological disorder. Finally, the patient must be informed of feasible alternatives such as hospice care and pain control [24].

From 1998 to 2004, 208 patients died using PAS, representing slightly more than one-tenth of one percent (0.10 percent) of deaths occurring in Oregon [24]. The majority of patients suffered from cancer; the median age was sixty-nine. Additional patients obtained prescriptions but did not use them; a total of 326 prescriptions were written in the seven-year reporting period.

In November 2001, then U.S. Attorney General John Ashcroft announced that doctors were prohibited from prescribing controlled substances, such as barbiturates, for use in PAS, under the provisions of the federal Controlled Substances Act [24]. Ashcroft indicated his intention to prosecute physicians in Oregon who prescribed lethal drugs for patients. Oregon sought a court order to block the injunction, and in April 2002, the U.S. District Court reaffirmed the Death with Dignity Act. Ashcroft then appealed the ruling to the 9th U.S. Circuit Court of Appeals. In May 2004, the court ruled against Ashcroft, asserting that states may make laws regarding the right to die [25]. In February 2005, the U.S. Supreme Court agreed to hear the case [26]. On January 17, 2006, the Supreme Court upheld Oregon's Death with Dignity Act and ruled that the administration had overstepped its authority in regulating state medical practice (*Gonzales v. Oregon*, 04–623).

The Netherlands Experience with Euthanasia

The Netherlands decriminalized euthanasia in 1999. Euthanasia was illegal yet tolerated prior to this decision. Physicians may assist suicides and also provide active euthanasia under certain conditions. Patients must be incurably ill, although no time limit is set for the definition. Patients must make repeated and explicit requests to have their lives ended, and must be mentally competent. A second physician must verify the patient's request. About 2 to 3 percent of deaths in the Netherlands are associated with PAS or euthanasia [27]. Critics charge that comatose patients are euthanized without their consent, and that physicians consistently fail to report instances of euthanasia [28]. Persistent concerns about abuses of the system led to a further tightening of the regulations. The Netherlands policy remains highly controversial, but served as a model for the Oregon Death with Dignity Act [29].

Ethical Issues

Both ethicists and physicians find PAS and euthanasia troublesome. Even strong proponents, such as Quill, urge that such practices be tightly regulated [19]. For many physicians, hastening a patient's death violates the oath to do no harm.

Nevertheless, the desire to relieve suffering and pain is strong. A central concern is the risk of abuse: the coercion of dying patients by family or physicians (for example, to cut medical costs), the involuntary euthanasia of unconscious or severely disabled patients, and the premature use of euthanasia, rather than as a last-ditch action [30].

There is strong support for improving end-of-life palliative care and pain management to provide dying patients with alternatives to PAS. The growing hospice movement, in which dying patients are given "comfort care" but no other treatments, is a reflection of the wish of patients to die a more "natural" death. Hospice care is more widely available and more frequently covered by insurance. Support for some form of PAS, however, remains stable in the U.S. population. It is likely that some degree of "backdoor" PAS will continue.

Lessons from the Schiavo Case

Medical decisions involving the withdrawal of life-sustaining treatment are made every day in U.S. hospitals. The Schiavo case has threatened to turn settled ethics and law topsy-turvy. Why did Terri Schiavo become front-page news? The disagreement between her husband and her parents, the lengthy litigation process, and the involvement of right-to-life groups more typically focused on abortion issues pushed a private family matter into the public eye. The sensational media coverage encouraged, ironically enough, death threats against Mr. Schiavo and others. In the aftermath of Ms. Schiavo's death, we can only hope that the publicity increased awareness of the need for everyone to prepare advance directives.

References

1. The Terri Schiavo information page. *Abstract Appeal*. October 22, 2003. Available at <http://abstractappeal.com/schiavo/infopage.html> (accessed December 3, 2003).

2. *Schiavo v. Bush*. Case no. 03–008212–CI–20, Sixth U.S. Circuit Court, 2004.

3. A bill for the relief of the parents of Teresa Marie Schiavo, Pub. L. 109–3, 2005.

4. *Theresa Marie Schindler Schiavo, Incapacitated ex rel., et al. v. Michael Schiavo, et al.* Case No. 8:05–CV–530–T–27TBM, U.S. District Court, Middle District of Florida, 2005.

5. *Theresa Maria Schindler Schiavo, Incapacitated ex rel., et al. V. Michael Schiavo et al.* No. 05–11556, U.S. Court of Appeals, 11th Circuit Court, 2005.

6. Cerminara, K., and K. Goodman. Key events in the case of Theresa Marie Schiavo. April 22, 2005. Available at <http://www.miami.edu/ethics/schiavo/timeline.htm> (accessed April 24, 2005).

7. *Schloendorff v. Society of N.Y. Hospital.* 211 N.Y. 125 (1914).

8. Meisel, A., and K. Cerminara. *The right to die: The law of end-of-life decisionmaking.* 3rd ed. New York: Aspen Law and Business, 2004.

9. Walker, R. M. Ethical issues in end-of-life care. *Cancer Control Journal* 6, no. 2 (March–April 1999): 162–167.

10. Merz, J. F. The right to die. *Nature Medicine* 2 (January 1996): 96–98.

11. *In re Quinlan*, 70 N.J. 10, 355 A. 2d. 647, 1976.

12. *Cruzan v. Director, Missouri Dept. of Health*, 497 U.S. 261, 279, 1990.

13. Emanuel, E. J. Euthanasia and physician-assisted suicide: A view of the empirical data from the United States. *Archives of Internal Medicine* 162 (2002): 142–152.

14. Orentlicher, D. The Supreme Court and terminal sedation: Rejecting assisted suicide, embracing euthanasia. *Hastings Constitutional Law Quarterly* 24 (1997): 947–968.

15. Quill, T. E., R. Dresser, and D. Brock. The rule of double effect: A critique of its role in end-of-life decision making. *New England Journal of Medicine* 337 (1997): 1768–1771.

16. Meier, D. E., C.-A. Emmons, S. Wallenstein, T. E. Quill, R. S. Morrison, and C. K. Cassell. A national survey of physician-assisted suicide and euthanasia in the United States. *New England Journal of Medicine* 338, no. 17 (1998): 1193–1201.

17. Solomon, A. A death of one's own. *New Yorker*, May 22, 1995, 54 ff.

18. Quill, T. E. Death and dignity: A case of individualized decision making. *New England Journal of Medicine* 324, no. 10 (1991): 691–694.

19. Quill, T. E. *Death and dignity: Making choices and taking charge.* New York: W. W. Norton and Company, 1993.

20. *Washington v. Glucksburg*, 117 S. Ct. 2258, 1997.

21. *Vacco v. Quill*, 117 S. Ct. 2293, 1997.

22. Capron, A. M. Death and the court. *Hastings Center Report* 27, no. 5 (1997): 25–29.

23. Oregon's Death with Dignity Act, 127.800–127.995, 1997.

24. Oregon Department of Human Services. *Seventh annual report on Oregon's Death with Dignity Act.* Portland: Department of Human Services. March 2005.

25. *State of Oregon v. Ashcroft*, No. 02–35587, U.S. Court of Appeals for the Ninth Circuit, 2004.

26. Greenhouse, L. Justices accept Oregon case weighing assisted suicide. *New York Times*, February 23, 2005, A1.

27. van der Maas, P. J., G. van der Wal, I. Haverkate et al. Euthanasia, physician-assisted suicide, and other medical practices involving the end of life in the Netherlands, 1990–1995. *New England Journal of Medicine* 335, no. 22 (1996): 1699–1705.

28. Hendin, H., C. Rutenfrans, and Z. Zylicz. Physician-assisted suicide and euthanasia in the Netherlands. *JAMA* 277, no. 21 (1997): 1720–1722.

29. Orentlicher, D., and L. Snyder. Can assisted suicide be regulated? In *Assisted suicide: Finding common ground*, ed. L. Snyder and A. Caplan, 55–69, Bloomington: Indiana University Press, 2002.

30. Emanuel, E. J. What is the great benefit of legalizing euthanasia or physician-assisted suicide? *Ethics* 109 (1999): 629–642.

Further Reading

Meisel, A., and K. L. Cerminara. *The right to die: The law of end-of-life decisionmaking.* 3rd ed. New York: Aspen Law and Business, 2004.

• A comprehensive review of U.S. law on the right to die

Quill, T. *Death and dignity: Making choices and taking charge.* New York: W. W. Norton and Company, 1993.

• A reasoned argument in favor of PAS, with suggested guidelines

Snyder, L., and A. Caplan. Assisted suicide: Finding common ground. Bloomington: Indiana University Press, 2002.

• A collection of essays addressing the different ethical and policy aspects of PAS

13

Concluding Remarks: The Challenges of Science Policy

The foregoing chapters reflect the advanced scientific and technological society in which we live and the perceived need for sound public policies to manage these technologies. Since the Industrial Revolution, bigger, faster, cheaper, more efficient, more powerful, and more broadly available technologies affecting every part of our society have been developed. Many of these advances present concomitant risks and hazards, and raise societal concerns. From the Luddites who were (perhaps rightfully) concerned about the loss of their jobs from automation to large numbers of Americans worried about commercial nuclear power plants (and other unsightly, unpleasant, or risky activities) in their backyards, society has seen fit to regulate scientific and technological advances, including many in the biological and biomedical areas. Society, of course, is the ultimate benefactor of technological innovations, but it also pays any prices (such as the job dislocations the Luddites feared) that these developments carry.

The regulation of science and technology naturally raises the many key questions that have been examined in the foregoing chapters. What leads a society to regulate a scientific or technological domain? What types of regulatory controls are available and appropriate? What is the role of science and individual scientists in formulating and shaping policies? How do political considerations influence and shape policies? What are the roles played in policymaking by the public, special-interest groups and advocates, academics, regulated entities, and corporate interests? Comparing and contrasting the cases presented here gives some indication of the breadth of potential policy responses to scientific and technological advances as well as how policy decisions are reached. As we've seen, the process of policymaking is too often like making sausage.

Why Regulate?

From the cases presented earlier we can glean various factors that lead to the formal regulation of (or decisions not to regulate) science and technology. These relate to characteristics of the technology and its development; commercial, political, and religious interests; and attributes of the society itself. For example, what was it that led many municipalities in the 1970s (such as Berkeley, California, and Cambridge, Massachusetts) to enact bans on recombinant DNA technologies, but permits broad use across the United States of a wide array of assisted reproductive technologies with neither thorough long-term studies of the outcomes nor formal regulation of the methods (in stark contrast to most other developed countries)? Why does the federal government refuse to pay for research involving human embryos, but not ban it outright (as have some states and other governments)?

Typically, regulations are adopted because of a perception that the risks of a new or evolving science or technology are too great, and are not being adequately policed by the scientists, engineers, universities, or companies involved. Thus, we have seen the evolution of regulations in the areas of drugs and medical devices following the elixir of sulfanilamide, thalidomide, and the Dalkon Shield cases. In many areas, such as in the use of humans and animals in research, we saw that the government is highly reactive to these types of seminal events. This may not always be rational, because single events may or may not be indicative of a greater problem, and because the regulatory fix may not narrowly address the perceived problem. For example, in response to the thalidomide tragedy that unfolded in Europe, the Federal Food, Drug, and Cosmetic Act was amended to require the testing and proof of efficacy, but the hazards of thalidomide could have been (and essentially were in the United States) avoided by adequate safety testing, as required by the 1938 act.

An additional complication is that the risk perception by the public and its representatives (including the media) often does not match the scientific assessment of a new technology. Perceived risk is affected by many factors, such as the voluntary nature of the risky activity (e.g., local pollution hazards versus automobiles or smoking), the magnitude of the potential consequences (e.g., a rare chance of mass death—nuclear power or genetically modified foods—versus frequent but highly differentiated deaths—automobiles), and the belief that the individual has control over the activity (e.g., automobiles). Further, the uncertainties surrounding new technologies may make the risks speculative or difficult to ascertain, and policies may reasonably reflect avoidance of unknown or unknowable risks.

If the public perceives high risk in an activity, there might be calls for regulation that are felt to be inconsistent with the scientific assessment of risk. The perceived risks of a technology may induce the public to cry out for strict regulations, as with genetically modified foods, even though these products carry essentially the same risks as unmodified crops. Of course, the fact that humans may be viewed as having interfered with the food in a new way can evoke various sentiments about nature and humanity's role in manipulating it, about unknowable risks, and any risks that are acknowledged would be seen as morally less acceptable because they arise from errors of commission (as opposed to omission). In the aforementioned case of recombinant technology, the bans enacted by numerous municipalities across the country led prominent scientists to meet (at Asilomar, California, in 1975) and call for a self-imposed moratorium on the technology until the safety issues could be addressed. Misperceptions of risk may also produce outcomes in the courts, such as tort lawsuits for silicone breast implants, which are in stark contrast to the accumulating scientific evidence regarding the safety of the product.

In contrast, sometimes the affected public may perceive that regulations are overly protective. Thus, AIDS activists were primarily responsible for changes to FDA regulations that enabled fast-tracking of drugs for life-threatening diseases, where they perceived the risks of drugs that were perhaps less well understood because of limited testing to be quite acceptable, given the potential benefits. Different constituencies will often judge the risks and potential benefits of technologies in disparate ways, and it is important to understand and incorporate the views and values of different affected groups in arriving at good policies. The current outcry (and spate of lawsuits) regarding the safety of new drugs such as Vioxx is a stark example: because of the small increased risk of heart attack associated with antiinflammatory drugs such as Vioxx, millions of patients suffering from the pain of arthritis and other diseases lost access to a medication that helped relieve their symptoms. Of course, the perception of the risk in the Vioxx case may be shaped by the revelations that Merck's scientists suppressed data about the risk, undermining the ability of the regulators, physicians, and the public to properly assess whether the small risk was justified by the marginal benefit in pain management offered by the drug.

Government may shy away from regulating products, such as tobacco, that play significant roles in the national economy. In this case, a long tradition of free enterprise runs counter to the federal role of protecting the safety of the populace, despite overwhelming evidence of the risks of tobacco consumption. Tobacco regulation has been slow in coming, and still avoids controlling tobacco as a drug. The United States also does not tightly regulate alcohol, an agent that arguably has greater social and medical impacts than tobacco. The attempt to regulate alcohol by the passage of the Eighteenth Amendment in 1919 led to a flourishing black market. The motivation for banning alcohol stemmed more from social disapproval led by the temperance movement than medical concerns, but the experiment in "legislating morality" largely failed.

The risks of scientific and technological advances need not be physical ones but may be based on the perception that scientists are doing something that conflicts with strongly felt social mores, as with human cloning and embryonic stem cells, or euthanasia. Note that while a vocal religious constituency has called for bans on embryonic stem cell research because it necessarily requires the destruction of human embryos, another (and arguably overlapping) constituency—couples requiring in vitro fertilization in order to conceive and people who wish to promote technology that lets couples have their own children—has intervened to prevent regulations that could (as they have in Italy) limit the numbers of embryos created and ergo destroyed in the pursuit of making babies. End-of-life decisions, generally viewed as the purview of individuals and families, have taken a highly public trajectory, in which many parties feel compelled to judge the morality of decisions.

Finally, there are fundamental social norms that may be implicated by regulation, such as those regarding freedom and liberty from unwarranted governmental intrusion. In the United States, these values may underlie the refusal to regulate in vitro fertilization, which is shrouded in the sanctity of the physician-patient relationship. AIDS activists believed they had a fundamental right to try any drugs that could help their desperate plight, and their special case helped expand the regulatory model practiced by the FDA. As a general rule, there is a suspicion in the United States about regulations that act as prior restraints on our freedoms to act, perhaps leading to a higher threshold of perceived risk before regulating a technology (as with in vitro fertilization and genetically modified foods).

The Role of Science

Scientists are key to science policymaking; good science is a necessary but not sufficient part of the policymaking process. It is important to recognize that while the technical and methodological skills and substantive knowledge of scientists are critical to good policy decisions, there are limits on scientists' specific competence.

Scientists are not specially empowered to speak to issues of values held by the broader society. Certainly, those with a public voice, such as movie actors, may find themselves called on to offer value judgments on public policy matters, but scientists, in our opinion, should avoid the temptation to use platforms provided to them *as scientists* for the advocacy or espousal of matters beyond their specific scientific competence. Scientists must be careful to discriminate between their science and their values, or risk that their science will be discounted or ignored by those who hold contrary values. Nonetheless, scientists are members of society and should participate in public discourse as citizens.

Scientists (and science organizations such as the American Association for the Advancement of Science) must acknowledge that they are not necessarily disinterested or objective participants in the policymaking process. Science is a highly competitive social enterprise. What and how science is done, who gets funded, and who gets credit are all affected by power and influence within the scientific community. Moreover, science is a societal activity, and necessarily is subject to explicit and implicit societal controls. Explicit controls include decisions about funding priorities as well as bans or other restrictions placed on unacceptable activities (cloning and stem cells, genetically modified organisms, and nuclear technologies). Implicit controls arise from unstated restrictions on what research might be funded, or research that might stimulate unwanted attention from special-interest groups that may not want to know what science (typically the social sciences) has to offer.

Money raises perhaps the most vexing problem for science. Increasingly, science follows the money, and government policies and commercial interests have large effects on what research is pursued, and how. Government has used its power of the purse to promote research on desired technologies, especially those having military or dual-use (military and nonmilitary) applications (such as high-density television), or those that might arguably affect national security, such as research into biological weapons. Technology-forcing methods include push mechanisms (paying directly for research and development on a desired technology) and pull mechanisms (ensuring a market for products that are developed via private funding, and even mandates—as with California's unsuccessful 1990 attempt to mandate that 2 percent of automobiles sold in the state had to be electric by 1998). This raises the fundamental question, What should be the role of government-funded research?

The government should fund research that serves a unique public interest, such as weapons and defense; research that has the potential to contribute to social welfare but that might not be done or done fast enough, or that might not be made broadly available, if left to commercial interests; and research devoted to building or developing a domestic industrial or scientific capacity. For example, the government has at times spent substantial sums on basic big science activities like the Superconducting Super Collider and other large physics experiments, NASA's trips to the moon and exploration of the solar system and beyond, and the human genome project. These may be attributed to the aspirational goals of a society rich enough to undertake such research because it's challenging, it will not be funded by commercial interests as there is no profit to be made, and it can demonstrate and promote technical, scientific, and military prowess and superiority (recall the space race of the early 1960s).

In addition, the growth of commercial ties between academics and firms in the biomedical sciences has created substantial conflicts of interest, particularly in the biomedical sciences. Increased media coverage of recent cases of fraud and misconduct may damage the public's perception of and trust in scientists. Commercialization also adds to the competitive environment of science and arguably changes the culture of science, which prizes the open exchange of information.

At bottom, the pursuit of governmental or private support may create an incentive on the part of scientists to oversell the potential benefits of their work, and to attempt to prioritize their activities for funding over competing social needs. We believe that scientists and scientific organizations must avoid these instincts. They should be as objective as possible, acknowledge uncertainties and the limits of knowledge expressly, and address threats to objectivity explicitly.

Conclusion

The case studies presented in this book provide examples of the possible trajectories and the many challenges facing policymakers (legislators and regulators) as they attempt to develop cogent policies for or using science. The future will generate new issues that will raise new challenges. Scientists, policymakers, and the public have roles in the upcoming debates. We all should strive for knowledge, flexibility, and awareness of the limitations of the policy process.

Index